Green Lands for White Men

SCIENCE · CULTURE
A series edited by Adrian Johns and Joanna Radin

Green Lands for White Men

Desert Dystopias and the Environmental Origins of Apartheid

MEREDITH MCKITTRICK

The University of Chicago Press Chicago and London

The University of Chicago Press, Chicago 60637
The University of Chicago Press, Ltd., London
© 2024 by Meredith McKittrick
All rights reserved. No part of this book may be used or reproduced in any manner whatsoever without written permission, except in the case of brief quotations in critical articles and reviews. For more information, contact the University of Chicago Press, 1427 E. 60th St., Chicago, IL 60637.
Published 2024
Printed in the United States of America

33 32 31 30 29 28 27 26 25 24   1 2 3 4 5

ISBN-13: 978-0-226-83180-0 (cloth)
ISBN-13: 978-0-226-83469-6 (paper)
ISBN-13: 978-0-226-83468-9 (e-book)
DOI: https://doi.org/10.7208/chicago/9780226834689.001.0001

Library of Congress Cataloging-in-Publication Data

Names: McKittrick, Meredith, author.
Title: Green lands for white men : desert dystopias and the environmental origins of apartheid / Meredith McKittrick.
Other titles: Science.culture.
Description: Chicago : The University of Chicago Press, 2024. | Series: Science.culture | Includes bibliographical references and index.
Identifiers: LCCN 2023052478 | ISBN 9780226831800 (cloth) | ISBN 9780226834696 (paperback) | ISBN 9780226834689 (ebook)
Subjects: LCSH: Schwarz, E. H. L. (Ernest Hubert Lewis), 1873–1928. | Reclamation of land—South Africa—History—20th century. | Reclamation of land—Political aspects—South Africa. | Irrigation farming—South Africa—History—20th century. | Climatic changes—South Africa—History—20th century. | White people—South Africa—Economic conditions—20th century. | Climatology—Political aspects—South Africa. | Water diversion—South Africa—History—20th century. | South Africa—Race relations. | Kalahari Desert.
Classification: LCC S616.S6 M35 2024 | DDC 631.5/87096—dc23/eng/20240112
LC record available at https://lccn.loc.gov/2023052478

♾ This paper meets the requirements of ANSI/NISO Z39.48-1992 (Permanence of Paper).

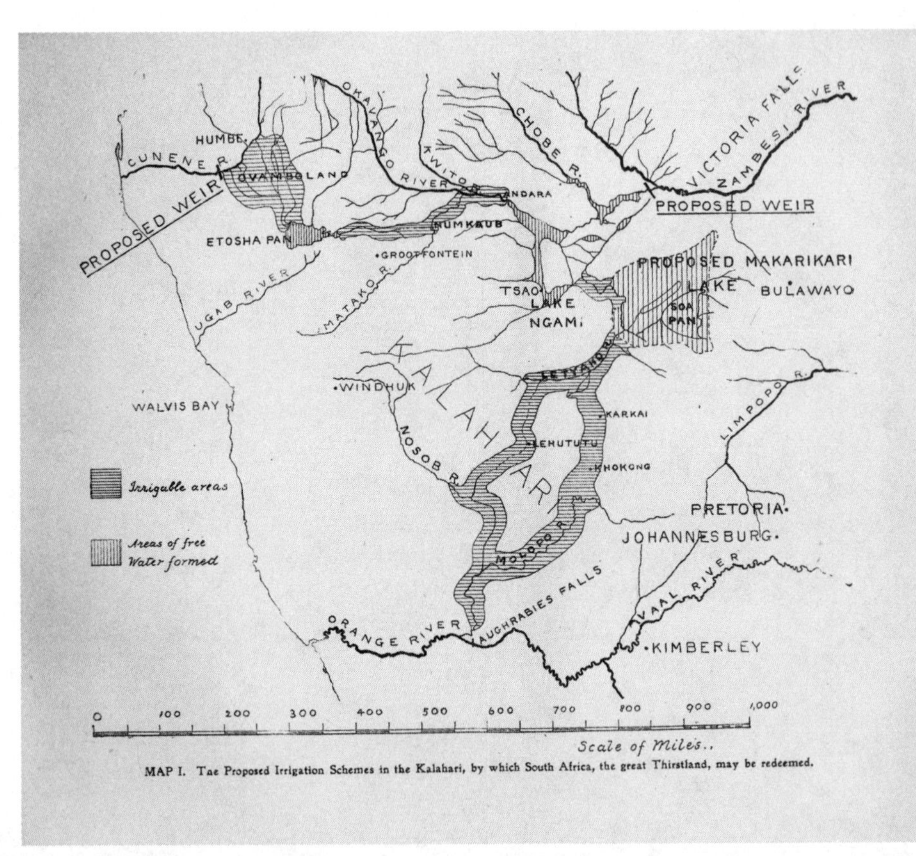

MAP I. Tae Proposed Irrigation Schemes in the Kalahari, by which South Africa, the great Thirstland, may be redeemed.

Figure 0.0 Schwarz's scheme. Source: E. H. L. Schwarz, *The Kalahari; or, Thirstland Redemption.*

CONTENTS

Introduction 1

1 Lost Lakes and Vanished Rivers 23

2 The Origins of Rain 56

3 The Invading Desert 80

4 White Men's Fears 103

5 Watering the White Man's Land 136

6 "The Kalahari Dream" 166

7 Redemption Reimagined 195

8 Afterlives 229

Epilogue 245

Acknowledgments 251
Notes 255
Bibliography 291
Index 309

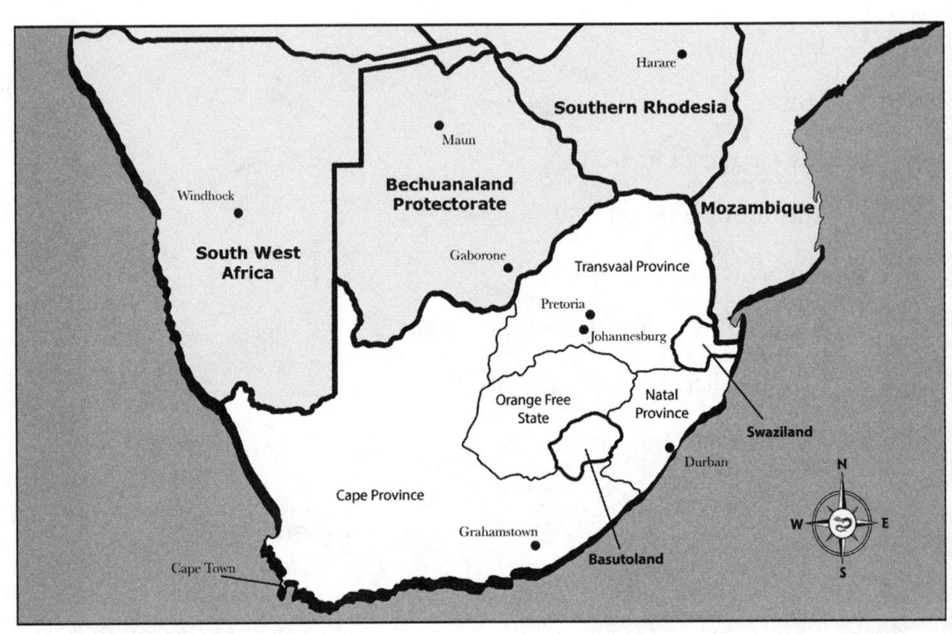

Map 1 Colonial Southern Africa

INTRODUCTION

On January 30, 1918, an article in the Johannesburg *Star* issued an ominous warning: South Africa was drying up. This aspiring white man's land at the end of the so-called Dark Continent had once resembled North America, a fertile land "clothed with forests and prairie vegetation." But now the Kalahari's "sun-scorched wastes" generated hot winds that drove back the ocean air whose moisture produced rain, and aridity intensified every year. The Union of South Africa had been forged just eight years earlier, in the aftermath of war between the Boer republics and the British Empire. Its 1.5 million whites, one-fifth of the population, had laid claim to more than 90 percent of the land and all its mineral wealth. But without human intervention, the *Star* argued, it was fated to become as uninhabitable as the Sahara.[1]

That southern Africa was drying up was not news to the *Star*'s readers. For a century, white farmers had watched springs and wells disappear. Explorers and missionaries who ventured into the Kalahari and the coastal Namib Desert had reported dry riverbeds, smoothed boulders, and even seashells—all evidence of a much wetter past. Professional and amateur scientists, farmers, and interested observers all debated the causes of the apparent disappearance of southern Africa's water. But by the early twentieth century, government experts insisted that there was no process of progressive desiccation. Periodic droughts were a normal feature of southern Africa. The degradation of land and water resources and the mounting economic losses from drought were the fault of white farmers and their backward agricultural practices. Many—perhaps most—white South Africans rejected the experts' message. They insisted that the changes in the land had natural, nonhuman causes. The rains were failing and the desert was expanding, even in places untouched by white settlement. The question was why.

The article offered an answer. It absolved whites of responsibility by arguing that the land itself was to blame for the unfolding apocalypse. Natural processes of erosion, unleashed by the continent's topography, had caused the rivers that had once fed inland lakes to turn toward the coast. Deprived of their water supply, the lakes vanished—and with them, the atmospheric humidity that had generated abundant rains. The desert was born, and a self-reinforcing cycle began. White farmers were therefore correct when they said rainfall was declining and that it was not their fault. But there was a solution. Past environmental conditions could be restored by diverting two major rivers that lay outside South Africa's borders into the subcontinent's interior, where they would refill the Kalahari's large, shallow basins. Drought would no longer threaten white survival; instead, a vast and newly greened land would be opened for white settlement.[2]

The article had no byline. But the language and ideas were those of Ernest Schwarz, professor of geology at Rhodes University in Grahamstown, in the Eastern Cape. When Schwarz repeated his message before the country's scientific elite six months later at the annual meeting of the South African Association for the Advancement of Science, his colleagues responded with polite skepticism.[3] But the British-born geologist had tapped a deep reservoir of white fears and aspirations. Members of Parliament, native commissioners, writers, professors, engineers, and white farmers in South Africa and the countries now known as Namibia, Botswana, and Zimbabwe threw their support behind what became known as the Kalahari Thirstland Redemption Scheme. The country's largest munitions manufacturer, American by birth, presented an anonymous donation—from a British industrialist—to fund the costs of a government investigation into its feasibility. Schwarz died in 1928 while researching river basins in West Africa. But support for his scheme survived. More investigations followed, and scientists grew increasingly vocal in their rejection of Schwarz's ideas. This simply generated a backlash, revealing the depth of popular mistrust of the new and expanding scientific elite. At farmers' meetings and in letters to newspaper editors, white men insisted that such experts were completely out of touch with the realities of South African life.

The Kalahari Thirstland Redemption Scheme had counterparts around the world. White South Africans' fears for their environmental and racial future coincided with the rise of a global politics of whiteness, and a sense that European settler colonialism was hitting ecological limits. In the first half of the twentieth century, engineers, architects, and other "visioneers" proposed transforming climates and racial regimes in North Africa, Brazil, and Australia by moving and holding water on dry land.[4] None of these schemes ever came to pass, but the intellectual currents they reflected and nurtured are with us today.

This book excavates the popular and populist environmental and racial ideas that circulated around the settler colonial world and informed such schemes—ideas that blended local experiences and transnational conversations, but which diverged from a coalescing body of official knowledge. It is a study of white fears, white aspirations, and the construction of white identity during a time noted for the proliferation of imagined dystopian and utopian futures.[5] It seeks to understand the quotidian ideas and everyday worldviews that were crucial to creating and sustaining white supremacist societies beyond the realms of statesmen and ritualized racial violence. And it argues that the environment—in both its physical and its imagined forms—has been central to the construction of the "global color line," and the "white man's lands" that line was designed to protect.[6] The perceived threats posed by both aridity and dark-skinned peoples fueled popular fears of white extinction, and spawned ideas about climatic and racial futures that challenged emerging expert knowledge.

The Schwarz scheme became part of a larger debate over how to secure white power and white prosperity in South Africa, and over what constituted rational (and, by implication, white) knowledge. In the first decades of the twentieth century, South Africa's leaders saw industrial capitalism, environmental conservation, and scientific agriculture as the path toward a secure future. Schwarz's supporters imagined a different path: an engineered climate and a vast country of independent white farmers that extended beyond South Africa's existing borders. Both visions rested on segregation, but the role of white farming differed in each.

These decades were characterized by the increasing displacement and disenfranchisement of Black people, even as they began to advocate more forcefully for political rights. They were also characterized by increasing paranoia on the part of white South Africans, who argued that existing policies were inadequate to mitigate the forces threatening "white civilization." Reading these "archives of the visionary and expectant"—the ideas and predictions that were written off as uninformed or irrational by experts—reveals forms of white environmental knowledge and racialist thought that were often at odds with experts, but circulated widely.[7]

The lack of historical attention to the Kalahari Scheme and its counterparts is the result of the methods and assumptions of much recent scholarship on environmental and scientific knowledge creation, white nationalism, and "high modernist" engineering projects. This literature, which has contributed enormously to our understanding of how apparently national stories have transnational dimensions, tends to rely on official archives and the writings of mainstream intellectuals, thus inherently privileging statist and elite narratives. It also tends to focus on the origins of contemporary ideas and infrastructure, thus resulting in history being read backwards from a known endpoint. And it shares an assumption that popular knowledge is local, even insular knowledge, and that it has been gradually subsumed by global expert knowledge.[8] Conversely, recent writing on white racial ideas has suggested a certain homogeneity until the mid-twentieth century, when popular white nativist movements around the world diverged from political and economic elites who were making their peace with a decolonized world.[9]

This is an oddly racialized view of intellectual history, in which Indigenous knowledge systems continue to flourish among nonwhite communities but not among white ones. Indeed, histories of imperial knowledge often tacitly presume a binary between colonists and Indigenous people.[10] But neither category was in itself homogenous, and there was significant variability in what different whites "knew." Historians' traditional focus on scientific and intellectual elites, on ideas that dominate our present, and on schemes that were built rather than merely imagined obscures important aspects of settler colonial history—not least how race and environment have been fused in the popular imaginaries of white communities. This book suggests that both environmental and white nationalist ideas have long and heterogeneous histories.

Global schemes to transform society through transforming the climate were envisioned and popularized at a time when scientific understandings of both climate and arid lands were in their infancy. It was also a time characterized by a sense of existential crisis brought on by world war and economic upheaval, as well as racial anxieties among whites about their future on the planet. Our own time is, in some respects, not so different, marked as it is by fears of climate change, ecological limits, demographic shifts, and economic dislocation, as well as popular skepticism about experts. Moreover, a populist nationalism—often with a dose of virulent racism—has become increasingly visible in many of the same settler societies that embraced this vision of green lands for white men a century ago.

We need to understand the circulation of these other bodies of knowledge, because they continue to shape people's receptivity to a wide range of messengers, including science skeptics, anti-elitists, and white nationalists. In *Merchants of Doubt*, Naomi Oreskes and Erik M. Conway's powerful examination of science denialism, the authors draw connections between multiple attempts to discredit emerging scientific consensus, from tobacco's health effects to human-induced climate change. But they cannot explain why people seem inherently willing to mistrust experts and why, as Ronald Doel put it, "lone dissenters" and "merchants of doubt" find such an eager audience. Doel suggests that those who accept the claims of skeptics may "embrace a consistent but distinct worldview (or environmental view) about which academics know little."[11] The same is true of those who today embrace white nationalist causes and who question the authority of elites and experts. These groups, of course, have transnational networks of their own. This is not a new phenomenon: we see it among the white men who for three decades advocated for Schwarz's vision of a future South Africa.

Transnational Networks of Knowledge

The proposal to inundate the Kalahari marked a moment when a growing cadre of scientific experts in South Africa was forced to confront widespread skepticism of their authority, not only among those whom they ruled without consent but also among those who were their supposed political equals. These experts, many of them trained overseas, lamented that their attempts to inform the white public about the flaws in Schwarz's plan had the reverse effect of further entrenching popular enthusiasm. In their frustration, they characterized their opponents as misinformed, unscientific, and wedded to irrational ways of knowing the world. But popular ideas were not simply the remnant of some pre-scientific body of folk knowledge, and they were not parochial. Like scientific experts themselves, Schwarz's supporters were embedded in a complex web of local and transnational systems of knowledge.

Over the past quarter century, scholars have explored the centrality of global networks forged by imperialism to the creation of a whole range of scientific subfields, including ecology, botany, agronomy, hydrology, medicine, climatology, and meteorology.[12] More recently, the spatial form of those networks has been imagined in more diverse ways, to include connections between colonies and beyond the formal bounds of empire.[13] Yet the focus remains on the circulation of particular kinds of people, ideas, and institutions: those centered on the state and those that are the precursors of contemporary equivalents, such as today's botanical gardens and game parks, or models of range ecology and climate change. The actors are primarily professional scientists, usually in government employ, and other types of experts and state officials.[14]

This literature has offered important insights into how colonial science functioned as an instrument of power. But opposing a transnational world of primarily white experts who birthed today's scientific practices and ideas to a local one of primarily Indigenous intellectuals creates a binary that is simultaneously spatial and racialized. It situates global networks within a progressive vision of history, overlooking those networks that are "ephemeral and even fleeting."[15] Historians who have broadened their lens to include networks of trade and consumerism, migration, and anti-colonial resistance have been able to incorporate the stories of actors who operate outside or in opposition to the machinery of imperialism. But our conception of networks of scientific, environmental, and racial knowledge among whites is still tied very closely to the imperial state.

Global white solidarity and the quest to secure "white man's lands" emerged contemporaneously with modern science, and in much the same context. A century ago, W. E. B. Du Bois described the global nature of what he called the "new religion of whiteness," defining it as "the ownership of the earth forever and ever. Amen."[16] Recently, historians have explored the creation of the "global color line" and the unspoken racial dimensions of ostensibly nonracial ideas. Tyler Stovall argues that nineteenth-century imperialism helped entrench the idea of "freedom" as a specifically white entitlement. Where Europeans had previously been divided into responsible people with property and poor "revolutionary savages," imperialism placed all white men in opposition to colonized subjects, thereby racializing the idea of liberty. Marilyn Lake's study of progressivism highlights the importance of a global politics of whiteness in forging new ideas about the role of the state in securing the common good. Miles Powell demonstrates how the American conservation movement emerged out of perceived threats to white racial dominance.[17]

But like the histories of imperial knowledge, histories of transnational whiteness have an elitist bent to them. They are stories of how prominent intellectuals, reformers, and political leaders shaped policies that created and reinforced the color line in "white man's lands" around the world, in the spheres of labor, immigration, citizenship, and voting rights.[18] To be sure, historians have grappled creatively with their sources to identify the contributions of nonwhite assistants in the development of global scientific knowledge in the late nineteenth and early twentieth centuries. These studies perform an important recuperative function. But they broaden the picture without fundamentally changing it, leaving intact the equation of global scientific knowledge and white colonial knowledge. There has been less curiosity about the white knowledge systems that deviated from that of an increasingly professionalized class of scientists and experts. As Powell notes in *Vanishing America*, "Many—perhaps most—Americans held environmental and racial views that differed radically from those of elite white men."[19] We know surprisingly little about those differences in the United States or anywhere else, and we know even less about where they originated and how they were shared, sustained, and changed over time and space. The important works that do explore white popular knowledge almost universally end their story before the late nineteenth century, leaving intact a teleology that has white knowledge, but not Indigenous knowledge, converging with scientific orthodoxy by the twentieth century.[20]

In the early twentieth century, popular and expert systems of knowledge might indeed differ radically, as Powell suggests, but they were more entangled than exclusive. Sandra Swart has shown how rural white intellectual worlds, even in communities historically regarded as isolated, were shaped by both global and Indigenous networks of knowledge.[21] Around the settler colonial world, farmers used forked sticks to find underground water, as had their ancestors in Europe, but they now insisted that science could explain their success. Several groups of South African farmers tried to recruit the California rainmaker Charles Hatfield to come to South Africa, but also wrote the US Weather Bureau and a Berkeley meteorology professor to check his credentials, while a white sheep farmer in the arid Karoo, trained as an engineer, built his own rainmaking apparatus based on what he had read about experiments in the United States and Australia.[22] The German farmer who proposed the precursor of Schwarz's climate-engineering scheme read John Wesley Powell's report on the arid lands of the United States, traveled to Egypt to study irrigation, and used the research of German and Russian scientists to argue for his proposal's feasibility. These engagements with a wider world all centered on a quest to secure water, whether in the form of rain, rivers, or groundwater, as a means of making white settlement more secure. It was a quest that took on renewed urgency in the late nineteenth and early twentieth centuries, as settler colonialism expanded into the world's arid and semiarid lands.

Dry Lands, White Man's Lands

The environment has been a central if often only tacitly acknowledged factor in the project of empire, settler colonialism, and white supremacy. What Alfred Crosby termed "neo-Europes" were largely defined by their climate and how would-be settlers perceived their suitability for creating new societies modeled on their homelands.[23] But by the mid-nineteenth century, the relationship of settler colonialism and environment was changing. Many of the places that resembled "home" had been claimed and occupied. The settler colonial "explosion" that James Belich vividly describes in *Replenishing the Earth* took place in new kinds of environments—places that were, on the whole, drier than earlier zones of expansion.[24] There is both a racial and an environmental dimension to this story, but the two have generally been treated separately.

Settlers are not like other migrants. They carry their sovereignty with them and seek to create self-sustaining societies in new lands. The ultimate aim of settler colonialism is not the exploitation of the Indigenous population, but its effacement and replacement.[25] Government policies were designed to effect this outcome even where Indigenous populations remained large, as in South Africa. Effacement could be discursive as well as physical—a failure to register the presence of Indigenous people, or an act of counting and mapping them out of existence.[26] This is certainly the case in South Africa, where whites were always a minority amid a Black majority. In the nineteenth century, white settlers had expected Indigenous peoples to fall away and vanish before the onslaught of white civilization, as had supposedly happened in North America and Australia. Even as it became apparent that demographic realities would be something quite different, the hope that South Africa could become a white man's land like Australia or North America survived and eventually formed the basis of "Grand Apartheid."

Making the settler presence appear natural and perpetual implied certain gender and generational as well as racial relations. A "white man's land" required not just white men, but white women and children. As Lorenzo Veracini notes, it goes almost without saying that the archetypical "pioneer," whether in North America, Australia, or southern Africa, was a white man with a white wife and white offspring to whom he could bequeath the land he claimed for himself.[27] A "white man's land" required whites *on* the land. This remained true even as cities swelled and "pioneers" began to be seen as part of national pasts rather than the present. In 1927, the American geographer (and future Johns Hopkins University president) Isaiah Bowman suggested that the white world remained interested in "the land question" because, among other things, "There is . . . the feeling that our kind of people ought to occupy the land of which we are possessed."[28]

But by the late nineteenth century, much of the land available for settlement posed a problem for this possession-by-occupation. These new lands were dry—drier than the lands settlers had come from. Aridity was simultaneously a blessing and a curse. It meant lower population densities of Indigenous peoples, and it meant that even at subtropical latitudes settlers could escape the ravages of some lethal diseases. But aridity also imposed harsh limits on white settlement. These so-called drylands are classified today using a ratio of mean annual precipitation to mean annual potential evapotranspiration. But their other major feature, which would have been most noticeable to people at a time when it was difficult to measure evapotranspiration, is the extreme variability of precipitation. The inherent ecological variability of drylands, now understood through the concept of "disequilibrium ecology," posed a challenge for white settlers who were accustomed to more predictable climates. It also posed a challenge for an emerging class of scientific authorities whose expertise rested on "repetition, standardization, and predictability."[29]

Settler colonial concepts of what lands were or were not suitable for settlement were fluid and shifting.[30] Extensive propaganda and exuberant optimism supported early waves of expansion into the drylands. The US Great Plains—the famous "Great American Desert" of Zebulon Pike's 1806 explorations—were transformed into "Nature's great flower garden where Eden might have been."[31] American land speculators shaped and reflected popular opinion when they insisted in the mid-nineteenth century that "rain follows the plow."[32] In Australia, a geographer who suggested in the 1910s that the continent was too arid to support extensive white settlement faced such a popular backlash that he eventually left the country.[33]

In Donald Worster's memorable phrasing, the world's arid lands were seen in the nineteenth century as an "instrument of world economic dominance." In the context of a settler society, this was a project to "induce settlement in an empty land, to fabricate an empire *de novo* out of yeoman farmers, miners, and manufacturers."[34] Worster's study is an American one, but it had variants in other aspiring white men's lands. Belich tells a global version of this story, though its environmental context is only implied by the repeated expansion and contraction of white settlement into marginal lands that followed wet and dry phases. Belich is not an environmental historian, and much of what drives his narrative are the material realities of gold reefs, export markets, British capital investment, and economic booms and busts. But the effects of what amount to many individual experiences—of drought, of losing one's home, of financial ruin—matter well beyond the aggregated economic data they generate. The stories people told about their experiences mattered. Economic busts and droughts did more than contract the zone of white settlement; they generated existential fears that were both climatic and racial in nature.

A fear for one's continued existence denotes something more emotionally powerful than a concern with competition from Black or Asian laborers, or the financial setbacks that result from drought or global recession. Such existential fears emerged as settler colonialism began to hit limits. In her transnational history of global white identity, Marilyn Lake quotes the alarm sounded by the liberal Australian politician Charles Pearson in the

1880s: that the white race, confined as it was to the temperate zones, was running out of room to expand, and would lose its dominant position as it was "thrust aside" by "the black and yellow races."[35] In his study of white supremacy in the British world, Bill Schwarz argues that ordinary white citizens expressed such fears differently from elites; they often voiced the unspeakable, articulating sentiments and ideas "on the anxious margins of the public domain."[36] Those sentiments and ideas took on a particular cast in the drylands. Robert Wooding, writing about built and unbuilt water engineering schemes in Australia, argues that white citizens' interest in such schemes rose and fell in tandem with drought conditions. Popular sentiment could swing from wild optimism about the potential of technology to conquer nature to "apocalyptic visions of decline and despair."[37] These "blueprints of distress" were racial and environmental, but they are rarely explored as both.[38]

The vision of an agrarian frontier that offered independence and prosperity to white men of modest means remained seductive even as it was proved false in one economic bust or catastrophic drought after another. The search for technological solutions that would push past environmental limits and secure white dominance has to be understood not just as a search for profit, but as a response to these existential fears. If irrigation, controlled by the state and other powerful actors, would usher in the economic dominance Worster writes about, Schwarz's scheme and its counterparts around the world promised something quite different: the transformation of the drylands by increasing and stabilizing their rainfall. Men did not have to appeal to the government for rain. They did not have to pay for it, or mortgage their land to a bank to get it. Rain was a democratic source of water that would allow all white men to prosper. Its leveling function was particularly important in South Africa, where drought had created a large class of "poor whites" whose low standard of living blurred racial hierarchies and undermined the myth of white superiority.

For settlers in southern Africa's dryland environments, precipitation was the most important feature of climate. And so the story of the quest to redeem the Kalahari is also a story of grappling with climate knowledge. Climatology and meteorology have lately garnered a lot of attention from historians. Much of this work has the express intent of excavating the origins and history of our contemporary understandings of climate and climate change, and it has been extremely important in this respect.[39] Historians who explore popular ideas about climate that lay outside the scientific mainstream tend to focus on time periods prior to the twentieth century, before new kinds of observation technologies offered scientists access to more accurate understandings of climate and its universal drivers.[40] Those who rejected this new form of knowledge are portrayed as "kooks and cranks," their ideas representing "detours and dead ends."[41] South Africa's scientific elite would have agreed. They labeled those who challenged their authority on the issue of climate change as backward, ignorant of science, or just plain stubborn.[42]

But establishing this knowledge was slow work. Paul Edwards has noted that understanding the workings of the global climate is "one of the hardest challenges science has ever tackled."[43] Scientists understood the general pattern of global atmospheric circulation by the mid-nineteenth century. But it took another hundred years to connect that understanding to weather patterns on the ground. And indeed, it is striking to read the accounts of South Africa's meteorologists, irrigation engineers, and agronomists in the early twentieth century and to realize how little they understood about the drivers of the country's weather and climate. In 1914 the chief meteorologist told a commission on drought and rainfall that the country's rain came entirely from the Indian Ocean. It was a popular novelist and former magistrate who suggested that north-south shifts in the Intertropical Convergence Zone also played a role, reflecting a major component of our current understandings of climatic seasonality. But, he added, he also suspected that telegraph poles and lightning conductors were one cause of reduced rainfall because they caused "a leakage upwards of the electricity stored in the earth."[44]

South Africa's experts might have disdained the scientific pretensions of such popular intellectuals. But they could not tell farmers why the rains failed, or predict when it might happen again. They could not even offer an accurate picture of past rainy seasons. The government's meteorological stations used methods of recordkeeping suited to Northern Europe rather than to the drylands of the Southern Hemisphere. Their annual records began in January, reflecting the Gregorian calendar year rather than an austral summer rain cycle that commenced in September or October. The figures were given as monthly totals, though the entire month's rain might have fallen in just three hours. There were just a handful of stations scattered over an enormous country where just a couple of miles could separate a location that got no rain and one that received a flood-inducing deluge. Many farmers had better rainfall records than their government experts did. In short, official expertise hit its limits when faced with the variability of arid landscapes. There was nothing predictable about South Africa's climate. But here, as elsewhere, nineteenth-century scientists had labored mightily to identify patterns in its rainfall.[45]

Diana Davis has demonstrated how our contemporary discourses about arid lands—particularly ideas about "desertification" and its purported links to "deforestation"—have colonial roots. But from the start, state knowledge of arid lands was contested. In the absence of useful expert knowledge, white and Black farmers alike fell back on their own ways of understanding the world and its weather. As we will see, these ideas were not exclusively local nor global; they were entanglements of both. White farmers drew on personal experience and the experiences of those around them—including Black farmers, although it is extraordinarily difficult to trace these influences in the sources. But they also found inspiration in their fellow whites colonizing arid lands elsewhere in the world, repeating what they had heard about California, Australia, or even Soviet expansion into Central Asia. The more educated among them engaged with the world of scientists, selectively drawing from climate theories that matched local knowledge derived from experience. Nineteenth-century climate science was dynamic, and offered a buffet of possible theories. In his early work, the geographer Alexander von Humboldt had proposed that lakes helped to generate rainfall. Almost a century later, another German geographer, Eduard Brückner, sought to understand the relationship between land-based moisture and rain. Prominent US foresters suggested well into the twentieth century that forests might increase rainfall by releasing vapor into the air. These were people whom South Africa's scientific elites and government technocrats also read. But educated farmers and government experts drew differently on their work. They did so, interestingly, in the "multiscalar" ways that Deborah Coen has identified as typical of the origins of modern climatology, but which have been overlooked by many scholars: by combining spatial scales that were global and local, in which scientists and others with "wide-ranging claims to climate expertise" interacted.[46]

Transforming an "empty" arid land into a greened land for white men required technology, in southern Africa and elsewhere in the world. The question was what kind of technology, and what kind of transformation. State visions tended to focus on the centralized control of water, in the form of storage dams and irrigation schemes.[47] But in the early twentieth century, popular visions were rather different. The North American idea that "rain follows the plow"—that white agriculture on the Great Plains was improving its climate—appealed because it promised white farmers prosperity with autonomy, a way to avoid becoming ensnared in debt and reliance on the state. The idea of increasing rainfall was a seductive one. The geographer Bowman, assessing the limits settler expansion was facing by the 1920s, lamented the willingness of Australia's politicians and public to believe that arid lands could be densely settled by white farmers. "The hard fact remains that no amount of political ardor can increase the rainfall. The semi-arid and arid interior of Australia will not yield to aspiration merely. Its climate takes no account of votes. . . . Money cannot invoke clouds and rain!"[48]

Aspirations cannot generate rain. But aspirations matter to history. The dream that technology could supersede the limits imposed by climate and demography, could enable the expansion of white men's lands into the world's deserts, and could engineer a society where white poverty was unknown generated racial-environmental imaginaries that had lasting consequences. David McDermott Hughes writes about white farmers in postindependence Zimbabwe who built dam reservoirs and other water features on their farms as a means of asserting belonging and ownership in a land dominated by Black Zimbabweans. But he also argues that this focus on changing the landscape allowed them to "imagine the natives away." Similarly, Jeremy Foster argues that in the years after the formation of the Union of South Africa, whites used the landscape to imagine an all-white territory, thereby engaging in the "imaginative erasure" of Black South Africans. By the time Schwarz's scheme was being debated, Foster argues that this vision of South Africa as "white" despite its Black majority "had become an integral part of the white worldview."[49]

Histories of the Future

The Kalahari Thirstland Redemption Scheme was about futures both feared and desired. Historians don't spend a lot of time thinking about how their subjects imagined the time the historians themselves inhabit. To use Reinhart Koselleck's terminology, our work tends to prioritize "spaces of experience" over "horizons of expectation."[50] Even when we acknowledge the contingent nature of historical change, this focus on experience over expectation builds a kind of teleology into our stories. It privileges the past expectations of the powerful and, especially, of the state and its agents: those historical actors who had the greatest capacity to transform their expectations into experience, to bring forth the future they imagined.[51] Turning our lens toward the futures that failed to materialize allows us to see what otherwise remains hidden. People's fears and desires come into focus, as does the spectrum of the possible as they understood it. This necessarily shifts our understanding of how they understood their present.[52] It also shifts our understanding of our own present, which looks less natural and inevitable when we recapture the diversity of historical people's expectations for their future.

The past is littered with these alternative futures, the apparent dead ends of history. But their historical effects can be difficult to identify, in part because the act of imagining an undesirable future can itself set in motion a series of actions designed to avoid its realization. For this reason, as one essay on future scenarios observes, "Predictions that to-day appear implausible may . . . have been the most important of all."[53] Visions that were taken seriously by people at the time can seem highly improbable in hindsight. Ideas and proposals that were written off by experts as the work of isolated "kooks and cranks"—and which often appear that way to us today— may have had relevance and popular support at the time.

Focusing on these forms of future-making shifts the framing of high modernist schemes away from the elitist and institutional perspectives that dominate the study of so many built projects. Pivoting toward proj-ects that were imagined but not built—the many "unrealized utopian projects of high modernism," in Philipp Lehmann's phrasing—expands our conceptual field to include those past "horizons of expectation."[54] The fact that the Kalahari Scheme was not proposed by a government agency or employee and was never built allows us access to worlds obscured in the stories of the state-sponsored schemes that were constructed. The public enthusiasm for engineering the climate and "redeeming" the Ka-lahari reveals white citizens' fears and aspirations for South Africa in the decades between the creation of the Union of South Africa in 1910 and the consolidation of apartheid half a century later. There were multiple imagined paths toward a "white man's land." As a result, the eventual outcome of a society structured around a particular kind of segregation looks less inevitable, and the grand ambitions of apartheid in the 1960s and 1970s become more comprehensible.

Redeeming South Africa's White Minority

In 1920, Ernest Schwarz published a book outlining his scheme. Its title—*The Kalahari; or, Thirstland Redemption*—reflected more than his penchant for dramatic flourish.[55] It rooted his high-modernist project firmly within older racial and environmental imaginaries. "Thirstland"—a direct trans-lation of the Afrikaans "Dorsland"—was a local term for the arid lands that stretched north of the early zones of white settlement.[56] Schwarz rejected the term "reclamation"—used by the US government to describe its aspirations for arid lands—in favor of "redemption." The term had religious undertones, but it also had both an environmental and a ra-cial meaning in the United States. "Redemption" referred to restoring fertility to exhausted or waste land, and it continued to be used in the US South even after "reclamation" came into common usage in the late nineteenth century.[57] It was also used by American whites to describe the restoration of white rule in the postbellum South. In Schwarz's imagined future, both the dry lands of southern Africa and its white population would be redeemed.

Historians have written extensively about the danger white poverty posed to racial hierarchies in South Africa. "Poor whites" jeopardized the supposed prestige of whites in the eyes of nonwhites, and the class differences they revealed undermined the myth of the unity of the *Volk*, or Afrikaner nation. Schwarz tapped into the fears of many farmers—less acknowledged by historians who have written about white poverty—that they were themselves one drought away from becoming poor whites. "There is hardly a farm in South Africa which is secure," he wrote, arguing that most white farmers were kept solvent through artificial pricing and other government interventions that were in turn funded by the profits from the previous century's mineral discoveries. This was a false independence and a precarious prosperity. "South Africa cannot go on living on the mines, as we are doing today," Schwarz insisted—invoking a concern, widely discussed in the 1920s, that the diamond pipes and gold seams would be exhausted in the not so distant future.[58]

The image of whites abandoning or being driven from the countryside reinforced a sense that South Africa's status as a "white man's land" was tenuous. White landlessness, whether seen as a cause or as an effect of poverty, portended a time when whites would no longer, in Bowman's phrasing, "occupy the land of which we are possessed." The vast historiography that seeks to explain the origins of segregation and apartheid has not sufficiently acknowledged the importance of a white countryside to a "white man's land." This is the result of an economic and urban bias in the scholarship. A liberal British interpretation that laid responsibility for apartheid at the feet of racist Afrikaner nationalists gave way in the 1970s to an economic argument that linked both segregation and apartheid to modern capitalism. Later, a cautionary note was sounded: if race could constrain as well as empower the actions of capitalists—as it surely did—it could not simply have been a tool wielded by the economic elite. Jeremy Krikler, in his call to incorporate the "primacy of the politics of white supremacy" into explanations for segregation and apartheid, encourages us to see the world as whites saw it in the first half of the twentieth century, however odious that perspective might be to us today.

> For whites, racial supremacy in South Africa—unlike in the South of the USA—was always challenged by the facts of demography: whites were a

> minority in a land conquered from black people. And we should not underestimate the power of this sense of being in the minority to animate policy and responses to developments.[59]

White workers, Krikler notes, framed their opposition to employers' policies in terms of how those policies would affect the country's white population. They spoke of "the right of existence of the White Population of South Africa."[60] Invoking perceived threats to "white civilization" was a strategy to get the attention of the powerful, but it also reveals that people understood racial demographics as an existential problem.

In the context of South Africa, "history from below" has meant an effort to recover the most marginalized and silenced voices: those of Black South Africans. The result, as Neil Roos notes, is that the white community has been treated as monolithic and that, with very few exceptions, elite voices have continued to stand in for everyone's voices. Roos calls for greater attention to "the culture and history of ordinary white people in a society where power and society were racialized, and for whom 'being white' was central to identity and everyday experience." Whites were not divided into those who supported state efforts to create a racial state and those vanishingly few who opposed it. Rather, there were multiple ways in which "ordinary whites related to the production, organization, and maintenance of a racist society." As Roos notes, white South Africans could be "part of the rural poor, the 'army of the unemployed,' or even the 'aristocrats of labor,'" and yet could simultaneously be "elites, bound to segregated society by the privileges of whiteness, however contested its terms often were."[61] To the extent that historians have looked at how such ordinary whites helped to create and perpetuate a racist state, they have largely focused on urban whites.[62] But rural whites had their own relationship to their racist society. They lived in a world saturated by anxiety and fear—of demographic "replacement" or "swamping," but also—and relatedly—of climatic apocalypse. For them, a countryside emptied of white farmers represented the death of civilization itself.[63]

After white minority rule ended in 1994, historians began to reassess how whites developed racial thinking and practiced racism. Two puzzles emerged from this reassessment. The first is that the racial theories of experts were largely irrelevant to policies or popular views. The second is that outside of a handful of intellectual elites—many of them, as Keith Breckenridge notes, virtual pariahs in settler society—whites talked surprisingly little about race. In short, for a society engineered so thoroughly around race, the engineers' views seemed to matter little, and race seemed almost to not require discussion. Saul Dubow suggests that this is because "racist assumptions were so prevalent in the common-sense thinking" of the time.[64] Yet by virtue of its absence in the historical record, there has been little exploration of this "common-sense thinking." How precisely did race figure into the quotidian perceptions of white South Africans?

Cognitive dissonance was part of everyday life and language in South Africa. As J. M. Coetzee put it, "Blindness to the color black is built into the South African pastoral."[65] Whites wrote about "farmers" and meant only white farmers, or about "people' and meant only white people, despite the Black majority around them, often even on their own farms. In the gallons of ink devoted to discussing and debating Schwarz's scheme in letters to editors and government officials, in articles in newspapers and farming journals—many of which referenced the views of whites who were not the sort to write for publication—and in government reports and self-published pamphlets and books, the existence of Africans is scarcely acknowledged. This is not unique to the sources around Schwarz; it was built into the everyday linguistic conventions of white South Africans. But the structure of the Schwarz archive, built as it is around whites' fears of their own annihilation and their aspiration to live in a country of white men, systematically erases not just the voices of Black people but their very existence. How does one deal with this erasure? How does one responsibly write a book about it?

This book takes discursive erasure as both a problem to be investigated and a feature of whites' horizon of expectation. It asks how a place like southern Africa—where 99 percent of the land is classified as drylands or hyperarid,[66] and where Black people outnumbered whites by a ratio of three to one—could conceivably be imagined as a lushly greened land of white yeoman farmers. It is a book about white people's ideas, but those ideas are not divorced from politics and economics. The future is not a neutral space: imagining it is a way of testing, apprehending, and wielding one's own power.

Toward a History of Popular Racial-Environmental Imaginaries

The Kalahari Thirstland Redemption was not just a river engineering scheme. It was a path to a future that looked radically different from the present: a humid climate instead of a dry one, an economy dominated by agriculture instead of mining, a white population that was predominantly rural rather than urban, a country whose territory extended hundreds of miles beyond its present borders instead of being confined within them, and a society whose demographic balance looked more like the settler colonies of North America and Australia than like the tropical African colonies to the north.

Schwarz's particular genius was his ability to read popular sentiment—to understand the deepest fears and aspirations of white society in all their complexity. British South Africans insisted that the minority status of the white population was an existential threat, and that the country needed to draw hundreds of thousands of immigrants from Northern Europe. Afrikaner nationalists argued that the problem of white poverty was the true existential threat, and demanded that it be solved before the borders were opened to white immigration. Farmers and their sympathizers insisted that agriculture had to remain the "backbone" of the country and its economy, even as capitalists invested in mining and engaged in large-scale land speculation, and manufacturing and cities boomed. Techno-enthusiasts embraced large irrigation schemes, while others claimed that they were economic boondoggles that trapped white farmers in systems of debt and state surveillance, or even that they were contrary to the will of God. A handful of paternalistic liberals insisted that "natives" needed protection and opportunities to prosper, while most whites clamored for more cheap labor and insisted that Black South Africans had competitive advantages that whites lacked. Some farmers embraced the "modern" farming methods promoted by agricultural experts, while others insisted that the methods were too costly to be economic, and challenged expert claims that traditional farming practices caused environmental harm.

Remarkably, Schwarz took these divergent social imaginaries and forged them into a coherent whole. His scheme would solve the problems of white poverty and white minority status. It would restore agriculture to its rightful place and produce wealth to diversify the economy. It would radically reshape the landscape, but would do so using simple technology to restore a past equilibrium that had been lost through geological happenstance; at one point Schwarz suggested that a mere pile of logs would be sufficient to turn the Chobe River inland. It would render white farms profitable without compromising Black subsistence. It would create a white countryside without depriving farmers of their cheap Black labor force, which would remain conveniently available yet not an integral part of the white nation. And it would allow a modernization of farming that protected white farmers' independence from the forces of capitalist exploitation. Most of Schwarz's supporters did not embrace every aspect of his scheme; they picked and chose from this package based on what most spoke to their concerns or to their assessment of the problem. And no one asked too many questions about the place of Black labor in a white man's country. Indeed, the uncertainty over the project's feasibility was its strength, lending it a "mobile and mutable" quality.[67]

Overarching everything was a shared sense that the racial order would have to be based on some form of racial separation that resulted in a country that was white. As historians have noted, segregation was a fuzzy concept in the first half of the twentieth century, meaning different things to different people. But, like the apartheid system that grew out of and superseded it, segregation is a spatial concept. Space is racialized in a segregated society. White leaders and intellectuals across the political spectrum recognized that segregation, meant to make white supremacy a reality, required more land: more for Africans, who sought to leave reserves that could not sustain them; more for "land hungry" whites—decommissioned soldiers, young people—who wanted to farm but found the cost of entry prohibitive; and more for "poor whites" who lacked skills to fill jobs in the cities.[68] When Schwarz suggested creating conditions for denser agrarian settlement in South Africa and opening new lands to white settlement beyond South Africa's borders, he was not completely out of step with mainstream thinking. Politicians and agronomists alike advocated for "closer settlement" of whites in rural areas. South West Africa (now Namibia), granted to South Africa as a class C mandated territory in 1919, was seen by successive South African governments as a possible solution to its problem of poor and landless whites.[69] Prior to 1923, many hoped that Southern Rhodesia (now Zimbabwe), also a white settler colony, would be incorporated into South Africa. And from the earliest negotiations over creating the Union of South Africa until the 1940s, the possibility was left on the table that the "high commission territories" of the Bechuanaland Protectorate, Swaziland, and Basutoland—modern-day Botswana, Eswatini, and Lesotho—would be incorporated into South Africa.

Schwarz suggested that the Kalahari was the perfect laboratory for segregation, a place where "natives" and whites could remain apart. "The country is so vast," he assured readers, that southern Africa's Black residents "need not come in contact with the white settlements at all."[70] Schwarz was not just promising his fellow white South Africans a world in which their position as the dominant race was beyond question; he was conjuring a future in which whites could simply ignore the existence of the Indigenous majority. In short, South African whites would enjoy the same luxury as many of their counterparts in North America and Australia. They would get the kind of settler society that seemed to have faded from historical possibility by the twentieth century. But it would require a wetter climate to make this world a reality.

Outline of the Book

Schwarz was born and educated in London amid two revolutions. The first was a transformation in scientific understandings of the planet's geological and climatic past; the second was a "settler revolution" that drew large numbers of Europeans and their descendants into the world's arid lands. Chapters 1 and 2 situate the Kalahari Thirstland Redemption Scheme in this global context. Chapter 1 explores how these two revolutions came together to shape Schwarz's life and career. Late nineteenth-century explorers and would-be farmers encountered a variety of arid landscapes containing dry river and lake beds, signs of water-based erosion, and marine fossils. This evidence of previously wetter conditions raised urgent questions about the climatic past and future. Chapter 2 explores the formation of a cosmopolitan narrative about the importance of surface water in the regulation of climate. By the time Schwarz arrived in South Africa in 1896, white settlers and some geographers had married emerging scientific ideas to their own experiential knowledge. They argued that the world's arid environments were desiccated ones whose water had drained away, and that restoring that surface water would also restore the rainfall. In South West Africa, German colonist farmers and officials, influenced by these ideas, argued that creating a viable settler colony would require river diversion and climate engineering. They proposed the precursors to Schwarz's scheme.

Chapters 3 and 4 examine the local dimension of these transnational ideas about aridity and rain. Public support for Schwarz's scheme rested primarily on its promise of increased rainfall. Chapter 3 looks at the vernacular climate ideas of white South Africans, who generally believed that rainfall was declining, and asks why experts devoted so much energy to trying to refute this belief. Conversations about climate are also conversations about the future—our own continued existence in the world—and about morality and responsibility. They reflect larger concerns about the nature of social and political orders. Chapter 4 focuses on how white fears for their continued existence in a majority-Black country intersected with the climate-change fears discussed in chapter 3. The specter of a countryside emptied of white people generated a "back to the land" movement that sought to increase white immigration and "redeem" the poor white population. But the quest to place large numbers of whites on the land foundered on the economic and ecological limitations imposed by aridity and rainfall variability.

Chapter 5 brings together the environmental and racial ideas explored in earlier chapters by examining the role of water in engineering a white man's country. Public enthusiasm for the Kalahari scheme reflected not just vernacular environmental knowledge but also a pervasive faith in the power of science and technology to solve any problem, and an assumption that it was the job of the state to secure the prosperity of white farmers. White citizens flooded government offices with ambitious and occasionally fantastical schemes to move water across the landscape and engineer white agrarian prosperity—demonstrating not just faith in technology but a belief that the natural world was inherently hostile to the project of settler colonialism, and that a radical remaking of the environment was the only means to secure white safety and power.

Chapter 6 turns to the debate over the Kalahari Thirstland Redemption Scheme in the 1920s. It weaves together the threads explored in the previous chapters—fantasies about the Kalahari and its hinterland, popular beliefs about arid environments and climate change, fear of the African majority, and a utopian faith in technology—to show why there was such deep and lasting support for Schwarz's scheme. Public calls for a government investigation of the scheme were answered in 1925; but a highly critical report, and Schwarz's sudden death shortly thereafter, did little to dampen public enthusiasm. Chapter 7 considers the reasons for this sustained enthusiasm in the two decades after the government's initial investigation. It shows how popular constructions of white innocence and popular ideas about the requirements of a white man's land were partially incorporated into expert thinking and government policy as the country moved toward more radical forms of segregation.

Schwarz's scheme was never built. But by the 1960s, some aspects of the world he had promised his supporters had become reality. Chapter 8 concludes the book by linking the social and environmental engineering projects of "Grand Apartheid" both within and beyond South Africa to the popular ideas about racial and environmental futures that had been mobilized under the banner of Kalahari Redemption. In the epilogue, I consider what new evidence has concluded about the climate of the early twentieth century. And I examine the parallels between Schwarz's South Africa and the rise of climate skepticism, geoengineering enthusiasm, and resurgent white nationalism today. The story of Schwarz's unbuilt scheme is a story of how popular ideas and populist demands reoriented political and scientific elites' understanding of possible and desirable futures. It is a story of how fears of racial "replacement" and suspicion of experts resulted in concessions to rural whites at the expense of the Black majority—and about the origins of popular support for racial partition that took distinct forms under Grand Apartheid in the 1960s. And it is a story of how an increasingly racialist state came to embrace technology as a solution to ecological problems that had their origins in political and economic inequalities—and how, in the process, it further entrenched those inequalities. It is a cautionary tale for our time.

1 * Lost Lakes and Vanished Rivers

The Kalahari Thirstland Redemption Scheme could not have been imagined outside two revolutions that shaped Ernest Schwarz's life. The transformation in scientific understandings of the earth's history offered him an intellectual framework for thinking about the planet's geological and climatic past. And the dramatic nineteenth-century expansion of Europeans into the world's drylands propelled him to southern Africa.

Ernest Hubert Ludwig Schwarz was born in London in 1873 to parents who had emigrated from Germany in the 1850s. The family had settled among the German community in the fashionable district of Sydenham and baptized their ten children in the nonconformist Hamburg Lutheran Church, choosing fellow German immigrants as godparents. But the family had also sunk roots deeply in Britain. Friedrich, Schwarz's merchant father, had anglicized his name to Frederick and become a naturalized British subject. Later in life, Ernest would anglicize his own middle name, trading Ludwig for Lewis. The family grew increasingly prosperous over the course of Ernest's childhood, and Schwarz attended the elite Westminster School. He then studied paleontology and geology at the Royal College of Science, with a brief stint at the Camborne School of Mines.[1] If mining and metallurgy were the practical applications of Schwarz's education, the young man's choice of mentor—the renowned field geologist John Wesley Judd—indicates more purely scientific interests. Judd, whose expertise included paleontology and evolution, remained an influence beyond Schwarz's years of formal education; Schwarz wrote him about his experiences in South Africa, and dedicated his first book to him.[2]

That a curious student would be drawn to geology and paleontology in the late nineteenth century is not surprising. These were dynamic fields full of intellectual excitement generated in part by the 1859 publication of Charles Darwin's *The Origin of Species*. In a critical introduction to one of Darwin's books, Judd claimed the famous naturalist for his own field, arguing that "had Darwin not been a geologist, the 'Origin of Species' could never have been written by him."[3] Darwin's seminal work, in turn, gave new significance to geology as well as paleontology and the young field of climatology (which at the time was linked to geology).[4] The fossils embedded in the upper layers of the earth's crust took on meaning beyond indicators of past life; now they also revealed the process by which such life had emerged and changed. They offered a chronology not only of evolution but of the changing climates, landforms, and waterways that had shaped evolution. New ideas about geology, climate, and biology also invigorated debates over the age of the earth. In the century before Schwarz's birth, scientists had begun using geological evidence to question long-accepted biblical chronologies. By the time Schwarz was in school, some were suggesting that the planet was hundreds of millions, or even billions, of years old.[5] These ideas about a geologically ancient planet emerged alongside the realization that it had undergone dramatic change. In 1837, a Swiss zoologist named Louis Agassiz controversially argued that glaciers had in fact once covered much of the planet's Northern Hemisphere. By the end of the century, scientists were beginning to understand that glaciation had happened multiple times, and some were suggesting that it had also affected the Southern Hemisphere. These challenges to the perception of an unchanging planet formed one context for Schwarz's ideas about southern Africa's past.

In 1895, a year after his graduation from the Royal College of Science with honors in paleontology,[6] Schwarz embarked for the Cape Colony. He spent his first month in South Africa exploring the Swartberg, the Western Cape's tallest mountains. Famed for their fossils and rock art, the mountains form the boundary between South Africa's two largest semiarid regions: the Little Karoo, to the south, and the Great Karoo, to the north. That the first thing Schwarz did upon arriving in South Africa was take an extended fossil-hunting trip suggests that his interests in paleontology and geology had drawn him to southern Africa. He clipped an advertisement for an assistant geologist at the newly created Geological Commission of the Cape of Good Hope, and applied for the job, which he did not get. Instead, he traveled inland and joined thousands of other Britons in the gold mining boomtown of Johannesburg.

Johannesburg was located in the South African Republic, also known as the Transvaal. The republic was governed by Afrikaners, descendants of Dutch and French Huguenot settlers who had migrated away from the British-controlled Cape of Good Hope in the mid-nineteenth century. The discovery of diamonds on the border of the Cape and the Afrikaner-governed Orange Free State in the 1860s, and of gold in the heart of the South African Republic in the 1880s, caused radical changes in what had been a poor and marginalized agrarian society. By 1899 the Transvaal produced 27 percent of the world's gold supply; by 1913 the figure was

40 percent.[7] British money and expertise poured into the region, to the point where foreigners, or *Uitlanders*, numerically dominated the white population—a fact that would eventually pave the way for war between the British Empire and the Afrikaner republics.

Much of the Transvaal's landmass was populated by semiautonomous African polities and Afrikaner pastoralists who claimed huge tracks of land. Johannesburg was an anomaly amid this agrarian landscape. It had not existed a mere decade before Schwarz's arrival. But by 1895 it was a raucous, chaotic boomtown of a hundred thousand residents, all there because of the gold-bearing reef that ran through the center of the city. Johannesburg was a magnet for British capitalists, mining professionals, and laborers from Australia, the United States, Germany, France, and Italy—and tens of thousands of African migrant workers from around the subcontinent who were tasked with the dangerous job of extracting ore from ever-greater depths beneath the earth's surface. A visitor in that year wrote, "Though it is rapidly passing from the stage of shanties and corrugated iron into that of handsome streets lined with tall brick houses, it is still rough and irregular, ill paved, ill lighted, with unbuilt spaces scattered about and good houses set down among hovels." He compared it to the mining cities of the American West, "a busy, eager, restless, pleasure-loving town, making money fast and spending it lavishly, filled from end to end with the fever of mining speculation."[8] It was also a cauldron of political intrigue, as British mine owners and imperialists sought to gain political control of what the British under-secretary for the colonies called "the richest spot on earth."[9] Schwarz arrived just months before the under-secretary, in league with the politician and mining tycoon Cecil Rhodes, staged an unsuccessful coup.

Schwarz had come to Johannesburg not to work for the gold mines—which might have been expected, given his education—but to edit *Scientific African*, a popular science magazine. He joined the South African Philosophical Society, whose creation in the late 1870s spoke to an emerging "South African" identity rooted in the twinned creeds of colonial nationalism and scientific rationality.[10] But his tenure as magazine editor was brief. *Scientific African* closed its doors after publishing just five issues. Schwarz was hired by the Cape Geological Survey to write a report on whether there were coal deposits in the Swartberg—the very mountains he had explored the previous July. The job was temporary, but a few months later he was hired permanently. He began working as an assistant geologist alongside Arthur Rogers, the man who had gotten the job he had wanted the previous year.[11] The discovery of diamonds and gold had caused a surge of interest in the subcontinent's geology. George Corstorphine, the Cape's first professor of geology, was hired to coordinate the

Cape survey, but his joint positions at the South African Museum and the newly established South African College (the future University of Cape Town) left him little time for fieldwork. The two assistant geologists became the primary surveyors in the field.[12]

Corstorphine, Rogers, and Schwarz were products of the second revolution that shaped Schwarz's trajectory—what the historian James Belich has called the Settler Revolution. Born in the United Kingdom within five years of each other, they had all studied geology at British universities. Their education and subsequent careers owed much to the revolution in scientific understandings of the earth and its history. But their arrivals in South Africa, all in the same year, were part of the mass migration of millions of Europeans and Euro-Americans from temperate, humid environments to subtropical drylands. In 1788 the British had landed shiploads of convicts in Australia; seven years later, they had seized control of the Cape Colony. By the early nineteenth century, British officials were planning for the large-scale settlement of British subjects in the Cape. During this time, Anglos and Anglo-Americans explored and surveyed the lands of the modern US West and Canada. Belich provides a sense of the scale of this "Anglo explosion," which intensified in the second half of the nineteenth century. The population of California grew twenty-five-fold between 1848 and 1860, while that of Texas grew twentyfold from 1836 to 1860. The United States added ten new Western states between 1864 and 1890. Western Australia saw a sixfold increase in its population between 1891 and 1911, and Western Canada's provinces experienced similar growth. In South Africa, thousands of British settlers arrived in the Cape Colony and Natal. A few decades later, the gold-mining boom drew large numbers of British citizens—including, for a short time, Schwarz himself—to South Africa's dryland interior.[13] Further north, British settlers encouraged by Cecil Rhodes moved into Southern Rhodesia.

Although Belich focuses on the English-speaking world, other Europeans also were on the move. In the first half of the nineteenth century, Afrikaners unhappy with British rule in the Cape and unable to afford land there colonized southern Africa's interior. Brazilian ranchers moved into the dry plains of the *sertão*, followed by farmers lured by high cotton prices and government policies supporting agricultural settlement.[14] In 1830 France seized Algiers; by 1896 there were 578,000 *colons*, or French settlers, in Algeria—the second-most populous African settler colony after South Africa itself. In the late nineteenth century, Germany imagined its new colony of South West Africa as a new *Heimat*, or homeland, and colonization societies encouraged German settlement there.

In almost every place where it happened, this late nineteenth-century settler "explosion" was largely a process of confronting, claiming, occupying, and seeking to understand dryland environments. In the eighteenth century, the exotic environment most familiar to Northern Europeans was that of the humid tropics, via trade in the Indian Ocean and, especially, tropical plantation colonies in the West and East Indies.[15] But in the nineteenth century, the Settler Revolution and growing scientific interest in the planet's natural history combined to bring unprecedented numbers of Europeans into direct contact with the planet's dryland ecosystems.

Forty percent of the earth's land surface is drylands, which are technically defined by the ratio of precipitation to potential evapotranspiration (map 2). Drylands have a water deficit; all rain that falls is potentially lost to evaporation or transpiration (though in reality, some of it seeps into groundwater reserves or runs off in watercourses). For this reason, they tend to lack permanent lakes or rivers. The most arid drylands often have internal drainage systems in which runoff, rather than flowing into the ocean, empties into interior lakes or pans where it evaporates. These bodies of water are often brackish or saline, and their size fluctuates over a single year and from year to year.[16]

When northwestern Europeans and their eastern North American counterparts encountered the world's driest lands, techniques to measure evaporation and transpiration were rudimentary and very few people even thought to try. But they understood that water and land had a different

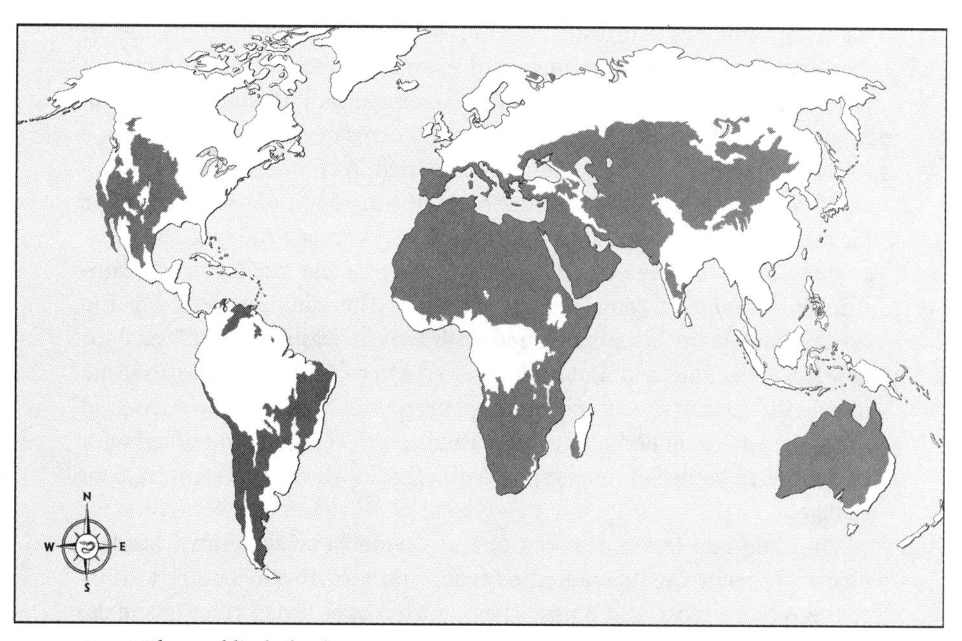

Map 2 The world's drylands

relationship in these places than in those they had come from. Rainfall was relatively low, and it was highly variable in its distribution over time and space. Surface water was rare and ephemeral. It was dawning on these newcomers that arid environments occupied a far larger proportion of the earth than they had imagined. The driest of these places contained undulating sand dunes, stony plains, deep canyons, or flat-topped mesas and buttes. Others, with slightly more rainfall, were savannas that stretched as far as the eye could see. The new arrivals sought to settle the grasslands of the Kalahari, the southern African Highveld, the Brazilian *sertão*, and central North America, as well as the deserts of Australia's interior and the southwestern United States, often wresting these spaces from indigenous pastoralists and hunter-gatherers. They moved into drier areas in years of good rainfall, past Goyder's line in Australia or the hundredth meridian in the United States, and found themselves forced back to rainier climes when drought struck.

In 1803, the Louisiana Purchase nearly doubled the size of the United States—and added an enormous expanse of drylands to a country that had been dominated by humid environments. In 1806, the leader of the expedition surveying the plains at the heart of the new territory pronounced them "a sterile waste like the sandy deserts of Africa."[17] The 1820 Stephen Long Expedition confirmed this view. The map illustrating that expedition's official report famously labeled the Great Plains as the "Great American Desert," and the text compared it to the Sahara or Central Asia, some of the driest places Europeans then knew.[18] By the 1860s, the Great Plains had been reinvented as an agricultural Eden and people were pouring in, drawn by higher-than-average rainfall and by the 1862 Homestead Act, which promised up to 160 acres of free land to settlers.

Not all lands could be so easily reimagined. As pioneers staked homestead claims in the Plains, an American doctor found himself traversing the much drier Colorado Desert, more than a thousand miles to the west. Joseph Widney marveled at what he described as "the parched and death-stricken remains of some ancient world."[19] The early surveyors of the Great Plains were little concerned with why it was a "desert," or with how it had become arid. But Widney faced a very different landscape, one filled with aquatic fossils and stands of dead trees, bordered by a sculpted shoreline where no body of water existed. These features, combined with local tales of vanished human populations, told a story of recent, radical change.

The Colorado Desert was not unique. Evidence of water in waterless places is a recurring theme in the accounts of nineteenth-century visitors to the world's most arid lands. They puzzled over broad river channels, shallow lake beds, marine fossils, and rocks worn smooth by erosion, as well as the occasional evidence of once-dense populations. Visited in the right month of the right year, some of these places did not look dry at all, as rain filled stream beds and shallow lakes and nourished carpets of vegetation. Visited a short time later, they seemed waterless and devoid of life.

Those who explored these drylands came from Europe or from the well-watered and humid zones of existing settler colonies—places like the eastern United States and Canada, or the coastal zones of the Cape Province, Brazil, New South Wales, and Victoria. They came from a world that, in the words of Jamie Linton, "normalizes copious volumes of liquid surface water." For a people who had a bedrock faith "in the universality of humidity," these new environments presented conceptual challenges.[20] Indeed, travelers' responses were rooted in preexisting ideas and expectations as much as in the physical realities of the places themselves. The British geologist John Walter Gregory later recalled that the first Australian explorers fervently hoped—even assumed—that "beyond the waterless wastes . . . rose a cooler, better land, with well-watered valleys, timbered hills, and turfed steppes." In an example of what Patricia Limerick calls "wish-fulfillment geography," many believed they would find a large river flowing northwest, which would link British settlements in the south to the northern part of the colony.[21] Such a river does not exist. But in 1846, Thomas Mitchell believed he had stumbled upon it: "There I found then at last the realization of my long-cherished hope—an interior river falling to the north west in the heart of an open Country extending also in that direction." Mitchell was convinced that this river, now known as the Barcoo, would cross the continent. Its existence was therefore "typical of God's Providence of affording access to extensive regions beyond."[22] But Mitchell's faith proved misplaced. The Barcoo does not continue flowing northwest. It turns southward and ultimately vanishes into Australia's "dead heart."

In the absence of such water in the present, travelers speculated about its existence in the past. In the Sahara, an English traveler in the late eighteenth century suggested that the vast, dry *chotts* or lowland pans were the semimythical Lake Tritonis. The geographer James Rennell elaborated on this idea, citing classical sources as evidence that a large lake in the Sahara had been connected to the Mediterranean by a river. Rennell suggested that the lake had dried out when a sand bar had formed between the lake and its water supply.[23] Decades later, one of the first European visitors to central Asia reached a similar conclusion, suggesting that the nature of gravel deposits on the valley floors of Persia indicated "that these extensive basins were formerly lakes." There was, William Thomas Blanford wrote in 1874, a "resemblance to an old sea- or lake-shore . . . the lower spurs of the hills, here formed of vertical shales and sandstones, being rounded in outline, as if worn by the sea."[24] Blanford and others imagined a lost waterscape: a united Aral and Caspian Sea, and a much larger Persian Gulf. Not everyone agreed with Blanford's theory. The Swedish explorer Sven Hedin, who traveled through the arid lands of northwestern China, insisted that the climate of Central Asia's arid regions had long been stable. But around the arid world, observers continued to link the peculiar geomorphology of arid places to a wetter past. In northern Senegal a French colonial official and surveyor observed, "There is no longer any circulation of surface water, but there remains a dead network. . . . Everything suggests that this network was powerful."[25] This debate continued well into the twentieth century.

Folk Hydrologies

In southern Africa, a specific narrative emerged about the geological and human history of the diverse drylands that dominated the western half of the subcontinent (map 3). Europeans had lived at the Cape since the 1650s, first as Dutch East India Company soldiers garrisoned in a fort and trading with the local Khoekhoe populations, and later as farmers raising crops and cattle in the lands close to what is now Cape Town. As their population grew, the newcomers slowly spread inland. By the time Schwarz was born, the indigenous pastoralist and hunter-gatherer populations of Cape Town's hinterland had been dispossessed—and in some cases exterminated—by Dutch pastoralists, or *trekboers*, who had slowly colonized the interior, their movements trending eastward toward lands with more abundant rainfall. The human costs of this colonization were immense, in terms of both individual lives and collective lifeways. Descendants of Khoekhoe pastoralists labored on white farms. Some foraging communities had been wiped out in what can fairly be termed a genocide, and entire languages had vanished.[26] Other Indigenous communities had fled deeper into the interior of the Kalahari, or across the Orange River into what is now Namibia. Yet, despite the longevity of Dutch settlement at the Cape, much of southwestern Africa's interior remained terra incognita to Europeans well into the nineteenth century. Its major human communities, geographical features, and river courses were unmapped and unknown, the subject of rumor and speculation. When Schwarz was a child, the rivers he would later propose to divert were drawn on European maps in a variety of ways, most of them inaccurate.[27]

Just beyond the winter rainfall region around Cape Town—often referred to as a "Mediterranean" environment—lay the Karoo, whose name originates in an indigenous word for aridity. Across the Orange/Gariep River, southern Namibia's rainfall averages less than two hundred millimeters

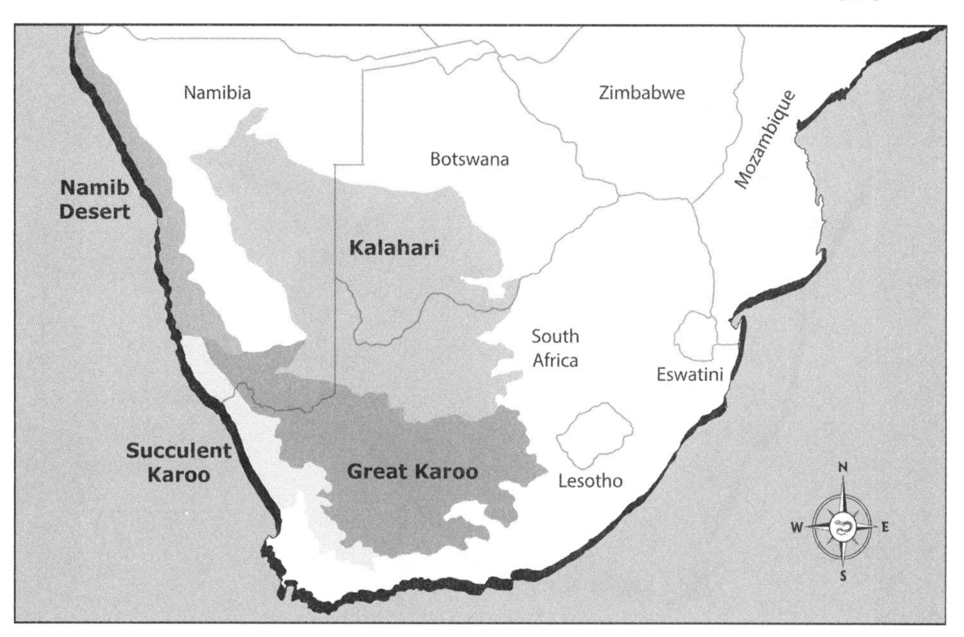

Map 3 Southern Africa's arid lands

(eight inches) a year. Conditions are even drier along the coast, where the Namib, one of the world's oldest deserts, runs north into Angola. Arid conditions extend to the east. Botswana is dominated by the Kalahari, an enormous expanse of semiarid savanna that stretches into eastern Namibia and northern South Africa. Rainfall gradually increases to the north and east. But to the north, rainfed agriculture only becomes marginally possible near the Angolan border. And in the east, South Africa's highveld, an inland plateau that covers much of the east-central interior, is also drought-prone. In all of these places, potential evaporation far exceeds annual rainfall, and surface water is rare (map 4).

As the Cape geologists recorded and analyzed the region's faults, folded mountains, and weathered geological formations, Schwarz and his two colleagues were immersed in an environment vastly different from that of their childhood home. The Karoo is a place of open spaces and harsh landscapes, with plant and animal life adapted to arid conditions. The temperature is hot by day and cold by night. Rainfall is seasonal and averages six to ten inches per year, but is highly variable: it may come in torrents, unleashing ephemeral but raging rivers, or it may not come at all. When it does rain, carpets of brilliant flowers or shoots of bright green grass appear seemingly overnight. Observers remarked on these transformations. The former missionary and Cape botanist John Croumbie Brown noted

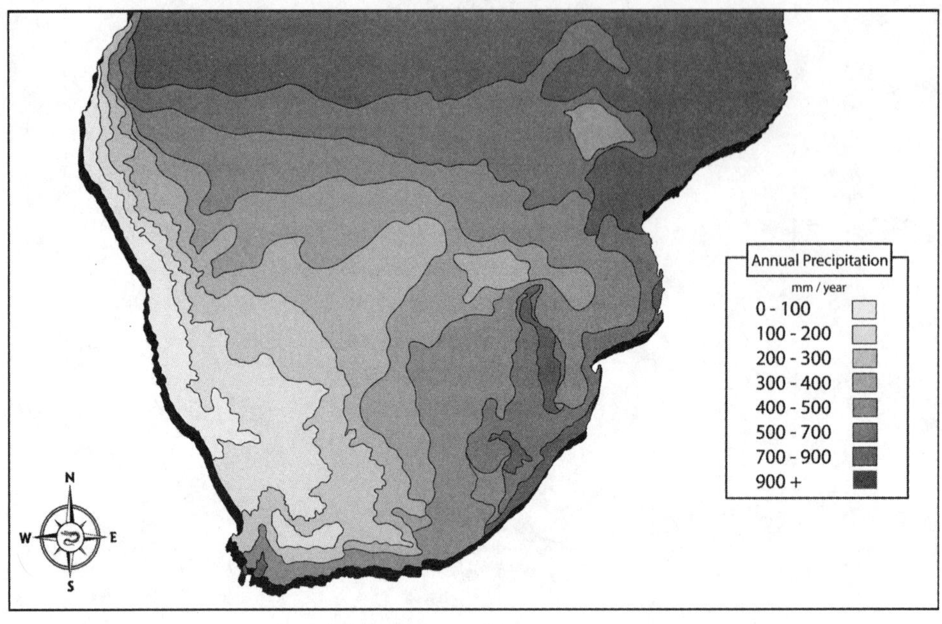

Map 4 Average annual rainfall in southern Africa

the "astounding facts" of dry rivers that could become violent torrents in the course of a day, rising by thirty feet or more.[28] In the early twentieth century, the newly arrived resident commissioner in Ovamboland, on the Namibian-Angolan border, observed the "most astonishing effect" that rainfall had on the area: "During the winter months one might be inclined to regard Ovamboland as almost desert country."[29] The jarring changes were later interpreted by some people not as evidence of flora adapted to a particular kind of rainfall regime, but as evidence that wetter conditions had once been the norm (figure 1.1).

The geologists' encounter with this parched world was a common feature of settler colonialism in the nineteenth and early twentieth centuries, and the sensory and emotive power that such environments exerted over men from temperate climates cannot be underestimated. The missionary Robert Moffat, establishing a station at Kuruman, on the Kalahari's margins, wrote at the onset of a severe drought in 1820, "As an inhabited country it is scarcely possible to conceive of one more destitute and miserable."[30] Seventy years later, a German geographer who traveled in what is now southern Namibia wrote, "One cannot think of anything more monotonous, more unattractive than this part of the 'dark continent,'" noting the empty stream beds, the "hostile thorns," and the dried grasses that made "the whole country" resemble "a vast stubble-field."[31] William

Figure 1.1 Water lilies bloom in a flooded channel of the seasonal Cuvelai River delta. The photo represents a scene that inspired outsiders to imagine the possibility of permanent surface water in a place that was devoid of it for much of the year. Source: Basler Afrika Bibliographien, Personal Archives Anneliese & Ernst Rudolf Scherz (PA.4), Hahn collection, S70 0019. Photo by C. H. L. "Cocky" Hahn.

Charles Scully, a former magistrate in South Africa's desert northwest and one of the country's most popular writers, wrote of the coastal mountains in South Africa's most arid region: "For stern, uncompromising aridity, for stark, grotesque, naked horror, these mountains stand probably unsurpassed on the face of the globe."[32] Schwarz, journeying to South West Africa in 1918, wrote in this tradition when he recorded his "dreadful dreary journey across the Karoo" in his diary.[33]

And yet, from the moment of the earliest encounters, amid all this waterless waste, there were rumors that conditions were very different far to the north—that "'the large sandy plateau' of the philosophers," as the missionary and doctor David Livingstone put it, had limits.[34] A man who hunted in the Kalahari in the early 1840s recounted the military defeat of a Tswana polity whose inhabitants had fled westward across the desert "and for some years located themselves on the borders of a vast inland lake."[35] In 1849, Livingstone was part of a small group that traveled to this lake, known as Ngami. Encountering a river flowing out of the Okavango Delta, Livingstone was told it came "from a country full of rivers—so many no one can tell their number—and full of large trees." Livingstone was captivated by the possibilities that water promised: "The prospect of a highway capable of being traversed by boats to an entirely unexplored and very populous region, grew from that time forward stronger and stronger in my mind." The arrival at Lake Ngami was anticlimactic. Livingstone wrote: "It is shallow. . . . It can never, therefore, be of much value as a commercial highway."[36]

Many others followed Livingstone—obsessed, as one scholar puts it, with Ngami's "assumed vast resources of fresh water in an otherwise arid wasteland, and with the idea of using its waterways for navigation to access remote areas."[37] Despite the difficulty of crossing the "Thirstland," at least eighty-one white visitors came to Ngami between 1849 and 1880. Half were hunters, but there were also traders, as well as "missionaries, naturalists, settlers, engineers, photographers, tourists, and even honeymooners."[38] They found a watery maze of channels cut through floating plains of papyrus and bordering innumerable islands. The Okavango Delta, of which Ngami is a part, is one of the world's largest inland deltas (figure 1.2).[39] Its water originates in the rain-rich highlands of central Angola. Summer rainfall swells the upper Kavango River and its tributary, the Kwito, from January through March; this water travels southeast for hundreds of miles until, in northwestern Botswana beginning in March, it fills the innumerable channels of the delta. Water only reaches the delta's southern margin in June, during the local dry season. At its peak, the delta is larger than the state of Connecticut and almost as large as the country of Eswatini. But all of its water evaporates or transpires in situ; none reaches the sea. It is a remarkable ecosystem that bears scant resemblance to what Livingstone surely imagined when he first heard about "a country full of rivers."

Foreign visitors venturing further north eventually found the perennial rivers at the northern edge of southwest Africa's arid lands: the Kavango, the Chobe—a tributary of the Zambezi—and the Kunene, to the west (map 5). They marveled at the contrast between the rivers, with their broad floodplains, and the arid lands that stretched for hundreds of miles to the south. And they imagined the kinds of futures such water resources might make possible. Two men from Natal, traveling in the region in the 1880s, wrote that the lands around the Kavango River "would support many thousands of Europeans, in absolute independence and even wealth."[40] Others were equally enthusiastic. Curt Von François, governor of South West Africa, reported in 1891 that the Kavango and its hinterland promised even poor settlers a secure livelihood.[41] Another German military officer wrote in 1907 that the drylands around the rivers could support millions of cattle. "An industrious people could raise up a second Argentina," he suggested, evoking another dryland that had recently attracted

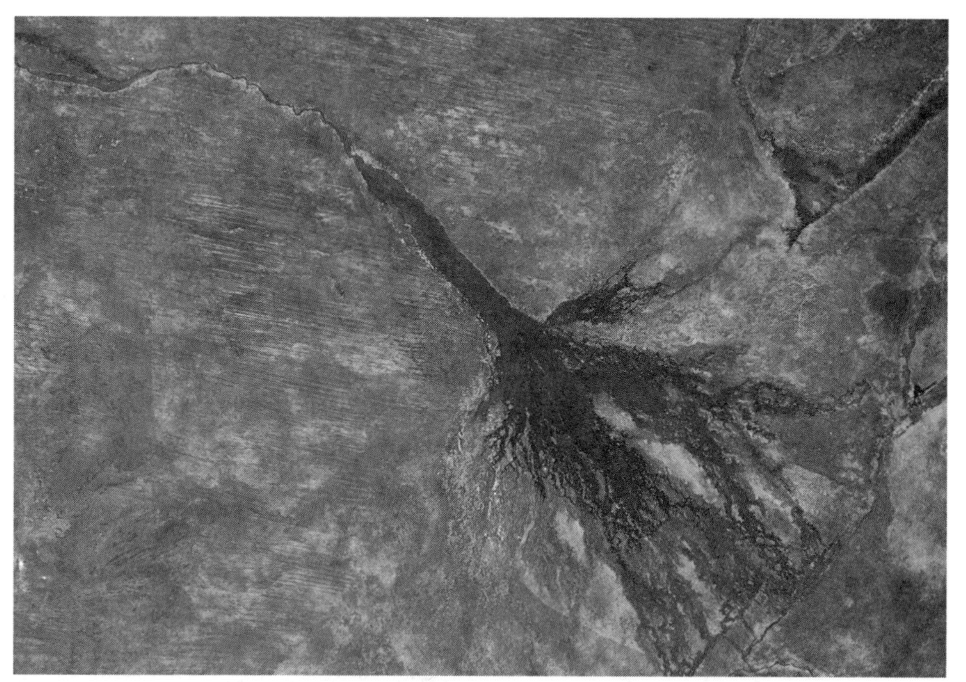

Figure 1.2 This satellite image shows the alluvial fan of the Kavango River. In the upper right corner, the Chobe River flows south and then makes a dramatic right turn. The dark thread between the Okavango Delta and the Chobe shows the interconnectedness of these river systems—connections that sparked the imaginations of men like Schwarz.

millions of European settlers.[42] Travelers acknowledged that dense populations lived along these waterways. Yet their accounts implicitly assumed that such populations would be replaced by white settlers who would make more productive use of the water resources.

These visitors faced puzzles similar to those encountered by explorers in arid lands elsewhere in the world. Livingstone, traversing the Kalahari in 1843, wondered at its dry watercourses and marine fossils. Following an empty channel that "must have been as broad as the Thames at Westminster," Livingstone came to "what must have been a large lake," where he found fossilized bones. His missionary colleagues made similar observations. Describing a valley in the Bamangwato kingdom in contemporary Botswana, the missionary Robert Moffat wrote in 1854 that the boulders had been smoothed by the force of flowing water, though the valley was now dry (figure 1.3).[43] John Croumbie Brown argued that the physical geography of South Africa's interior "tells of a time when what is now so

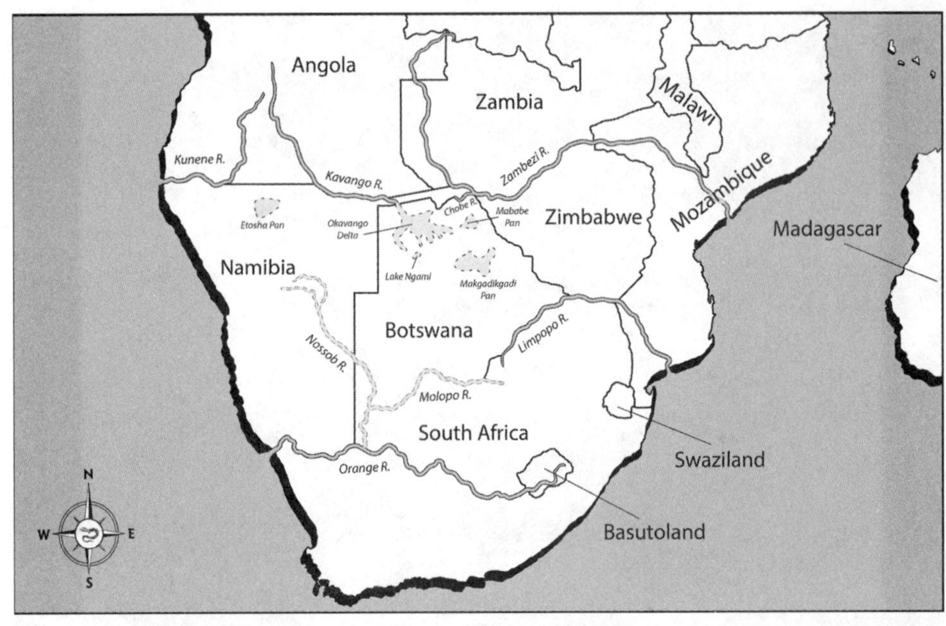

Map 5 Southern Africa's rivers and seasonal pans. Schwarz proposed to divert the Kunene toward Etosha, and the Chobe toward Ngami and Makgadikgadi in northern Botswana. He argued that the water would reach the Nossob and Molopo Rivers, which rarely flow.

arid and dry was once a land of lakes and torrents" (figure 1.4).[44] Other government employees, often some of the only literate Europeans in these places, concurred. The Irish-born Scully recalled seeing dry river courses "choked with mighty boulders weighing tons, as well as ordinary river drift, all worn as smooth as eggs. It was clear that the flow could not have been any sudden phenomenon, isolated in time; these river beds must have carried water continuously for long ages." Others noted how frequently they encountered ancient camelthorn trees that were dead and dying, indicating recent drying conditions.[45]

Lake Ngami also seemed to offer evidence of a wetter past. A distant shoreline indicated that it had once been much larger. Here, too, was "another of the proofs of desiccation met with so abundantly throughout the whole country." Describing the lake's outlet, Livingstone wrote, "The water supply of this part of the river system . . . takes place in channels prepared for a much more copious flow. It resembles a deserted Eastern garden, where all the embankments and canals for irrigation can be traced, but where the main dam and sluices having been allowed to get out of repair, only a small portion can be laid under water."[46]

Figure 1.3 Rounded boulders like these in the Namib Desert were regarded by some people as the result of water erosion, and proof that the desert had once contained abundant water. Photo by the author.

Figure 1.4 Dry riverbed in western Namibia. Photo by Olga Ernst. CC license.

Others agreed. Nearly five hundred miles to the west, the Swiss botanist Hans Schinz, who crossed what is now the Namibian-Angolan border region between 1884 and 1887, described the land between the Kunene River and the Okavango Delta as "the dry and sand-covered basin of a system of formerly extensive inland lakes, among whose remains we have to consider Ngami and the numerous salt pans, which are no longer far away from complete desiccation."[47] Siegfried Passarge, the first professional geographer to visit southern Africa, concluded in his pioneering survey of the Kalahari that "a wealth of observations not only indicates a decline in water in historical times, but deposits, river beds, old lakebeds, imply a very different climate and a much greater water supply."[48]

These accounts of lost lakes and vanished rivers, and of perennial rivers that lay more than a thousand miles from Cape Town, rippled through settler society. Most white South Africans had never seen the rivers Schwarz proposed to divert, or the empty channels whose water would be "restored." These lay far beyond the zones of white settlement. But the stories of those who had seen them circulated widely. Most South Africans who visited these areas did not write books, but wrote about their adventures in newspapers, invoked them years later in parliamentary speeches, or recounted them to entertain friends and neighbors. In a 1918 Senate debate—the first of many—on Schwarz's scheme, a South African senator told his colleagues, "I have traveled in my younger days through these inland dried up river beds and there I have seen everything complete, everything just as nature seems to have left it except the water." His colleague recalled traveling in German South West Africa and noticing "the dried up rivers that must have flown [*sic*] at very recent date and supplied Ngami and other lakes."[49]

Meanwhile, the stories published in travel narratives were recounted until they were well known even to those who had never read the books—and, indeed, to those who could not read at all. When in 1920 Schwarz published a book outlining his scheme more fully and wrote, "I am stating no new idea of my own, but rather developing the facts recognized by actual travelers," his fellow white South Africans knew which travelers he meant, and they shared his idea about the authority the words of such travelers carried.[50] Livingstone's was the best known of these canonical accounts. His books were widely available and his ideas were invoked by virtually everyone, scientists and popular authors alike, who later wrote about the region. Livingstone's portrait hung in the halls of the South African Parliament, and when the assembly discussed Schwarz's scheme in 1925, an Afrikaans member pointed to it when he stated, "There is no doubt that the three lakes were filled by this river."[51]

The dry riverbeds seen throughout southern Africa were understood not as a normal feature of arid environments, but as a sign of dramatic changes. Two of the most famous were technically tributaries of the Orange River: the Molopo, a large channel that forms much of the border between South Africa and Botswana; and its tributary the Nossob, which runs from eastern Namibia into Botswana.[52] Both flowed very infrequently; and when they did flow, sand dunes in their beds prevented water from reaching the Orange. Their physical features invited those who saw them to imagine a different past. A resident along the South Africa–Bechuanaland border noted that parts of the Molopo were eight hundred yards wide, and that it once "must have carried a very great volume of water."[53] A member of Parliament whose district included a large part of the Molopo expressed the sense of a river that had vanished: "One thing is certain, that in the past a large river ran which is today no longer there."[54] A contributor to the government agricultural journal, writing in 1910, recalled visiting both the Molopo and the Nossop. "Now, none of these rivers ever run now, not even in the rainy season," he wrote, before recalling a rare flood in the early 1890s that had poured water into a tributary of the Molopo. It was, he argued, proof of an "old Boer's story" that the Okavango River "at one time nourished the present dry and waterless Kalahari."[55] Schwarz echoed these stories when he described the boulders that had accumulated in the riverbed and which served as evidence of once powerfully flowing water.[56]

That Molopo flood made a powerful impression on observers. The agronomist William MacDonald, advocating "the conquest of the desert" in 1913, was told by a local man that millions of fish had appeared in the river, which had not run in anyone's living memory: "And what we want to know is, where did these fish come from?"[57] The river flooded again in 1917, creating a shallow lake. A resident of nearby Gordonia wrote in 1923, "If I had not seen the accumulated water there, I would not have thought it possible. It appeared like a little blue lake, a vast expanse of clear blue water, with millions of fish in it. It was a grand sight indeed, and well worthy of seeing in the Kalahari."[58] Schwarz, describing the same flood, told a drought symposium in 1923 that "considering the flat, sandy nature of the country and the enormous volume that came down this river, there can be little doubt but that the water had come through the Kalahari from the Ngami region." This suggested that "the old river can be restored in its entirety."[59]

Such stories were part of a folk hydrology, one that saw the world connected by water in all its forms and dimensions: the vapor that moved through the air to create rain, the rivers that had once crisscrossed the land, and underground water channels that mirrored those supposedly vanished rivers. A man suggested that the springs in the small farming community of Rietfontein were fed in some way by Ngami and Etosha, more than a thousand kilometers away. Schwarz evoked a similar idea of water moving along invisible channels when he explained the presence of water in the Molopo. "The only explanation that appears feasible is that the Zambezi, Chobe, and Okavango waters have filtered through underground, along the course of their former surface flow, and have issued where the river-bed is still open."[60] Others agreed. One farmer wrote, "A few years back the water did overflow from the Zambesi basin and made its way down to as far as Abiqua Puts [a pan in the Molopo riverbed], and as a remarkable coincidence we all experienced a splendid season." The flowing river not only affected the year's rainfall; it also proved that "it is certainly not going to cost very much to construct a weir that would make this a regular occurrence."[61]

Others found evidence that the water from a vastly larger Kavango drainage system had once flowed not only into the Molopo and the Orange, but from Makgadikgadi Pan toward the Limpopo River. Livingstone's companion William Cotton Oswell wondered whether the Botletle River, flowing out of Ngami, might eventually "unite with the Limpopo" hundreds of miles away.[62] Aurel Schulz and August Hammar, traveling through the Kalahari and along the Kavango and Chobe rivers in the mid-1880s, described a dry riverbed that seemed to link the Makgadikgadi with the Limpopo. "At the time the Okavango flowed into the Limpopo, this river must have been an imposing stream, and one of the longest in South Africa."[63] Taken together, these stories of the drylands to the north combined with direct encounters with dry riverbeds to create two settler imaginaries: one of a well-watered land of rushing rivers and lakes to the north, and another of an expanding ghost landscape of dry river channels and lake beds to the south. This folk hydrology nurtured white fantasies that these rivers had the potential to remake the region's economy and its demography. But it also fed fears that this water was disappearing, jeopardizing the future of whites in the subcontinent.

Climate Instability and Dystopian Futures

Around the world, nineteenth-century travelers interpreted signs of abundant water in waterless places as evidence of climate change. As Blanford put it, "For inland seas and lakes to have occupied the interior of Persia, and for large deposits to have formed in them, it is evident that the climate must have been much damper than at present."[64] The geologist and botanist Ralph Tate similarly linked Australia's geomorphology to its climatic past. "A vastly increased rainfall over what is now the arid region of Australia in former times is demanded by the extinct rivers and lakes and the former existence of large herbivores," he wrote.[65] These ideas were possible partly because of new ideas about the age of the earth and its dynamic climatic history. In the absence of any persuasive theories behind the mechanisms of climate change, a story of the planet's past that included glaciers causing mass extinction of life also made other, similarly dramatic—and catastrophic—changes seem plausible. It suggested that deserts also had histories. Passarge, for example, argued that the Kalahari's past contained both drier and wetter periods. The "Pluvialzeit," or rainy time, had been preceded by a long "Wüstenperiod," or desert period.[66] The apparent recent drying trend, then, was not necessarily progressive; it could be cyclical. Such shifts would have been less concerning if they could be located in the remote geological past. But in many places the evidence seemed to point to climate changes in historical time. For Widney, who stood on the shoreline of the then dry Salton Sea in the Colorado Desert, observing stands of dead trees, geological time was of little use in the face of evidence that seemed to indicate recent and dramatic change.[67] In Brazil, too, settlers in the *sertão* constructed a narrative of a recently-vanished pluvial past. As they struggled to cultivate cotton and food crops in the unpredictable climate, the settlers concluded that the *sertão* had once had thick forests and a wetter climate.[68]

North Africa's changes proved especially difficult to reconcile with geological time scales. Historically the region was part of Europe's known world, and the apparent evidence of its recent desiccation was well established. By the nineteenth century, popular opinion in France held that North Africa had once been among the most fertile places on earth, "the granary of Rome." Some colonial officials viewed its environmental decline as the result of invading Arabs who had deforested and overgrazed the land.[69] But others suggested that the real problem was a loss of surface water. The British engineer Donald Mackenzie wrote, "Arab traditions point out that several depressions in the Sahara were covered with water in A.D. 681, but since the year 1200 the water gradually disappeared."[70] The French speleologist Édouard-Alfred Martel thought the changes were even more recent. Not only had parts of Tunisia and Algeria dried up since Roman times, but travelers reported that in some places, chotts had vanished within the past fifty years, displacing nomadic peoples whose wells had dried up.[71]

All of these theories placed Saharan desiccation in historical time rather than geological time. Southern Africa's ghost landscapes seemed to be of similarly recent origin. Livingstone wrote to a colleague in London, "Perhaps you are aware that this country is becoming warmer, the fountains are drying up and there is the strongest evidence that at one period, this instead of being as it is now, chiefly a sterile waste, was one of the finest watered countries in the world." The missionary Moffat, stationed in the southern Kalahari, wrote, "The water was, not a very long time since, much more abundant than it is now," and suggested that the country might eventually become uninhabitable.[72] The hunter James Chapman wrote in 1862 that "within the knowledge of white men still living everywhere fountains have been drying up." In the ten years he had been coming to the land around Ngami, water levels in the region's large pans had fallen, "owing no doubt to the general desiccation going on."[73]

Indigenous inhabitants confirmed these accounts. "Native tradition," as represented by travelers, seemed unanimous that the Kalahari's climate was becoming drier. Chapman claimed that "older Bushmen" described a lake to the east of Ngami that had been twice Ngami's size just forty years earlier. A British hunter who traveled along the Chobe River in 1899 wrote, "The natives unanimously agreed that the inundations were decreasing in volume, and were not nearly so large as formerly." He had seen "disused dams erected to catch fish, and also quantities of shells of water-snails far away from the river, where I was told the water never came now."[74] Livingstone reported that local people insisted that dry streambeds had once watered rich herds of cattle.[75] Observed physical features and local knowledge seemed to converge on a similar story of change. A former magistrate in the Cape wrote about the region between the Kavango River and Ngami, using a ubiquitous racial slur for Africans:

> There is every indication that this country is drying up. Fountains that gave out fine springs of water, so the old Kaffirs told me, in their fathers' time, have not been known to flow for many years. This is a common remark all over the country, and there is evidence that it is so. Extensive pans, some more than a mile in circumference and 100 feet deep, with rocks or cliffs generally on the north-east side, with sandy bottoms, are now without water, when evidently they must have been full at some time.[76]

The appearance of so many places that seemed to have become arid in historical, not geological, times led some people to ask whether this was a global phenomenon. Scientists had begun to ask whether the earth was drying out in the eighteenth century—long before anyone imagined a planetary ice age, and before the extent of the world's drylands was understood—in an attempt to make sense of evidence that the earth had once been covered with much more water than at present. Mountaintops contained the fossils of shells, and shorelines indicated formerly higher sea levels. The water level of some European lakes had declined in historical times.[77] By the nineteenth century, the heavens seemed to lend credence to the possibility that Earth was losing its water. Improved telescopes showed landscape features of Mars and the Moon that suggested a wetter past: craters resembling lakebeds on the Moon and channels or canals on Mars. The nebular hypothesis, then the dominant theory of the solar system's formation, also supported the idea that desiccation was part of a natural cycle. The theory held that cosmic bodies went through a sort of evolution, from gaseous clouds to solid masses whose atmosphere gradually disappeared. Mars and the Moon were further along this evolutionary path and had already lost their moisture; the same waterless future awaited our own planet.[78] While astronomers looked to the heavens to understand earthly desiccation, others cast their gaze downward. Martel, the speleologist, argued that water was escaping underground through caves and fissures in the earth's crust.[79] This "constant and progressive burial of the waters" was turning surface hydrological systems into subterranean ones. The result would be "the inevitable desiccation of our globe."[80]

This question of planetary desiccation took on new relevance when Europeans encountered arid environments that showed signs of having once contained significant quantities of surface water. Central Asia seems to have made a particular impression on explorers and those who read their accounts. The sight of apparently abandoned agricultural lands and once densely populated communities, now covered with sand, suggested cause for alarm about what the planet's future might hold. In 1904 a group of geographers and geologists assembled at the Royal Geographical Society in London to discuss whether the world was losing its water. The Russian anarchist and geographer Peter Kropotkin told members of the society that evidence from Central Asia revealed "that the whole of that wide region is now, and has been since the beginning of the historic record, in a state of rapid desiccation." Kropotkin added that in eastern Russia, lakes, rivers, and marshes had dried up over the past three hundred years. And in southwest Siberia, a series of shallow lakes had vanished in the course of a century. Indeed, the entire Northern Hemisphere seemed endangered. "It is a geological epoch of desiccation," Kropotkin concluded, "independent of the will of man." Those who commented on Kropotkin's presentation challenged him to explain the mechanism of such desiccation, but they also described the apparent disappearance of water across Central Asia's lowlands and highlands, Great Britain, and North and South America.[81]

A decade later, British geologist John Walter Gregory took on the desiccationists. Gregory had spent five years as a professor in Australia, and had traveled in the interior and around Lake Eyre. In a two-part article, he argued that historical records demonstrated a stable climate around the Mediterranean and Central Asia, though it was possible that some places had enjoyed past "pluvial periods" followed by desiccation in "prehistoric times."[82] But Gregory did not have the last word, and other scientists continued to insist that planetary desiccation was real. It was a conversation that was ongoing when Schwarz began to publicize his scheme.

How to Explain a Lost Lake

The most dramatic evidence that southern Africa was drying up on a human, not geological, timescale was a remarkable process that travelers witnessed over the course of the second half of the nineteenth century: Lake Ngami was shrinking. In the late 1890s, it dried up completely. The famed colonial administrator Frederick Lugard, working for a company prospecting for diamonds in northwestern Botswana between 1896 and 1898, observed that "Lake Ngami has ceased to be a lake and is now practically dry."[83] Passarge declared it "dead" in 1896. The Tawana rulers who lived near its the shores moved their capital upstream—a process repeated several times as the water receded. A body of water that had assumed semimythical status among white southern Africans in the mid-nineteenth century, and which had symbolized the potential for commercial activity and settlement beyond the Thirstland, was gone in just a few decades. In the early 1920s, a local magistrate reported that the bed of the former Lake Ngami was a broad, grass-covered plain. It had become indistinguishable from the rest of the Kalahari.[84]

Ngami's history was more complicated than the story of its progressive nineteenth-century disappearance allowed. A Yeyi man born around 1820 reported that when he was a boy, old men had told him there had been no lake during their childhood, and that they had played along a river where the lake later formed. The local magistrate accepted this story in the 1920s, pointing out that white travelers had observed the remains of trees in the lake when it receded. Local informants also emphasized the dynamism of the channels that flowed into and around the lake, reporting the creation of new channels, particularly through the activity of hippo, and the shutting down of old ones.[85]

But it was Ngami's disappearance, not its initial appearance decades earlier, that assumed an iconic place in white imaginaries, because Ngami was only one water source of many that seemed to have disappeared. A hunter wrote in 1912, "Some years ago Lake Ngami was a wide expanse of clean water, but, as is the case in so many other parts of Africa, it gradually dried up till it reached its present depressing condition, being a mere parody of a lake."[86] Few white South Africans had seen the place with their own eyes, but the suddenness of its demise raised questions about the cause of such dramatic change—questions that were linked to transnational conversations about the origins of aridity. If arid environments were relics of well-watered ones, what had caused them to fall into a spiral of desiccation—and what did it mean for the future?

Again, Livingstone was the source of one theory: that geological changes had radically reshaped southern Africa's hydroscape. In the mid-nineteenth century, he suggested that much of the northern Kalahari had once been covered in lakes and that the water had escaped through a "fissure" or "crack" in the earth's surface. Victoria Falls, which appears as just such a crack, was the most likely culprit: "The fissure made at the Victoria Falls let out the water of this great valley, and left a small patch in what was probably its deepest portion, and is now called Lake Ngami." Livingstone suggested that parallel processes had happened along the Congo, Orange, and upper Zambezi Rivers.[87] German officials who explored the region in the early 1900s echoed Livingstone's idea about the Zambezi River "breaking through at Victoria Falls" and shifting its course toward the Indian Ocean.[88] The botanist Brown applied a version of Livingstone's theory to the dry lands of the Cape's interior, whose fossil evidence indicated that they had once been underwater. He suggested that a geological process of uplift, or rising land, in the interior had rerouted rivers toward the coast. These had carved new courses through the coastal mountain ranges that had formerly trapped the water in the interior. As these rivers acquired an outlet at the ocean, it caused "drainage on a stupendous scale."[89] Schwarz's entire scheme was based on the idea that changes in river courses had drained the continent's water into the oceans.

For those who believed that geological changes were causing a decrease in the region's surface water, the disappearance of Ngami was seen as the latest step in a process that was visible elsewhere. Only a few miles from the source of the Kavango River, in the highlands of Angola, the Kunene also has its headwaters. From their common place of origin, the two rivers flow away from each other. The Kavango runs southeast into Botswana, terminating in the Okavango Delta; the Kunene flows southwest to Ruacana Falls, where it turns westward and forms the border between Angola and Namibia. Although its journey through the hyperarid Namib often depletes its water before it reaches the ocean, the Kunene's course terminates at the sea, and in years of heavy rainfall its water drains into the Atlantic. Between the Kavango and the Kunene rivers is a large, flat triangle of land that is home to agropastoralist communities and was known to colonial administrators as Ovamboland. Rainfall here is just sufficient for agriculture and, in comparison to the arid lands to the south, this region is densely populated. In the nineteenth century it contained a number of centralized kingdoms, and for much of the twentieth it was as the main source of cheap labor for South West Africa's colonial economy.

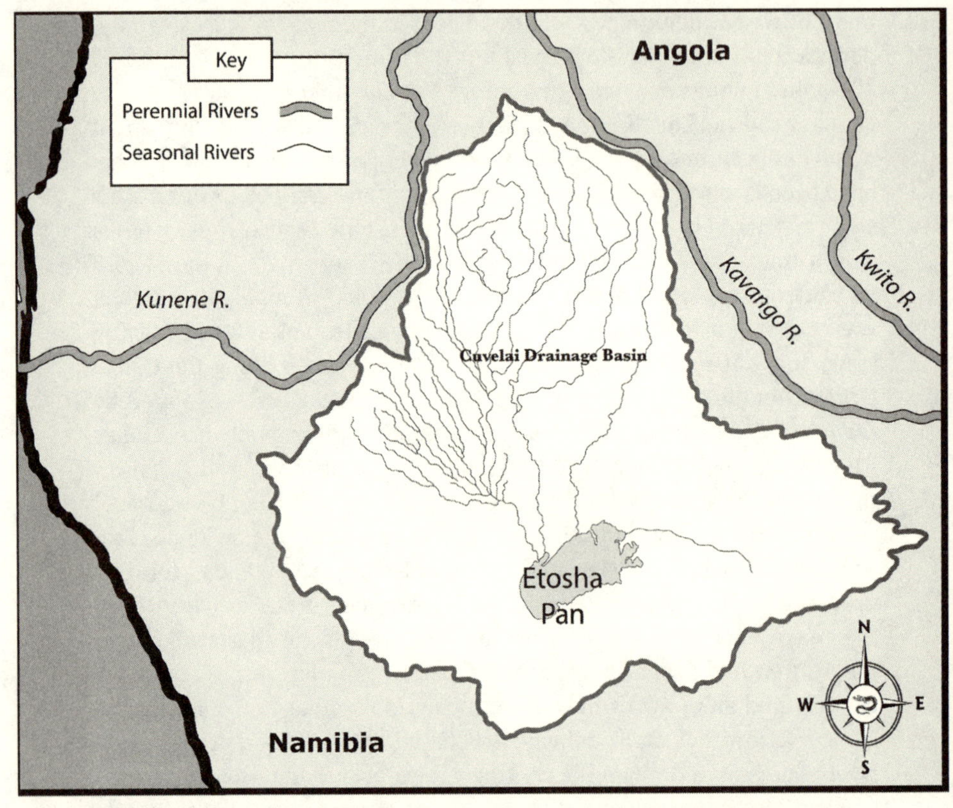

Map 6 The Cuvelai drainage basin. The Namibian side was known in the colonial period as Ovamboland.

This featureless plain had no perennial river to spark the imagination of would-be European settlers. But an annual flood, known as the *efundja*, fills shallow channels and transforms the entire region; this drainage system is now known as the Cuvelai Basin (map 6). It replenishes the wells of local farmers, and turns a dry and sandy landscape into one of shallow lakes rich with fish and frogs and dotted with water lilies. At the southern edge of the plain is a saline depression, seventy-five miles across, known as Etosha Pan, now part of a national park. In years of heavy flooding, water from the *efundja* flows into Etosha. The sense of a transformed land is reinforced by the arrival of groups of flamingos.

Until the 1940s, most outside visitors, traders, resident missionaries, and colonial administrators believed that these floods were caused by the Kunene overtopping its banks in the west.[90] A series of channels that originate near the Kunene and flow southeast into Etosha seemed to support this view. But the headwaters of these channels are filled with sand.

In a variation on the regional declensionist narrative, outsiders argued that these channels were silting up, thereby severing the plain from the Kunene and threatening the future of the *efundja*. Here as elsewhere, the theory was strengthened by local reports that the frequency and volume of the annual floods had declined within people's lifetimes. The potential loss of this annual flood was a grim prospect. But it also seemed to bode ill for the future of the Kavango River system in the east. In the early twentieth century the Kunene and the Kavango came to be seen, in the words of the irrigation engineer Francis Kanthack, as "sister" rivers.[91] The German geographer Georg Nitsche wrote, "Lake Ngami corresponds to the Etosha Pan," while "the marshland of the delta-like, divided lower Okavango" was the equivalent of the Cuvelai, or Ovambo, plain, with its multiple mazelike channels known as *iishana*.[92] The idea that the two systems worked the same way made it seem feasible that the Kavango system was suffering the fate of the Ovambo plain, progressively being cut off from the water supplied by perennial rivers.

Southwestern Africa therefore contained multiple "lost lakes." East of Etosha and Ngami were two other sizeable basins: Mababe Pan and Makgadikgadi, both showing evidence of once having held substantial quantities of water. More than a decade later, Schwarz—who quoted Nitsche at length and in the original German in his 1918 journal—described the salt crust on the surface of the Makgadikgadi Pan and wrote, "Conditions are entirely the same in Etosha."[93] Dry channels ran between these pans and the adjacent rivers, suggesting a single hydrological system—one in which the survival or death of a single body of water was linked to the survival or death of the others.

All of this evidence of a pluvial past could conceivably have been millions of years old. But Ngami's fate seemed to reinforce the continued relevance of a biblical time scale that operated in centuries and millennia. Schwarz's published textbooks referenced the geological epochs that were his profession's accepted chronological framework. But when he wrote about southern Africa's desiccation, he shifted to historical time, situating major geological changes in the recent past. He suggested, for example, that the creation of Victoria Falls might have postdated the arrival of Portuguese missionaries to East Africa. At other times he speculated that the falls had formed about five hundred years earlier, when the upper Zambezi had eroded a new path, thereby joining with a smaller coastal river.[94] These widely known stories of vanished lakes and rivers help us understand how Schwarz's readers, including men who accepted that the earth was millions or billions of years old, could situate such dramatic geological changes within historical time frames.

Others insisted that such changes had to have human, not geological, causes. A range of explanations emerged for Ngami's disappearance, rooted in different timescales and divergent understandings of its larger significance. Stories that centered human action came to exist alongside those that focused on impersonal geological forces. One of the most well-known proponents of human responsibility for desiccation was James Fox Wilson, whose observations about South Africa have been widely cited by environmental historians, though very little is known about him. In 1865, Wilson presented a paper at the Royal Geographical Society in London, before an audience that included David Livingstone. He pointed out the disagreements over the timescales in which events were unfolding. He also agreed with Livingstone and others that southern Africa's landscape contained evidence of a wetter past, and that geological changes that had drained ancient lakes accounted for the onset of aridity long ago. But the disappearance of water sources noted by so many South Africans was a more recent and localized phenomenon that could only be explained by human agency. In South Africa, Wilson argued, "It is not so much to the waywardness of nature as to the willfulness of man that we must assign the recent extension of the Kalahari Desert."[95] It was quite clear to him which men were causing such damage. He primarily blamed Africans for the state of affairs, writing in capital letters: "THE NATIVES HAVE FOR AGES BEEN ACCUSTOMED TO BURN THE PLAINS AND TO DESTROY THE TIMBER AND ANCIENT FORESTS."[96]

Wilson situated his view of African culpability within transnational conversations. He accepted that many of the world's deserts and drylands were places that had lost their water. And he suggested that this had happened due to the actions of the people who lived in these places. In North Africa, Southern Europe, the Near East, Central Asia, and the Western United States, Wilson argued, "the folly or neglect of rulers or subjects" had transformed "fruitful gardens into wastes." Northern Europeans and their descendants in eastern North America had also cut down forests, of course. But when white men felled forests, they *improved* the climate of these cold, damp places by reducing humidity and rainfall and raising temperatures. Removing trees therefore had "operated to the improvement of the soil, increasing its producing capabilities, and occasioning it to be better fitted for the residence of man."[97] The story of the world's forests and climates was, for Wilson, one defined by race. While the white race had felled forests in a way that improved the environment, the world's darker races had unleashed environmental devastation by engaging in the same activity. The drylands of the world, then, were almost uniformly the product of "inferior" races.

In the context of southern Africa, this was a relatively new argument. The idea that white settlers had disrupted nature's equilibrium had a long intellectual heritage. Europeans recognized that their plantation agriculture and the deforestation that accompanied it had damaged environments around the world. Prior to 1820, Dutch and British residents in the Cape also assumed that environmental changes and even drought were tied to European activities on the land.[98] But by the time Wilson wrote, the question of racial culpability for environmental destruction was a more open one. In southern Africa, missionaries were the vanguard of these new ideas about race, rain, and responsibility. As they moved inland and encountered the agropastoralist societies dwelling on the Kalahari's margins, missionaries linked the supposedly degraded state of dryland environments to the supposedly degraded state of African societies.[99] It was a short step to holding Africans responsible for the region's aridity. And, because it was generally accepted by white settlers that South Africa's Black farmers had only arrived in the seventeenth or eighteenth century—approximately the same time that Europeans arrived in the Cape—it was fairly easy to shift responsibility from white settlers to Black ones.

The missionary turned botanist Brown merged these older and newer narratives, arguing that deforestation and pasture burning had been practiced by African farmers since time immemorial, but that Europeans had continued and expanded them.[100] These ideas were restated and elaborated in the years before Schwarz wrote his own book. In 1916, for example, the Eastern Cape forester James Sim reiterated that while the Khoikhoi herders in the Cape had done little to damage the environment, "the Bantu races coming from the North-East . . . began the systematic destruction of the forests and the ruination of the grass velds by the burning of the grass. The Dutch and English settlers followed both practices."[101]

It is worth pausing to ask, as some historians have done, how many people outside the rarified world of British geographical society members and government scientists paid attention to the writings of people like Wilson, Brown, and Sim. But these authors did not write in an intellectual or political vacuum. Their ideas reflected broader intellectual currents in white rural society. And they influenced popular writers—including Schwarz—who then served as conduits back to the general public. As the twentieth century unfolded in South Africa, these two narratives—of environmental harm unleashed by the hands of either Europeans or Africans—coexisted uneasily. Colonial regimes across the continent regarded Africans as poor environmental stewards, and blamed them for environmental degradation. But in South Africa, the concept of white responsibility remained an important part of both scientific and popular discourse, filtered through the lens of class and ethnicity. Indeed, as chapter 3 will discuss, when the government of the new Union of South Africa investigated problems of agrarian crisis and drought, it was precisely the actions of white farmers—particularly poor and "backward" white farmers and Afrikaner pastoralists—that it and many of its witnesses highlighted. A unitary white identity was still very much under construction, and scientific experts moved seamlessly between blaming Africans and uneducated whites for the country's environmental ills.

But Ngami, far from centers of white settlement, and changing on human, not geological, timescales, seemed to invite explanations rooted in African misdeeds. At almost the moment that the lake became dry, stories circulated that local people had caused the change by engaging in the same kinds of "folly" that James Fox Wilson had cataloged for non-European societies around the world. One story recounted how abandoned papyrus rafts had cut off Ngami's water supply when the papyrus stalks rooted in the channel that fed the lake, forming massive floating islands of vegetation. A man who had traveled periodically in the Kalahari and its environs in the late nineteenth century shared one version of this story, told to him by white hunters:

> Thousands of natives used to do a great trade in ivory, grain, etc., with a big chief living on the Lake N'Gami, and used the "Okovango" as a water way, making huge reed rafts and floating themselves and produce down the river. Near the Lake was an estuary where the rafts lodged, and were deserted. Thousands of rafts collected and ultimately formed a weir or barrage across the river, diverting its flow in a northerly direction into great swamps, finally finding a new channel into the "Chobe," and thence down the Zambesi.[102]

Another version of this story blamed a local ruler for blocking the lake. Carl Weidner, an Orange River farmer who claimed to have traveled to Ngami, reiterated the popular view that there had once been "three great seas" in the northern Kalahari, "along the banks of which many natives, tillers of the soil, lived in opulence from the fruits of their agricultural enterprise." But he rejected the idea that geological or climate changes had caused these lakes to vanish. In a typical settler telescoping of African history, Weidner argued that a mere forty years earlier, "marauding Bechuanas" had arrived at Ngami's shores and begun seizing local residents to sell to "Arab slave hunters." At the same moment, profligate Boer hunters showed up and decimated the region's game. (As will be discussed in chapter 5, in the minds of many white South Africans, mobile "Boers" managed their environment nearly as thoughtlessly as did Africans.) Moremi, "chief of the Natives north of Lake Ngami"—Weidner appears unaware that Moremi himself was one of the "marauding Bechuanas"—sought to preserve his sovereignty by driving out the interlopers through a process of ecological warfare. Moremi had trees felled in order to dam the Kavango River above Ngami, thereby cutting off the water supply. The chief succeeded in his goal of driving outsiders away, but he also inadvertently flooded his own kingdom, turning "a once flourishing and productive countryside into what is today an unhealthy and miserable swamp."[103]

The stories that whites told about African misdeeds were more than simply racist fantasies. Many had their origins in African intellectual histories and political critiques. Almost invariably, white listeners were oblivious to these larger messages, even when they were the subject of the critique.[104] European ideas about rapid desiccation were reinforced by "native traditions" that described wetter conditions around the Kalahari a mere generation or two earlier.[105] The hunter Chapman was told by people along the Botletle River, flowing out of Ngami, that "the country is dead."[106] But, for the people living in the lands frequented by missionaries, hunters, and traders, it was the newcomers who were responsible for this state of affairs. Missionaries across southern Africa observed that their prayers, singing, and church bells were held responsible for reduced rainfall. People also argued that the activities associated with hunters and traders—the movement of wagons, the firing of guns, the killing of certain animals, and the importation of strange products into the country—caused droughts.[107] The stories that purported to explain Ngami's disappearance contained similar critiques. They targeted an increasingly hierarchical African polity, and possibly the white hunters and mixed-race slave traders who helped sustain its centralizing power. The Okavango Delta witnessed momentous political change in the nineteenth century—changes recognized, in muddled form, in Weidner's story. Tswana speakers known as the BaTawana began settling along Ngami's shores in the early nineteenth century. Between the 1840s and the 1880s, a small Tawana elite, fortified with firearms, expanded its control over the delta's fishers and farmers, reducing them to a subordinate status that many outsiders likened to serfdom. As Passarge recognized in 1896, the goods being transported to the chief on papyrus rafts were agricultural products demanded as tribute. Those who hauled the fruits of their harvest to the Tawana royal house abandoned the rafts because, of course, the Tawana gave them nothing in exchange to take home. The story that attributes the disappearance of Ngami to abandoned papyrus rafts thus revolves around enforced tribute and its larger consequences. The other version of the story, which claimed that the Tawana ruler had intentionally blocked the channels that fed the lake, also appears to question the legitimacy of Tawana rule. As outsiders to the region, and arrogant ones that that, the Tawana rulers destroyed a valuable ecological system that sustained human life.

When such stories moved into white settler society, their function changed. They reinforced the racial-ecological discourses that would form a steady undercurrent in the debate over the Schwarz scheme. Schwarz was not insulated from this settler intellectual world. In his book, he never suggested that people were responsible for Ngami's disappearance; like Livingstone and Chapman, among others, he attributed it to seismic or tectonic changes that had sent the water in another direction. But in a publication in 1925, he repeated one of these Moremi stories. His pivot may indicate how his interactions with others, as he sought support for his scheme, introduced him to new ideas; or perhaps it demonstrates his embrace of settler discourse in order to win popular support.

By the 1930s, popular knowledge had shaped expert discourse. In 1933 a man critical of the feasibility of Schwarz's scheme wrote to *Farmer's Weekly* that the country where the rivers would be diverted "must be very flat, and the rivers slow running, or else the Natives could not in the year 1886 have cut off the river which fed lake Ngami."[108] A man who participated in a 1937 survey of the Okavango Delta told the Pretoria Rotary Club, "The drying up of Lake Ngami was, he thought, largely due to the long-continued Native occupation of the country bordering its main feeder, the Taoge River, and the subsequent man-made drought conditions which had been induced."[109] Legend was becoming scientific fact. In the survey's 1938 report, the author—a civil engineer named John L. S. Jeffares—wrote, "Doubtless there is truth in the legends that former chiefs have made obstructions with a purpose."[110]

Stories of such distant places might initially have been traded among small groups of hunters and traders. But the fact that these stories were so widely known and accepted by the late 1930s indicates how a place most white South Africans had never seen became entrenched in both scientific and popular ideas about environmental change.[111] The stories found their way into print, and were filed away in the minds of readers. Later, they were spread through conversations among friends, neighbors, and acquaintances, including those who could not read. Some people later remembered the stories and quoted them in new public forums.

While evidence of the oral circulation of these ideas is elusive, the print sources sometimes reveal glimpses. The story about the papyrus rafts carrying tribute, which appeared in an agricultural journal with a relatively small circulation, was recounted fifteen years later by William Charles Scully in an article he published in the Kimberly *Diamond Fields Advertiser* on South African desiccation.[112] In the same year, a version of the story blaming an African leader for the loss of Ngami was repeated on the floor of the South African Parliament—but now with would-be white saviors added to the cast of characters. Andries de Jager, an

Afrikaner representative from Paarl, told his colleagues that Ngami had dried up because a chief had rejected the advice of two white men that he clear silted channels. The blocked channels created pools of water that attracted game, which made for easy hunting. But the chief paid a price for his focus on short-term benefits: "The water holes dried up, the game disappeared, and the king had troubles with drought."[113]

Life in the Field

While Schwarz was clearly familiar with local lore about Ngami, establishing the influence of foreign scholarship on his eventual scheme is difficult, because he was remarkably sloppy about citing his sources. But his reports for the Cape Geological Survey reveal that he read the work of scientists engaged in debates over the histories of arid lands. In one report, he cited work by the geologist John Gregory as well as that of Ferdinand von Richthofen, a mentor of Siegfried Passarge, the first professional geologist to explore the Kalahari. And he later wrote about theories of planetary formation, which had been important to the thinking about worldwide desiccation. But in Schwarz's first decade in South Africa, these scholarly debates were probably not at the forefront of his mind, even if he was aware of them. His job kept him in the field for long stretches of time. He sent letters and photographs of his Karoo adventures to his mentor, John Wesley Judd, back in London. Few of the letters survive, but the photos give a taste of Schwarz's life as a geological surveyor. In one, he reclines under a rock overhang (figure 1.5); in another, he excavates the fossilized remains of a Permian-era reptile—indicating that his interest in paleontology thrived despite the demands of his job.

Schwarz's first decade in South Africa proved a tumultuous one. As tensions rose between the Boer republics and their British residents, the rinderpest epizootic swept into southern Africa, killing perhaps 90 percent of the region's cattle between 1896 and 1898. Recurrent droughts compounded economic hardship. When Schwarz surveyed the Nuweveld, the region was in the grip of a drought so severe that Schwarz could not complete his work because the local farmers "were utterly unable to accommodate me and my horse."[114] As Schwarz conducted his geological surveys amid these environmental crises, evidence of an apparently wetter past caught his attention. In an article for the *Cape Times*, he wrote that the fossils he found in the Swartberg indicated that the area had once been a sea.[115] In an 1897 report from the eastern winelands region near Cape Town, he observed the remains of an ancient river that had once flowed westward into the interior.[116] In another report from the same year, Schwarz described the rivers of the Nuweveld Mountains, a range

Figure 1.5 Schwarz in the field. Source: Judd papers, Imperial College, London.

that runs parallel and to the north of the Swartberg. Schwarz was again supposed to be looking for coal, but it was water that caught his attention in the drought-ravaged land.

More than twenty years later, Schwarz recalled this trip to the Nuweveld Mountains as his initial inspiration for the Kalahari Scheme. It was, he said, the place where he had first imagined “a project to weir up all the rivers” that rushed down the mountainside into the plains to the south. In his book, he wrote that he had realized at the time that creating “an almost continuous line of great dams” would “create favorable seasons” and the same virtuous precipitation cycle that was central to his Kalahari Scheme.[117] But in his initial report on the area, Schwarz made no reference to a climate-engineering scheme—referring only to the possibility of trapping in reservoirs water that “runs to waste,” in order to irrigate the surrounding land. He wrote of the “grand sight” along the southern face of the Nuweveld Mountains, where “the rivers are seen running in deep narrow gorges, recalling the celebrated canons [*sic*] of Colorado.”[118] Schwarz had never seen those American canyons, but the image of water “running to waste” in the desert was, by the end of the nineteenth century, a shared one across much of the settler colonial world. Indeed, the year after Schwarz wrote this report, a resident of Graaff-Reinet, 150 miles to the east, had a remarkably similar vision. Observing the steep gradients of the rivers that flowed out of the Karoo's mountain ranges toward the coastal plain, English-born William Roe advocated "saving our rain from rushing seaward" by filling the gaps in the mountains formed by erosion in order to prevent "our country draining and drying into a desert."[119] Schwarz's and Roe's observations evoked the paradox posed by these drylands, with their parched landscapes amid evidence of abundant water. But they also contained a critique: that the indigenous inhabitants had failed to make these places productive, and that they were implicitly responsible for the desolation.

The two men insisted that an entirely different future was possible. Roe wrote that slowing the flow of the rivers would have the effect of "ameliorating our dry climate and changing the whole aspect of the surroundings."[120] Schwarz suggested that the vast plains of the interior "all discharge their water through one or two points which, if dammed up, would hold vast quantities of water, and be available for the irrigation of the rich phosphatic lands to the south."[121] The two men were surveying a dry landscape very different from their childhood home—one with the eye of a photographer, and the other with that of a field geologist. But both were dreaming of a subcontinent covered in water.

2 * The Origins of Rain

In the early twentieth century, competing sides in the debate over planetary desiccation operated from a shared understanding that aridity was not a permanent condition. The planet's climate, including its rainfall regimes, had changed significantly. But had such changes occurred in historical or geological time? Why had the drylands formed, and what had happened to their water—water that had once filled the empty lakes and channels, supported the life forms now preserved as fossils, and smoothed the rocks and shorelines? Could anything be done to restore prior conditions? These questions were rooted in emerging understandings of the hydrological cycle and the capacity of humans to change it—understandings that reflected both scientific and popular engagements with aridity. The answers are central to understanding why Schwarz's solution to South Africa's desiccation made sense to so many people.

Two years after Schwarz imagined damming the rivers of the Nuweveld, the South African War, also known as the Anglo-Boer War, began. In October 1899, just days before the war started, Schwarz wrote to Judd about a surveying trip to the border of the Boer-controlled Free State: "We had an exciting time of it towards the end of our trip as we were in sight of the Free State, and war was expected every day."[1] In a later letter to Judd, Schwarz recounted being injured in a crash while on a train filled with war refugees. But he remained otherwise removed from the main theaters of battle, and spent the three years of the war much as he'd spent the prior three, surveying the geology of the Karoo.

A South African View of the World

The war's end brought forth a unified South Africa, with the Boer republics of the Transvaal and the Orange Free State and the British colonies of the Cape and Natal melded into a single British dominion in 1910. An all-white franchise remained in all but the Cape Colony (where voting rights based on property and educational qualifications offered limited representation to Africans), disappointing many Africans who had fought on the side of Britain. Afrikaners were the majority of voters in the Union of South Africa, and they elected a Boer War general, Louis Botha, as the first prime minister. (Another Afrikaner general, Jan Smuts, became the second.) Schwarz now lived in an expanded polity that encompassed a larger variety of climates and geological features. It also included many more white farmers, although agriculture in the Transvaal and Orange Free State had been severely disrupted by British scorched-earth tactics. The years after the war were ones of contestation. Africans used the chaos unleashed by the war to reassert land claims, and were again disappointed when Britain took the side of white farmers, who gradually rebuilt their farms and reestablished dominance over a Black labor force.

The war's end also brought changes to the Geological Survey. Corstorphine left the Cape to become a consulting geologist on the gold mines in the Transvaal, and Rogers took over as head of the survey. A new geologist joined Schwarz in the field: Alexander Logie du Toit, who would later become one of the Kalahari Scheme's most vocal critics. Unlike the survey's other members, du Toit was an Afrikaner, born and educated in the Cape. While many of his ethnic compatriots were fighting the British during the South African War, however, du Toit studied mining engineering in Glasgow and geology at the Royal College of Science in London. He married an Englishwoman and in 1903 returned to the Cape, where he eventually won the respect of his colleagues around the world.[2] In the 1920s, when Schwarz was defending his Kalahari Scheme before a critical scientific establishment, the head of Harvard's geology department called du Toit "the world's greatest field geologist."[3]

Yet, like Schwarz, du Toit became a proponent of a geological theory that threatened to tarnish his professional reputation. The theory was continental drift, and it was controversial from the time it was proposed by the geophysicist Alfred Wegener in 1912 until after du Toit's death in 1948. Many American and European geologists ridiculed it, and Schwarz rejected it. But it proved to be du Toit's greatest legacy to geology.[4] The fact that du Toit's most important idea lay outside the mainstream throughout his career, despite his stature in the field, might prompt us to pause and look more sympathetically at Schwarz's ideas, which appear outlandish today. It was Schwarz's later intransigence in the face of mounting evidence against his scheme that marked him as a maverick within his profession. But he developed his initial theories at a time when no one had definitive answers to basic questions about southern Africa's climatic and geological past. And his insistence that Northern Hemispheric understandings of geology could not be blindly applied to the rest of the world resonated with colleagues like du Toit, who also struggled against the Northern bias of the field. These men were among the first generation of scientists who viewed the world from the vantage point of South Africa, rather than fitting South Africa into a world conceived in Europe and North America, and it caused them to challenge a range of accepted truths. Paleontologists who found early hominid remains in South Africa suggested that *Homo sapiens* had originated on the continent—an idea that, like continental drift, initially was ridiculed in Europe. The first South African geologists tried to fit local rock formations into European typologies. When this failed, they began coining distinct names for local formations, and began to realize that southern Africa had important contributions to make to the field. Southern African rock formations were far older than those found in Europe, and they bore similarities to those found in India, Australia, and South America.[5] Du Toit's theory, which aligned with that proposed by Wegener, was born out of his perspective as a geologist working in the Southern Hemisphere, including the work he did in the field with Schwarz.[6] Schwarz's career developed as the role of South Africa's scientific community in the world was changing from one in which scientific papers were "reports on South African marvels for European readers" to one in which published work became "a sincere effort to contribute to the universal body of science."[7]

Du Toit embraced life in the field, living out of a wagon with his wife and young son. But Schwarz soon tired of the peripatetic life. In 1904 he married Daisy O'Halloran, and shortly thereafter accepted a position teaching geology and geography at the newly formed Rhodes University College in Grahamstown, a post he held concurrently with his role as head of the Albany Museum's fossil collection.[8] Grahamstown, deep in the Eastern Cape, was a microcosm of the Anglo settler explosion. When Schwarz arrived in 1904, it was one of several towns scattered across the semiarid Eastern Cape interior, each established as a military post during the nineteenth-century "frontier wars" between the British and the amaXhosa in the east. Most of the towns bore the names of colonial administrators and soldiers who had driven the amaXhosa out of their land to make way for British settlement: Craddock, Somerset, Grahamstown, Fort Beaufort. This was a frontier that had only recently emerged from turmoil; the last in a series of wars between the British and the amaXhosa had taken place during Schwarz's childhood, and the last of the Xhosa territories had been annexed the year before he arrived in South Africa. By the turn of the century, these former frontier towns were the center of an expanding commercial agricultural sector focused on wool and ostrich products. Grahamstown was the largest of these Eastern Cape towns. A description of the city in 1912 noted its botanical garden, its tree-lined streets illuminated with gas lamps, and the expanding white residential areas creeping up the hillsides. It was, a historian notes, "a large and sophisticated town" by South African standards, with fourteen thousand white and Black residents.[9] These Eastern Cape towns were part of the Karoo, and they became centers of support for Schwarz's scheme, their residents' enthusiasm driven in part by farmers' vulnerability to both drought and the boom-bust cycle of the wool and ostrich markets. But Grahamstown had a more congenial climate for someone born in Britain. It received rainfall in all four seasons and, though in the twenty-first century it has experienced severe water scarcity, it was described a century ago as having comparatively dependable rainfall.[10]

It was from this well-watered perch at the edge of the Karoo, in the comfort of a university appointment in a town with many of the trappings of British life, that Schwarz would come to focus his attention on how to fend off climatic and racial dystopia. But there is little evidence that Schwarz dwelled on such perils during his earliest years in Grahamstown. During his first decade at Rhodes, he used the relative freedom of his academic post to explore controversial theories within geology, archeology, ethnography, and cosmology. His choice of topics reveals a man of high energy, insatiable curiosity, and a willingness—to a fault, most of his colleagues agreed—to embrace unorthodox views. He argued that some of the region's African cultures had been influenced by Viking and Germanic peoples while others were descended from Malays or Chinese. He became an avid proponent of Thomas Chamberlin's now disproven planetesimal theory of solar system formation. He suggested that the Chinese could have built the archeological site of Great Zimbabwe. (At the time, many scholars and rank-and-file whites argued that the stone complex was the work of Arabs or Phoenicians.) He wrote a newspaper article opining on the unlikelihood of sentient life on Mars. He wrote in an American geology journal that there had once been a land bridge linking South America and Africa, and he sought to provide geological evidence that another land bridge lay submerged under the Indian Ocean—a theory linked to the hypothetical "lost land" of Lemuria, which had been popularized by enthusiasts of the occult.[11]

Schwarz combined these forays into the unconventional with more mainstream intellectual pursuits. He set up the University College's telescope in his yard, named his home Observatory House, and invited Grahamstown's residents to observe the heavens.[12] He helped prove that South Africa had experienced repeated glaciation.[13] And he weighed in on the proposed border agreement between South West Africa and Angola, informing the League of Nations that the Portuguese captured slaves from southern Angola to work on cocoa plantations on São Tomé and Príncipe—an accusation that drew an outraged response from the Portuguese and their South

African business allies but proved among the more correct of Schwarz's many theories.[14]

Other work prepared the intellectual ground for the Kalahari Thirstland Redemption Scheme. In 1906, Schwarz puzzled over the pattern of southern Africa's watersheds and postulated a previous hydrological system, when "the rivers were very differently arranged from what they now are."[15] He located the causes of this changed hydrography in South Africa's topography: its high interior and coastal mountain ranges. Rivers that flowed south had steep gradients as they plunged off the interior plateau toward the coast. The sheer force of their currents caused erosion in their headwaters, extending their courses upstream until they "captured" the rivers that sluggishly crossed the flat plains of the continent's interior.[16] In the future, similar river captures would continue to remake South Africa's watersheds, Schwarz argued. "The coast streams are stealthily creeping inland," he wrote, arguing that they would eventually incorporate the upper tributaries of the Orange River, draining still more of South Africa's center.[17]

Understanding the watersheds was crucial to the economic future of the new Union of South Africa, Schwarz argued. "The great want of South Africa is water, and the most obvious way of procuring a supply is to dam up the rivers so as to catch the flood waters," he wrote in 1906. But stopping the coastal rivers' water was not practical, given their steep gradients and narrow valleys. In the interior, it would be easier to build dams but harder to keep the water, thanks to "the enormous evaporation that takes place in our half-desert country."[18]

Schwarz's observations were part of a larger intellectual context: a shared white South African belief that the subcontinent's water was "running to waste." This phrase, repeated endlessly by those who lamented the ravages of drought and the unproductivity of agriculture, had its own imperial history. In a study of colonial hydrology in India, one historian argues that "the moral imperative of irrigation engineering was defined in relationship to the natural tendency of water, like energy, to run to waste."[19] In Australia, a man proposing white settlement in the arid interior wrote of its seasonal rivers, "Even in the driest years, millions of gallons flow to waste."[20] Proponents of a "white man's land" in various settler colonies argued that existing water resources were underutilized, an argument rooted in "normative Anglo assumptions of productive lands and watered environments."[21] These assumptions were shared by other European colonizing powers, who justified their exploits around the world in part by citing Indigenous people's supposed failure to make productive use of their lands. As hydro-engineering capabilities improved, the "waste" of water became part of this concept of resource neglect. "Waste" came to have two meanings. It described arid lands: in English, "waste" was an older term for desert (and it remains the dominant word for desert in Afrikaans and German). It also came to describe the natural flow of nonengineered rivers. Those charged with preventing the waste of water were members of a colonial network of engineers who were trained in the same institutions and moved around the imperial world. In the British Empire, India was the hub of this colonial hydrology, and many of the men who worked in South Africa's irrigation department had studied and worked there, including both directors of irrigation between 1908 and 1941.[22]

In South Africa, this ideology of wasted water confronted the physical reality of arid landscapes. The behavior of water in arid lands was a continual source of amazement to those who witnessed it. Written accounts are replete with astonished descriptions of dry channels suddenly transformed into rain-swollen rivers—and dismay at that water's equally rapid disappearance. Southern Africa appeared to have enormous water resources that were being squandered. A civil engineer called for the nationalization of the water supply, and argued that the amount of water "which runs to waste into the sea unused" was "incredible."[23] The treasurer of a church-run irrigation settlement told the 1920 Drought Distress Relief Committee of a river that "comes down in an enormous stream, and you cannot cross it sometimes for days. Yet all that water, or most of it, goes straight into the sea."[24] A Johannesburg man argued that the government "must find a way of making provision to conserve the rain-water that runs to the sea and is wasted." Lack of water reduced food production, and thus the wealth of the country: "We must stop this drainage of money."[25] A fellow resident of the city lamented the "millions of gallons of water" that flowed even through small gullies during summer storms, "allowed to escape to the ocean."[26]

Water—in the heavens, on the ground, and under the earth's surface—was a constant preoccupation for those who were claiming and settling lands seized from Indigenous communities. In 1912, Schwarz was hired by the Cape Geological Survey to write a report on the water resources of the town of Uitenhage, 130 kilometers southwest of Grahamstown. Schwarz cautioned that drilling for additional water might interfere with the flow of existing springs, a conclusion supported by his former boss Arthur Rogers, who emphasized that the commission could not support additional boreholes.[27] Rogers's cautious oversight may have restrained Schwarz's conclusions; it also seems to have left the town dissatisfied. Two years later, the town hired Schwarz directly, as a consulting geologist. Working independently, Schwarz was less cautious. While he and Rogers had earlier stated that all the water sources in the basin were linked, so that tapping one could deplete others, now he argued that the springs on this new piece of property could be exploited without jeopardizing the supply of neighboring farms. In a passage highlighted in pencil by a reader, Schwarz wrote, "In fact the more water is brought to the surface by boreholes the more will be the inflow into the area from the underground sources. Most of the underground water flows uselessly to the sea and escapes on to the floor of the ocean."[28] Such confidence about the workings of the area's subterranean hydrology was a marked departure from the report supervised by Rogers two years earlier that claimed that it was impossible to establish the path of underground water resources. It expanded the idea of water "running to waste" to include that which lay underground, and offered a glimpse into Schwarz's germinating idea about the potential for human intervention to augment, rather than diminish, the water supply.

Lakes and Rain

This idea of wasted water had a greater significance than the simple loss of a valuable resource. For many South Africans, part of a pan-European intellectual world, water on the surface of the land or even underground had a direct relationship to the quantity of water vapor in the atmosphere and therefore to the potential for rain to fall. The idea of water running unused into the sea came to intersect with larger ideas about where rain came from and what could explain its absence.

The experiences of those who were claiming and settling the world's drylands were framed in part by the intellectual traditions they brought from Europe. One of the most well-known is the belief that forests increased rainfall. This idea dates at least to the ancient Greeks, but it coalesced in its modern form in the mid-eighteenth century, as British, Dutch, and French colonists observed the apparent impact of colonial settlement on the climates of tropical plantation islands. Naturalists and scientists made a case for forest conservation based not only on the economic and medical utility of vanishing resources, but on the grounds that large-scale deforestation was causing a decline in rainfall.[29] It was a short step from seeing recent deforestation as responsible for desiccation to viewing arid environments as the product of past deforestation, particularly given European propensities to assume that forests were the natural state of most undisturbed land. The result was that deserts and drylands went from being "simple facts" to human-made aberrations, or "desiccated former forests that must be rescued and made forests again."[30]

In the Brazilian *sertão*, a period of relatively high precipitation coincided with the rush of settlers into the region. Drought followed in

1845, leading to a popular theory that deforestation had caused declining rainfall. A campaign to conserve forests followed, born of a belief that reforestation was "the route back to the mythical once-verdant *sertão*."[31] In the United States, people suggested that a lack of trees accounted for the aridity of the Western regions.[32] In the context of nineteenth-century imperialism and scientific racism, Indigenous populations were increasingly blamed for this state of affairs. John Wesley Powell, famed surveyor of the "arid lands," held Native Americans responsible for the absence of trees in parts of the West where he thought they should grow.[33] George Perkins Marsh, an early American conservationist, believed that forests had covered most of the "habitable regions" of the earth and had disappeared at the hands of humans.[34] This was true even of those arid lands most familiar to Europeans. "I am convinced," Marsh wrote, "that forests would soon cover many parts of the Arabian and African deserts, if man and domestic animals, especially the goat and the camel, were banished from them."[35] Such ideas shaped French colonial policy in North Africa, where officials argued that what had once been "the granary of Rome" had been destroyed by Indigenous residents. They created projects to plant trees, and passed laws to restrict indigenous Algerians' access to forest resources.[36] In West Africa, French and British colonial officials alike worried that the destruction of forests, particularly on the southern margins of the Sahara, was reducing rainfall and causing the expansion of the desert.[37]

Although they shaped colonial policies, such ideas were controversial. Even Marsh admitted, "The effect of the forest on precipitation . . . is not entirely free from doubt."[38] And Diana Davis herself observes, "It is notable that in the writings of the few natural historians, geographers, and serious explorers of these deserts, the declensionist narrative of deforestation and desiccation is not commonly found."[39] She argues, however, that dissenting voices "almost never had any influence on policy development or implementation."[40] But while elite discourses of forests and deserts might have shaped the aspirations of many colonial administrators, in most places the colonial state had far less power in practice than it had on paper. Conservation legislation was often unenforceable and sometimes fueled revolt among colonial subjects. In South Africa, white settlers' voting rights constrained unpopular conservation policies aimed at them. Indeed, the popularity of Schwarz's scheme, resting as it did on the idea that geological processes were the cause of declining rainfall, suggests widespread dissent over the idea that human-led deforestation was a primary driver of desiccation.

The decoupling of historical and geological time, as scientists came to accept that the earth was much older than once thought, opened up new possibilities for understanding the origins of aridity as something "independent of the will of man," as Kropotkin put it. The northern border of the Sahara, long associated with Roman glory in parts of Europe, might lend itself to the idea of a landscape desiccated due to human carelessness, but other contexts were less accommodating of this idea. Australia's interior, for example, seemed to offer little evidence of human impact on the landscape. And Blanford, suggesting that Persia had once sustained much larger populations, wrote, "Some alteration [in the climate] may be due to the extirpation of trees and bushes, the consequent destruction of soil and increased evaporation; but this alone will scarcely account for the change which has taken place."[41] Those who later argued for planetary desiccation embraced Blanford's reasoning and argued that, while deforestation could be a contributing factor, the scale of the changes demanded other explanations and time horizons.[42]

Conceiving of the earth's surface as dynamic opened new ways to explain the origins of deserts and drylands. While Davis focuses on French officials who insisted that the Sahara's aridity was the product of human action, others, as we saw in chapter 1, suggested that the chotts had dried out when blowing sand had filled the channels that linked them to the sea, thereby cutting off their water supply.[43] The half dozen men who commented on Kropotkin's paper in 1904, and who between them had visited most of the world's continents and a wide variety of its arid lands, broadly agreed that explanations for desiccation had to be sought in geological forces. The elevation of land was a popular theory—one proposed in 1873, the year of Schwarz's birth, by Blanford. He suggested that rising land in Central Asia had caused water trapped in interior basins to drain away toward the seas.[44] Thirty years later, two of the men commenting on Kropotkin's paper also suggested that the elevation of land could explain desiccation, either by creating a barrier to oceanic winds carrying moisture or by causing coastal rivers to flow faster and deepen their channels toward interior water sources, "by means of which the water escapes from the land, so that the land has become dried up." This was the explanation some in the audience offered for the shrinking of lake levels in the Andes and "profound" desiccation in western North America, and it was one that Schwarz echoed in his 1906 essay on South Africa's hydrology.[45]

Mountains were well known to be precipitators of moisture, and people recognized that their inland slopes often lay in rain shadows. Less clear was how the loss of interior water could lead to desiccation. Some scientists said it couldn't. Francis Kanthack, first director of South Africa's Irrigation Department, was one of them. Like Schwarz, Kanthack was the British-born child of German immigrants. Educated in England and Germany, he had worked as an irrigation engineer in India before coming to the Cape Colony in 1906. He told a committee examining the problem of drought in 1914 that any apparent desiccation that was occurring in South Africa was "caused indirectly by human agency"—primarily mismanagement of land and water resources. In response, the committee's members asked Kanthack to explain Ngami's disappearance. He responded that he had "not been able to form an opinion" about why the lake had dried up. But he insisted that Ngami was an isolated case, not a sign of impending doom.[46]

This response satisfied few people. In the context of the many dry riverbeds farther south, in the Kalahari and its margins, Ngami's disappearance fed into a growing sense of alarm at the apparent changes in the South African climate. The parliamentarian de Jager insisted that Ngami's disappearance had caused water holes to dry up, and that this somehow had unleashed drought conditions. The "lost lake" was not only representative of larger changes; it was a *driver* of those changes. But how could the draining away of a lake result in decreased rainfall?

The answer to this question is found in the same European intellectual tradition that linked deforestation to deserts. It held that bodies of water, like forests, generate rainfall through the water vapor they release into the air. Historians have paid far more attention to the idea that forests affect rainfall than to the idea that bodies of water do.[47] In part this is because the former lay within human control: Before the late nineteenth century, planting trees was a more practical intervention than was creating bodies of water. But the two ideas shared the idea that "vapors" influence climate—an understanding that was modified in the nineteenth century to accommodate new understandings of the role of evaporation and condensation in the water cycle. The idea that land-based water, like forests, played a role in local rainfall seems to have been common sense long before it was scientifically investigated. It was an assumption so basic almost no one felt the need to state it.

The German naturalist Alexander von Humboldt, the early nineteenth century's most famous scientist-explorer, is well known for offering a scientific rationale for the link between deforestation and desertification. But initially he focused on the role of surface water, not trees. The first edition of his most influential work, *Views of Nature*, attributed the existence of the Sahara to hot winds and "the absence of larger rivers, of lakes, and of high mountains."[48] Humboldt was surely influenced by geographers like Rennell, who suggested that the Sahara had once contained a large inland sea. But in his second edition and in subsequent English translations, Humboldt amended this sentence, replacing "lakes" with "forests that generate cold by exhaling aqueous vapour."[49] Yet the connection between lakes and rainfall remained common sense for a long time.

The idea that rainfall could be generated from land-based moisture may have originated in Europe, but it had little practical relevance there. Indeed, it was popularly believed that draining swamps and cutting down trees *improved* climate—a belief British settlers brought to eastern North America, and one that was rooted in the fact that many parts of Europe suffered from too much moisture, not too little. The possibility that rainfall could be derived from surface water took on new significance for those encountering empty river channels and lakebeds in the world's drylands. The idea that the loss of these lakes and rivers had reduced rainfall offered an origin story that made sense in places where there was little evidence of deforestation.

Standing near the Salton Trough in the 1860s, Joseph Widney described the desert as "a serious disturbing element in the climate of southern California," whose "withering blasts make forays upon more favored territories around." He imagined a very different past: The desert basin as a gulf, "its waters shallow and easily heated . . . a steaming cauldron, keeping the air-currents above constantly saturated with moisture."[50] In Australia, the geologist and botanist Ralph Tate also saw changes in the interior's surface water as a self-reinforcing cycle. "The replacement of the arid plains by freshwater seas, though a consequence of an amelioration of climate, must have had a reciprocal effect, conversely as the existence of a dry zone produces aridity and thereby intensifies the effect."[51] In Africa, Passarge argued that much of the continent's rainfall derived from evaporation in the massive Congo River basin. Diminishing rainfall due to global climatic changes of unspecified origin had caused surface waters in northern and southern Africa to begin to disappear, setting off a self-perpetuating cycle that would eventually afflict the Congo basin as well.[52]

Some of these men went one step farther. If the loss of surface water had changed the climate for the worse, they argued that restoring that water would change it for the better. Widney suggested that diverting the Colorado River to refill the Salton basin would cool the air, add twelve inches of rain per year over an area twice the size of Ohio, and allow for agriculture on the surrounding fertile soils. He saw this not as a radical intervention into natural processes, but rather as a restoration of the natural order.[53] The proposal was popular, but an 1874 report concluded that his claims for climate amelioration required further investigation.[54] Charles John Frémont, governor of the Arizona territory in the late 1870s, embraced Widney's plan, arguing that it "would cool off the entire American Southwest, changing its arid climate," and proposing to Mexican president Porfirio Díaz that they work together to create "an agricultural paradise."[55]

Widney was not the only North American to wonder about the influence of water on the rainfall of arid lands. The American conservationist

George Perkins Marsh was skeptical that forests affected rainfall, but he had no such doubts about bodies of water. In 1855, a sea captain named William Allan proposed cutting channels to link the Mediterranean and Red Seas via the Dead Sea. Allan imagined that shipping and trade would be the main benefits. But a decade later, Marsh wrote in his classic *Man and Nature* that the primary benefits would be the establishment of a precipitation cycle: "The creation of a new evaporable area, adding not less than 2,000 or perhaps 3,000 square miles to the present fluid surface of Syria, could not fail to produce important meteorological effects. The climate of Syria would be tempered, its precipitation and its fertility increased, the courses of its winds and the electrical condition of its atmosphere modified."[56] It seems an odd claim in a book now best known for its warning about the detrimental impact humans can have on the natural world. But it reminds us that Marsh wrote in the context of the late nineteenth century's burgeoning optimism about the possibility of engineering the climate.

In North Africa, both the French military surveyor François Roudaire and the British engineer Mackenzie suggested in the 1870s that the chotts could be filled with water diverted from the Mediterranean or Atlantic. "The retreat of the waters of the [Saharan] sea appears . . . to have profoundly modified the climate of these flourishing regions," Roudaire wrote.[57] Both men argued that restoring what Roudaire called the *mer intérieure* would recreate the past climate. (Marsh's later edition of *Man and Nature* included a section on the project, stating that "the climatic effects would doubtless be sensible through a considerable part of Northern Africa, and possibly even in Europe.")[58] Roudaire secured the support of Ferdinand de Lesseps, who had recently overseen the completion of the Suez Canal, and countered objections that a large lake in the Sahara would evaporate and become a massive salt pan by arguing that the reestablishment of prior rainfall patterns would prevent this.[59] The newly constructed Suez Canal had proven that bodies of water could shape the climate, he wrote, because rainfall had increased on the land around it. The implications for the Sahara were dramatic: If a narrow canal could increase rainfall, a "vast gulf" recreated in the chotts could have a much greater impact.[60]

A scheme to flood Lake Eyre in Australia by diverting water from Spencer Gulf was suggested in the 1870s and rejected by the Australian Parliament as too expensive. But the idea survived. A letter to the editor of a newspaper in 1892 suggested: "The whole climate and character of Australia from north to south may someday be destined to be thoroughly changed, turning what are arid deserts into a huge inland sea, surrounded by cultivated shores, and extending over hundreds of miles of now useless inhabited country."[61] And in northeastern Brazil a spate of reservoir building in the 1860s, motivated by the need for a secure water supply, was followed in the early 1870s by historic rainfall. In spite of the severe drought that followed several years later, many people associated those heavy rains with the construction of the reservoirs.[62] Although most of these schemes focused on creating lakes, any application of water to the land's surface could theoretically increase precipitation. John Wesley Powell, lauded now for his ecologically sound vision of the US West, predicted in 1888 that bringing irrigation to "lands that now present the desolation of deserts" would allow the evaporation of water that would otherwise run to the sea,

> and the humidity of the climate will be increased thereby. . . . As the general humidity is increased, the moister air, as it drifts eastward in great atmospheric currents, will discharge more copious rain. . . . As the lands gain more and more water from the heavens by rains, they will need less and less water from canals and reservoirs.[63]

In the latter decades of the nineteenth century, a scientific foundation for the concept of reprecipitation was laid—almost inadvertently—by two men who had no interest in drylands or desiccation. In 1887 the naturalist John Murray published an article that attempted to quantify what proportion of rain falling on land ran to the oceans, and what portion evaporated into the atmosphere. The Scottish-Canadian Murray was one of the founders of the field of oceanography.[64] He had been part of the 1870s *Challenger* expedition, a four-year endeavor to explore the world's oceans. In subsequent decades he continued his marine research from Scotland and oversaw the publication of a fifty-volume report on the expedition. Murray's interest in the question of rainfall derived from his interest in the oceans: he sought to "estimate what portion of the rain that falls on the land finds its way back again to the ocean by means of rivers," in part because he wanted to quantify the solids that rivers deposited on the ocean floor.[65]

Murray's data was certainly imperfect. He used a world rainfall map created by Elias Loomis, a Yale professor of natural philosophy with an interest in meteorology, in order to calculate mean rainfall in various river drainage basins, on continents, and over ten-degree bands of latitude. He also relied on existing measurements of river flow, data whose inaccuracies mirrored the unevenness of European knowledge of the world. As a result, Murray overstated the importance of the rivers Europeans knew best, and underestimated the flow of many that were less well known.[66] His deceptively precise conclusion—that 1/4.499 of the rain that fell on land ran to the oceans—was thus built on a numerical house of cards. But it carried with it a powerful implication: that virtually all the rest of the water was evaporated into the air.

In an article titled "On the Origins of Rain," the German scientist Eduard Brückner took Murray's argument a step farther. Brückner, known for his work on glaciers and the reconstruction of past climates, is acclaimed today as one of the first to think about the social and economic impacts of climate change.[67] He had proposed a thirty-five-year weather cycle, but found that it operated differently in Western and Eastern Europe; his investigation into the origins of rainfall was an attempt to explain the discrepancy.[68] Most scientists assumed that rainfall came almost entirely from the oceans, but a few geographers, including the Austrian Alexander Supan and the Russian Alexander Woeikov, thought land-based evaporation contributed to rainfall, mainly in the form of transpiration from trees. Scientists were unable to calculate how much water vapor the earth's land surfaces contributed to the atmosphere, because there was no reliable way to measure global evaporation. Murray's calculations, combining river flow and rainfall data, offered a way out of this predicament. Brückner, simplifying Murray's figure, argued that if only 22 percent, or two-ninths, of the rain that fell on land returned to the sea, it followed that the rest returned to the atmosphere via the land itself. He concluded:

> The role of the land's surface is not passive in the water cycle; to an enormous extent it contributes to the moisture content of the air. . . . A water particle, which came to the country from the ocean through the atmosphere, falls an average of three times as precipitation, before it returns to the bosom of the ocean.[69]

Murray's and Brückner's ideas about the land-based origins of rain rippled far beyond the concerns that had given rise to them. When Passarge suggested that the African continent contained an internal climatic circulation system, he cited Brückner as well as Supan and Woeikov. The Congo basin's moisture was recycled moisture derived "from the Congo basin itself," and desiccation was happening because the Nile River carried away much of the Congo's water.[70] And by providing a scientific basis for popular belief, Murray and Brückner offered explorers and settlers a way to make sense of the aridity of the lands they encountered, as well as the mystery of why so many waterless places contained evidence of former lakes and flowing water. In 1900 a German geographer named Johannes Walther had argued that the features found within arid lands did not require dramatic changes to explain them. The strong variability of desert conditions—what Walther called the "constantly changing climatic balance of a dry climate"—in which rivers dry for many years could suddenly erupt in flood, could account for features like water erosion just as easily as a past permanent shift from a wet to a dry climate.[71] But his idea gained little traction among colonizers of drylands, perhaps because it seemed to contradict their own experiences and demanded perception of timescales so far beyond the human.

Others saw in Brückner's ideas not just an explanation of present conditions, but the promise that humans could shape the climate. When Widney and Roudaire thought about bringing water into deserts to make it rain, they were working from an understanding of the origins of rain derived from long-standing folk knowledge. Starting in the twentieth century, their successors cited the research of respected scientists such as Murray, Brückner, and Supan when they argued for the feasibility of their climate-engineering schemes.[72] One of these was a Brazilian military engineer named Luiz Mariano de Barros Fournier. In the same month that Schwarz first suggested flooding the Kalahari, Fournier proposed his own climate-engineering scheme to Brazil's National Society for Agriculture. Fournier argued that damming rivers and creating a massive inland lake in the *sertão* would create a supply of water vapor that would condense over the surrounding mountains, resulting in predictable rainfall in an area notorious for its droughts. Like Schwarz, Fournier attempted to publicize his scheme in a 1920 book, although it never achieved the popularity of Schwarz's.[73] In it, Fournier drew on a combination of vernacular knowledge and scientific authority. Taking readers on a tour through the last five hundred million years, Fournier argued that the *sertão* had once been covered by what he called the "Great Lake," one of several massive lakes in what is now South America. He described the gradual diminishment of these lakes as the Andes Mountains formed and water drained away, referring to the process as an epic struggle between the continent and the sea, ultimately won by dry land.[74] Fournier cited "Brückner's law" on the continental origins of rain to explain the *sertão*'s current aridity. "The rain depends mainly on the evaporation of the waters from the surface of the earth," he wrote.[75]

Brückner in Southern Africa

Murray's and Brückner's ideas offered an alternative to understanding southern Africa's aridity as the product of foolish African chiefs or reckless white farmers. In a geology textbook dedicated to his mentor, John Wesley Judd, Schwarz repeated conventional wisdom when he focused on forests as the source of reprecipitated water vapor. But on the first page of the chapter on water, Schwarz cited Murray's calculations for the land-based origins of rain.[76] Later, he claimed that the idea for his scheme had grown out of the process of writing the textbook. Thinking about the river systems of Africa as a whole, he said, had revealed to him a common principle: "This was that the original drainage that at one time used to supply the interior of the continent, was being reft or had been reft, by the more vigorous streams on the coast."[77] But Schwarz also drew on identical schemes proposed by others, though as usual, he did not acknowledge his intellectual debts. The Kalahari Thirstland Redemption Scheme had ancestors in the form of German scientists, engineers, and farmers who were influenced by global ideas about aridity and rain, as well as by their own encounters with southern Africa's arid lands.

Passarge is the most well-known of the scientists. Trained in geography and geology, he argued for a connection between surface water and climate in his 1890s survey of the Kalahari. Most of Africa's rainfall was drawn not from the oceans but from the land's surface, he wrote, citing Brückner among others. But southern Africa's unique geomorphology mitigated against such reprecipitation cycles.

> The water supply from the pluvial period is not well preserved. . . . It is constantly diminishing. This process finds its expression especially in the noticeable decline of the Okavango marshes. But the other changes to springs and rivers could also be explained by it, even if deforestation may have accelerated the process considerably in some cases.[78]

While Passarge was conducting his survey of the Kalahari in the 1890s, another German man was familiarizing himself with what is colloquially called Great Namaqualand—the arid region of southern Namibia that is the historic home of the Nama people, from whom the name derives. Ferdinand Gessert had been born three years before Schwarz, and the two men arrived in Cape Town within months of each other. Gessert had studied physics in Berlin and, like Schwarz, had traveled to southern Africa after his graduation—but with a different goal. Upon arriving in Cape Town in 1894, he made his way north to the German colony of South West Africa. There he purchased fifty thousand hectares of land from Nama chief Paul Frederiks in the southern part of the colony. In subsequent years he bought other, equally large tracts of land nearby. Annual rainfall on his farms averaged 100 to 150 millimeters; in some years there was no rain at all. With no apparent farming experience in any environment, much less a desert, Gessert began experimenting with ranching and irrigated farming.[79]

The existing biographies of Gessert do not explain why someone with a university education chose to take up farming on the margins of the Namib Desert. South West Africa, more than twice the size of Germany, became part of Germany's overseas empire in 1884. Largely because of its dry climate, which was considered healthy for Europeans, the territory was imagined as a settler colony from the start. Historians have shown how Namibia's aridity and harshness were turned into virtues by early propagandists seeking to establish a new kind of *Heimat*, or homeland.[80] There was a powerful colonial lobby in Germany, and in 1891 the Deutsche Kolonialgesellschaft, or German Colonial Society, sponsored an essay contest on the topic "What prospects does German South West Africa offer German settlers?"[81] Military officers, agronomists, geologists, and geographers were divided on this question. Most insisted that South West Africa could never be anything other than a livestock-raising colony supporting perhaps a hundred thousand German settlers at most. But the colonization society was keenly interested in the possibility of making South West Africa into a land of white yeoman farmers, so it favored more optimistic reports.[82] One of these was written by an agronomist named Ernst Hermann in 1888, based on three journeys through the southern part of the territory. Hermann insisted that it was "anything but a desert" and could support dense agricultural settlement. He dreamed of a local cheese-making industry, tucked conifer seeds into the crevices of a cliff he passed, and opined that there were vast reserves of both groundwater and agricultural land that the local people had not bothered to exploit.[83]

Hermann may have been the inspiration for Gessert's journey to South West Africa. He highlighted the richness of the soils along the Konkiep River valley, where Gessert eventually bought his farms.[84] Upon arriving in the colony, Gessert retraced Hermann's route, following the Konkiep to its confluence with the Fish River and then to the Orange River, which formed the border between South West Africa and South Africa. Gessert was dismissive of the agricultural pursuits of the Nama people.[85] But his experience of the Namaqualand environment was beholden to their knowledge and infrastructure, from the trees and plants that indicated the presence of groundwater to the paths that led into the well-watered gorges. And unlike Hermann, Gessert's account makes clear that the sites with the highest agricultural potential were already known and occupied by Africans.

The most optimistic assessment of South West Africa's agricultural potential was offered by Edmund Vollmer, who compared South West Africa to the arid regions of Chile and Peru. On those grounds he suggested, in a report published in the year that Gessert arrived in the colony, that South West Africa could support four to five *million* Germans by using riverine irrigation for small-scale farming.[86] Few people thought Vollmer's predictions were realistic. But the German Colonial Society embraced the idea that South West Africa could be an African analogue to settler colonies that were prospering in other arid lands. In 1904 its members financed the study tour of the civil engineer Alexander Kuhn to the Southwestern United States, in order to ascertain what practices could be transplanted to the German colony.[87] Kuhn's time in the United States coincided with an African uprising against German rule and settlement, making the optimism inherent in the idea of turning South West Africa into an African California even more breathtaking. But it indicates the transnationality of a settler colonial mindset that was emerging in arid environments. Aridity was a shared condition, one to be conquered and remedied through the careful application of water resources that would allow the creation of the kinds of agricultural communities found in humid, temperate Europe.

Gessert was part of this band of colonial optimists, seeing the future of the colony to be in the production of fruits, vegetables, and grain. He purchased another farm, which he named Sandverhaar, upstream from his original farm and closer to the rail line that Germany was building to connect the coast and interior (figure 2.1). In 1902 he traveled to Egypt to

Figure 2.1 Dead camelthorn tree on Gessert's farm Sandverhaar, one of Gessert's farms. The presence of so many dead camelthorn trees was taken by many white settlers as evidence of a drying climate. In fact, their prevalence reflects how slowly wood decays in arid environments.

Figure 2.2 One of Gessert's failed dam sites

study irrigation, and with little understanding of South West Africa's rainfall patterns or geology, he returned there and began building dams.[88] But the Konkiep, like other ephemeral rivers, could rage with a violence the newcomers had not thought possible. Most of Gessert's dams failed in short order (figure 2.2), as did the railway line that ran by his farm, when swollen rivers tore away dam walls and bridge pilings.

By 1897, three years after his arrival in South West Africa, Gessert had become a regular correspondent to several publications in Germany, including the colonization society's journal and *Globus*, a geography and ethnology magazine. His early articles reflect his attempts to make sense of the climate of his new home. He noted the layers of clay that curled like dry leaves when the river's flood receded.[89] He painted a picture of the cloud bank that formed on the southwest horizon from spring through autumn, and the dry wind that blew from the cold ocean across the hot desert. Gessert described to readers in Germany how falling rain was absorbed by this dry wind before it could reach the ground. "The rainbow forms imperfectly," he wrote, "footless, no ladder between heaven and earth."[90] His imagery recalls the novelist J. M. Coetzee's claim that one of the commonest themes in white landscape writing in southern Africa is that "the earth and the heavens are separate and even sundered realms."[91]

Gessert soon understood that the infrequent rain that fell on his farm was shaped by weather patterns to the north. If it rained heavily in the densely populated agricultural societies that lived between the Kunene and Kavango Rivers on the colony's northern border, it also rained in central and southern South West Africa.[92] The colony's only perennial rivers lay on its borders, and some Germans had already suggested diverting the Kunene, on the northern border, to relieve water shortages in the Ovambo plain, where large populations promised a future labor supply.[93] But Gessert went a step further. By 1897 he was imagining how aridity itself might be abolished. A diverted Kunene would not just bring water to people, he argued; it would restore Etosha Pan to its previous grandeur, and create an evaporation reservoir that would cool the air and diminish the power of the rain-robbing west winds.[94] This would ultimately generate more rainfall across the colony.

The evidence Gessert marshaled to support this idea illustrates the range of literature he managed to acquire from his outpost on the margins of the Namib Desert. South West Africa had fewer than one thousand Germans living in it, mostly clustered in a few towns. Yet Gessert was aware that others were venturing into arid zones around the globe. He cited the accounts of visitors to the Gobi Desert—where, he argued, the disappearance of a large lake had reduced rainfall and caused advanced civilizations to collapse. Gessert contrasted the fate of the Gobi with that of East Africa, where the Great Lakes sustained a rain-rich climate and the promising German colony of Tanganyika. He also must have read Powell's report on the arid lands of the United States, because he summarized Grove Karl Gilbert's chapter in the report, which argued that water evaporating from the Great Salt Lake in North America condensed on the surrounding mountains, resulting in markedly higher rainfall around the lake's eastern shores. And he discussed, approvingly, Roudaire's plan to create an inland sea in the Sahara.[95]

Gessert cited Brückner when he argued that seven-ninths of the rain that fell was derived from water evaporated from land surfaces.[96] He echoed Passarge's claim that surface water was vanishing due to geomorphological changes. Livingstone had first suggested that the Zambezi was the conduit for water to escape to the sea; Gessert added the Kunene, which emptied into the Atlantic. Both rivers, he said, had once been endoreic, or inland-draining. Their present form was a Frankenstein creation, the result of short, rapidly flowing coastal rivers that had "gnawed" into the terrain at their headwaters, slowly extending their courses inland until they met and seized the slower inland-draining rivers, sending all their water to the oceans. Gessert was restating the views of the men commenting on Kropotkin's 1904 paper, which was published in the same month as his article—a reflection that these ideas were already familiar to those interested in the subject of desiccation. Combining the ideas of Brückner, Kropotkin, and southern Africa's most famous explorers, Gessert concluded that without river diversion, rainfall would continue to decline. The vast lake that would result from diverting the Kunene River inland would save southwestern Africa from becoming uninhabitable.[97] Gessert represented this not as a radical alteration but as a minor change, under the logic that it was a restoration of previously existing conditions. Like most other Europeans, he believed that a portion of the Kunene already drained into the Cuvelai plain at peak flood; he was simply proposing to increase this drainage to its previous extent.[98]

Gessert paired irrigation with climate engineering. Describing irrigation schemes in Egypt and the Western United States, particularly those of California's fruit and vegetable industries, he imagined a land transformed from waterless waste to a fertile breadbasket. Grazing cows and sheep could be imported from Argentina, he wrote, and pastured on land farther from the central water supplies. Cattle could be sold to feed African laborers in the colony's expanding mines, and sheep could be exported to European markets. Fruits and vegetables could feed local markets or be dried and exported.[99]

Gessert had never seen the Kunene or the Ovambo plain; he was imagining the present place as much as the future one. Indeed, there was little information about the region, and the location of the colony's northern boundary was unclear, thanks to ambiguity in the treaty between Germany and Portugal. The site for the dam he proposed to build was later determined to be north of the colonial border, in territory claimed by Portugal. Gessert admitted that the topography of the region was uncertain, as the area was so flat that miniscule errors could misrepresent the relative elevations of different places. But he nonetheless insisted that damming and diverting the Kunene River would create a system of navigable waterways linking the Kunene, the Kavango, and perhaps the Cuando and Zambezi Rivers.[100]

Schwarz's proposal more than a decade later was simply a geographically more expansive but otherwise nearly identical version of Gessert's. It proposed diverting two rivers rather than one. The western half of Schwarz's scheme was Gessert's, but the eastern half also had a precedent. In 1907 the German engineer Albert Schmidt suggested that the Chobe (or Linyanti) River, a tributary of the Zambezi, could be dammed where it briefly formed the border of South West Africa, creating a reservoir that would permit a large irrigation scheme in the northern part of the colony. Not only had Schmidt never seen the Chobe; he never even visited South West Africa, relying instead on Passarge's published work and German war maps. Like other Germans imagining South West Africa's future, Schmidt modeled his plan on irrigation schemes in Egypt and in Southern California. But he also sought to change the surrounding environment and climate. The resulting lake would make the environment healthier for European settlement by driving away the "miasmas," mosquitoes, and tsetse flies associated with the swamp. It also would cool and humidify the environment: "A large part of German South West Africa, which today is desert and waterless, would be covered with vegetation."[101]

White Survival in a Desiccating World

In 1904, as Schwarz was settling into his university post in Grahamstown, South West Africa was engulfed in a war between its Indigenous Herero population and the German farmers and ranchers who had claimed their land and exploited their labor. German soldiers arriving in the colony were ill prepared to respond to the tactics of guerrilla warfare, and in the early part of the conflict they suffered a series of defeats at the hands of an enemy they regarded as their racial inferior. Within a few months, according to historian Jon Bridgman, "the Hereros had made off with most of the cattle owned by whites and had destroyed virtually every farm" in central South West Africa. "By April the German troops were despondent, discouraged, and demoralized and their leaders were in despair. The greatest military machine in the world had ground to an inglorious halt, and it was unclear when and how it would be set in motion again." In May a typhoid outbreak added to German woes.[102]

In March, *Globus*'s lead article was on the war. But Gessert, secure for the moment on his farm in the southern part of the colony, away from the fighting, had other things on his mind. In the same issue, he urged afforestation in South West Africa, arguing that it would slow the wind, cool the soil, and reduce evaporation, thereby allowing plants to use more of the rain that fell. But this would not be sufficient, he argued: "If one strives for climate improvement of a detectable size, then one must resort to other means."[103] Military operations had ceased while German forces struggled to regroup. Reinforcements were sent from Germany, along with a new military commander, Lothar von Trotha, who had led campaigns in East Africa and China. He vowed to offer no quarter, and took to speaking of annihilating the enemy. By the end of June 1904 there were five thousand German soldiers in the colony—a number that would ultimately rise to almost twenty thousand by the war's end.[104] But Gessert was looking north, beyond the conflict raging in the center of the colony. He published a two-part article expanding on his plan and described what would happen if the Kunene were not diverted. "The country will dry out [*ausdörren*] at the accelerated pace it has embarked upon in recent decades, on the path to the desert." Citing decades' worth of traveler accounts, Gessert argued that there was unanimity about the reality of this desiccation.[105]

In August 1904, von Trotha issued an order forbidding German soldiers from accepting the surrender of any Herero, including women and children, and instructing that any Herero person found in the colony be shot on sight. When officials in Germany protested, the order was replaced by one designating all Herero as forced labor. The war now came to Gessert's part of the colony, as several Nama groups under the leadership of Hendrik Witbooi joined the war. Paul Frederiks, the Nama chief who had sold Gessert his farms, remained loyal to the Germans, but most of his subjects deserted him for the leadership of his cousin and rival, Cornelius Frederiks. Nama armies used the harsh terrain to their advantage, engaging German soldiers and then retreating into the Kalahari or into the mountains to fight another day. Most white farmers found themselves sequestered in one of the towns under the protection of the German military, but Gessert rejected this option. He took some of his livestock and fled across the Orange River into South Africa.

Gessert's published writing records the rigors of the journey and the scarcity of water. But he was more interested in the land's possibilities than in its limitations. He wrote about the geology, rainfall, and vegetation of the terrain he crossed and its potential for dam building, as well as the animals and crops that white settlers were raising successfully in the Northern Cape. At a place where the rainfall was the same as on his own farms, he marveled at the extensive grain fields there and imagined his own future: "Because one can easily catch the autumn rains in the vast floodplains of German Namaland through low dams, the cultivation of wheat and oats is theoretically as feasible as south of the Orange [River]."[106]

Gessert's optimism is mind-boggling in light of all that had just transpired, and was still transpiring, in South West Africa. After all, the very existence of a "German Namaland" was in doubt. Gessert saw the war as a failure of German administration, but he also suggested that the conflict had ecological roots. In June 1904 he argued that the uprising was "a result of overestimating the capacity of the country"—a carrying capacity that was declining as desiccation spread. The real threat was, therefore, not the Indigenous population but the changing climate. Hundreds of Germans "have now settled in the Windhoek mountainland, in the beautiful delusion that the landscape will remain rainy as usual," Gessert wrote. "With the natives one will cope. Worse will be the struggle for existence when the whites are quarreling with the whites about the desiccating water bodies." The diversion of the Kunene was necessary to avoid this future.[107]

A decade later, in 1914, Schwarz wrote his optimistic report on the inexhaustibility of Uitenhage's groundwater. The South African Senate convened a committee to look into drought and rainfall, and another German farmer in South West Africa introduced South Africans to the ideas about river diversion and climate change circulating in South West Africa. In the *South African Agricultural Journal*, Heinrich Paulsmeier suggested carrying out a scheme similar to that proposed by the engineer Schmidt for the Chobe River. Schmidt had made no detailed claims about climate change, though his statement that a reservoir would increase atmospheric moisture enough to cover a large part of the colony with vegetation implied greater precipitation. Paulsmeier made the connection explicit. In language redolent of Schwarz's a few years later, he wrote, "As there can be no doubt that the presence of a large lake in the Ngami basin will immensely improve the rainfall of the whole country, every South African should interest himself in this question." The way to recreate the lost lake, Paulsmeier argued, was to dam the Chobe and Zambezi rivers at their rapids.[108]

Paulsmeier, writing from a German colony on the eve of World War I, had no way of knowing that his own home would be invaded by South African forces the following year and then ruled by South Africa for the next seven decades. But he was already imagining a pan-settler southern Africa, one grounded in environmental interventions that would improve the prospects of whites everywhere. The issue, he argued, was as important to white South Africans as it was to settlers in South West Africa.

Ultimately, Schwarz did not so much invent a narrative about the climatic past and an engineering scheme to create a new climatic future as channel the ideas of many different people, both within southern Africa and around the world: Arid environments had once been wetter. Geological processes could account for the disappearance of water. The presence or absence of surface water affected rainfall. But Schwarz had none of the fatalism of the men in Europe offering dire warnings of planetary desiccation. Although he had been trained as a geologist, his intellectual orientation was more like that of the engineers who imagined the transformation of the Sahara, the *sertão*, and the Australian interior; and more like the German farmers colonizing the arid lands to his north. Humans could reshape natural processes, and the future of white civilization in Africa depended on their intervention.

3 * The Invading Desert

Schwarz's Kalahari Scheme was rooted in the idea that rainfall was declining in southern Africa and that as a result, the desert was encroaching into areas of white settlement. In 1920 he devoted the first fifty pages of his book to assessing the evidence for this proposition. But much of the public was already on his side. For more than a century, southern Africans had been speculating about the region's declining rainfall and desiccation—a process white settlers referred to as "drying out," "*opdroë*," or "*austrocknen*," and which was understood to include both declining rainfall and the expansion of the desert into fertile regions. These ideas spanned colonial borders, and cast change within Manichean terms. In 1920 the South West African newspaper *Allgemeine Zeitung* translated for its readers an article from the *South African Farmer's Advocate and Home* magazine that stated: "In the heart of Africa there is a desert, whose evil influence on the climate of the vicinity, and thereby on the rainfall of the Union, is beyond doubt. The Kalahari has been expanding its borders for centuries, casting its devastating spell over South Africa."[1]

Within the white community, popular understandings of climate emerged from the interplay of European intellectual traditions, African climate narratives, and lived experience. In the nineteenth century, men of science also had asked whether southern Africa was experiencing progressive desiccation and decreased rainfall. But by the early 1900s, an intellectual divergence was underway, part of a larger process of carving specialized scientific fields out of what had been a broad and inclusive natural science.

No single issue better illustrates this widening gulf between expert and popular thinking than the question of whether South Africa was "drying out." By the time the Union of South Africa came into being in 1910, government scientists and commissions were insisting that rainfall patterns had not markedly changed and that the ravages of drought had other, nonmeteorological causes. They agreed that a process of desiccation was taking place, but insisted that it was unrelated to rainfall patterns. Government commissions solidified this expert orthodoxy. A 1914 Senate committee concluded that "all available evidence goes to prove that there has been no definite diminution in the rainfall of South Africa, during historic times."[2] Another commission came to the same conclusion in 1920. But many, perhaps most, white South Africans continued to believe otherwise. They disdainfully characterized these "experts" as out of touch with the lived reality of "practical men." And they insisted that the new scientific orthodoxy rested on flawed evidence and failed to account for the actual behavior of water. This opposition to expert opinion united Afrikaner and British South Africans around the country, in urban and rural areas, and across lines of education and class.[3]

The arguments in favor of declining rainfall reveal the complexities of settler vernacular knowledge about the climate. The boundary between scientific and popular knowledge, one firmly drawn and reinforced by experts themselves, was unstable. All white South Africans found themselves stymied by a lack of clarity about where their rain originated, by the limited data that existed for South Africa's past climate, and by basic misperceptions of dryland ecologies. The public was more willing than the experts to admit these limitations—particularly the problems with how rainfall data was collected, and thus how drought was recorded. But both state-sanctioned and white vernacular knowledge were embedded in a larger context of racial power that rested on claims to rationality and modernity.[4]

Weather Stories

In the years after Union, the rules of the weather broke down in South Africa. Or so said a man writing from the Northern Cape under the pen name "A Farmer." "The weather seemed to get out of joint in November, 1911," he told readers of *Farmer's Weekly*. In that month, the traditional start of the rainy season in much of the country, his farm had received five times its usual rainfall. But then the rains had stopped. Such variability was not unusual, however, and "Farmer" was cheered by the subsequent arrival of southeast winds, which traditionally signaled that sustained rainfall was imminent.[5] His optimism was shared across the country: a Transvaal correspondent reported that while things had been very dry, "Farmers with long Transvaal experience are of opinion that we will get plenty of rain during the months of January and February in the new year"—generally the wettest months in the summer rainfall areas.[6]

But it was not to be. As "Farmer" noted more than a year later, "Day after day thunder clouds would form in the west and often look very promising . . . and toward sunset melt away, to go through a similar performance the following day."[7] On the outskirts of Johannesburg, and in Weenen in Natal, powerful thunderstorms produced hail that destroyed fruit crops.[8] But the regular rainfall that supported plant growth was elusive. A "prominent farmer" in the Karoo was reportedly spending the astronomical sum of ten pounds a day to feed his stock at a time when they normally would have been on pasture.[9] A missionary wrote that "many poor farmers are already ruined by the drought, and we do not know when He shall turn again his Mercy towards us."[10]

Hopes rose again in February 1912. *Farmer's Weekly* exulted, "Good rains have fallen over most of the Union, and we also hear over most of Rhodesia. The tension which in parts was becoming nerve-racking, has ceased, and we live and breathe again, and smile and go and look at the lands and speculate on possible eventualities."[11] But the rain did not last, and the season proved to be a poor one in many parts of the country.

The dry winter that characterizes most of southern Africa passed, and the austral spring of late 1912 began. White South Africans inundated the prime minister's office with requests that he declare a day of "humiliation and prayer." In the Orange Free State, a woman wrote at the request of her neighbors—perhaps because they were not literate—to ask for a day of prayer "so that the Almighty might grant us rain on this land and give food and water to the starving animals."[12] The day was duly declared in October 1912, but to no apparent effect. A farmer writing under the pen name Spes Bona wrote, months into the rainy season, "The present season, 1912–13, is ranking in our experience as the most extraordinary we have ever experienced. . . . The drought has turned everything topsy-turvy."[13] The season ended with below-average rainfall around the country. The next season was even worse, although it began on a promising note when good rains came early. Farmers sprang into action. "Probably more ground was cultivated than has ever been the case before in the history of South Africa. Visions of a record harvest and a record output of maize for export or home use floated before the eyes of the agricultural population." Then the rains stopped. "The veld became burned up, springs ceased to flow and feeding for stock soon became an acute problem to many." Locusts invaded the Free State.[14] By January 1914, *Farmer's Weekly* reported that South Africa was in the grip of what one Free State farmer called the worst drought in fifty years. Two million sheep owned by white farmers had already died, and farmers looked anxiously ahead to the long dry season, when they would have no feed for their livestock.[15] In March, as it became clear that farmers would again suffer huge losses, the South African Senate convened a commission to study rainfall, drought, and soil erosion.

A year later, *Farmer's Weekly* noted a transition in the sunspot cycle, and opined that this indicated a shift from dry to wet conditions. "We are almost certain to have a series of good seasons now," the editor wrote.[16] But in March 1915, the rains abruptly ceased. A farmer in Natal said that in thirty years of growing maize, he had never seen a drought set in so late in the season.[17] And a year later, conditions had not improved. A farmer in the Cape Midlands wrote, "We have been looking for a break-up of the drought, and in most cases looked in vain."[18] Indeed, by some official measures, the 1915–16 drought was the worst since the creation of the Union of South Africa.[19] Far to the north, in the Ovamboland plain, where Schwarz would propose to divert the Kunene three years later, 1915 marked the start of a catastrophic famine that is remembered even today for its high mortality rates. When Schwarz visited the region in 1918, the famine had ended only a year earlier. Human remains still littered the landscape and people's experiences doubtless shaped how they described the climate to the visiting geologist.

The dire tales of white farmers and *Farmer's Weekly* sit oddly with accounts of this period that rely on official government records. In 1941, a member of the irrigation department concluded that the five-year period of 1908–13 had the highest rainfall of any five-year period between 1904 and 1938—9.3 percent higher than the mean.[20] In contrast to the letters of farmers and the observations of the *Farmer's Weekly* editor that 1913 was a disastrous year, government records show no drought in the arid region of South Africa in 1913—true for fewer than half of the years recorded in the 1941 study—and relatively limited drought for the semiarid region. The year 1914, ranked as disastrous by many farmers, shows up in government records as fairly average.[21] Historians, too, see this period as one of prosperity, in which the ground was laid for "the longer-term triumph of [white] capitalist agriculture over [Black] peasant production."[22] This was due in part to the passing of legislation limiting the options of African participation in commercial agriculture. Most famously, the Natives Land Act in 1913 designated more than 90 percent of the country for white land ownership, and placed stringent—although poorly enforced—limits on African tenancy and independent farming on white-owned land. Yet in the pages of *Farmer's Weekly*, white farmers expressed little sense of triumph.

Government drought assessments were based on a region's deviation from its average yearly rainfall. There are two problems with this method of calculation. The first is that South African rainfall records were kept according to a calendar year, which aligned with the agricultural calendar of the Northern Hemisphere. But for the summer rainfall regions that constitute the majority of South Africa, where rains begin between September and November and end between April and June, such reporting lumped together the late rains of one season and the early rains of the next. The annual rainfall figures therefore measured parts of two growing seasons, rather than all of a single season. The second problem was that rainfall figures were compiled as monthly and yearly totals. Rainfall in arid environments is not only sparse but highly variable. The importance of measuring distribution over time therefore increases with aridity. High monthly or annual totals could obscure a disastrous season in which a few large storms, spaced weeks apart, accounted for most of the precipitation.

These problems are evident in the report of the 1914 Select Committee on Droughts, Rainfall, and Soil Erosion, which reprinted the entirety of South Africa's rainfall records. At first glance, this is an impressive set of data that includes twenty-one different locations. To some extent, even these statistics illustrate the problem of rainfall variability: the commission's data reveals that Middelburg, located in the summer rainfall region of the Karoo, received rainfall within one inch of the annual mean in only four of the thirty-five years on record. But a more careful examination reveals the weakness of sweeping claims about rainfall patterns. The oldest records in the Transvaal and Free State were from the 1880s, and had gaps due to the South African War. In Natal, coastal Durban's rainfall records went back to 1873, but record keeping in the interior began only in the 1890s. Cape Town's observations dated from 1841, but 80 percent of the Cape stations were established after 1877, and the oldest station was in the winter rainfall region close to the coast, where climate patterns were different from those in most of the country. Far less was known about historical rainfall in the arid interior. And all this data was organized around a calendar year that reflected the seasons of the Northern Hemisphere and presented rainfall totals as monthly figures.[23]

The statistics offered by the weather stations were therefore all but meaningless as a reflection of farmers' actual experiences. For example, for Middelburg—which later became a center of Schwarz support—the report showed a total rainfall of 21.51 inches for 1906 and 16.46 inches for 1907—both higher than the calculated mean rainfall of 14.25 inches per year. But if the monthly figures were calculated based on the actual farming calendar, from September to August, the picture would change. The 1905–6 farming season was a dry one. Between September 1905 and August 1906, only 12.47 inches of rain fell in Middelburg—below the recorded annual average, and less than 60 percent of the reported 1906 total. Most of the rain for 1906 fell at the end of the year, at the start of the 1906–7 growing season. The report also recorded only 16.46 inches of rain for 1907. But if the monthly totals from September 1906 to August 1907 were tallied, Middelburg actually got 26.67 inches of rain—nearly double its supposed average.

So did Middelburg farmers experience drought in 1905–6 and abundance in 1906–7? Not necessarily. 1905–6 was a drier season overall, but the rains were spaced more predictably. Some rain fell in November and December, and most fell in January and February, as expected. Farmers may have suffered from the early cessation of the rains, however, as very little rain fell in March through May. In 1906–7, although more than twice as much rain fell as in the previous year, farmers actually got *less* rain in January and February, while almost five inches fell in November, at the season's onset. There was more than twice as much rain in April and May, late in the growing season, as there had been in January and February, the peak season for pasture and crop growth. Livestock dominated the region's agricultural economy, and the uneven rainfall would have posed challenges for the management of both pastures and disease. To complicate matters even further, monthly totals fail to account for the distribution of rain over the course of a month. In arid and semiarid environments, rain often comes in cloudbursts—powerful thunderstorms that precipitate intensely for a short period of time. In short, none of the government's data can reliably indicate how farmers fared in those two years.

The narrative of "drying out" went beyond rainfall totals to consider *how* the rain fell: its distribution and intensity. White southern Africans had a variety of terms for rain, rendered in Afrikaans, German, and more rarely English. Afrikaans had the richest vocabulary. *Stofreën, stuifreën, motreën, kiesa,* and *jakkalsdou* were all terms for various forms of light rain or drizzle. *Voorjaarsreën* (spring rains) and *ploëreën* (plowing rains) marked the start of certain agricultural activities. Farmers spoke of *stortreën, plasreën, fontein reën,* and *waterloop reën,* all denoting some degree of heavy rainfall but distinguished by what they did to the land—recharging springs or causing ephemeral watercourses to run, for example. In South West Africa, farmers referred to brief downpours as *Platzregen,* localized showers as *Strichregen,* and widespread rains as *Landregen.*[24] Across South Africa, many white farmers insisted that the frequency of these various types of rains had changed—that rain now came in the strong downpours typical of 1907 rather than as gentle plowing rains. In South West Africa, some farmers argued that the *kleine Regenzeit*—the brief period of early rains that preceded the main rainy season—had vanished.

The Drought Commission acknowledged that across South Africa, "there is a very general tendency among farmers to believe not only that the annual fall has diminished, but also that the rain now usually comes later in the season and that the 'good old fashioned' gentle soaking rains are much less common." The commission's members admitted that they could not definitively refute the belief that the timing and intensity of the rains had changed, because they did not have such data.[25] Yet they insisted that droughts were primarily the result of white farmers' behavior, not climatic patterns. Farmers rejected these conclusions for two reasons.

As we'll see below, many disagreed with the commission about what constituted responsible farming. But they also pointed out deficiencies in official rainfall statistics that were apparent to anyone who worked the land. A Northern Cape farmer wrote in 1915, "It is clear . . . that rain-gauge records when the details are not minutely analyzed, and when only the annual totals are considered, must very frequently convey the erroneous idea of the actual amount of effective wetting such an area has received."[26] Another insisted that rainfall often came in "isolated spatters" that did farmers no good. "Nevertheless they help to swell the yearly total to an extent quite out of proportion to their real value to the country."[27] Such critiques were still relevant almost two decades later: "It is not a question of how much rain falls, but how it falls," one farmer pointed out. "What use is 20 inches if you get 10 of it in one day?"[28]

Sometimes there was no data at all. A new arrival to Lobatsi, in the Bechuanaland Protectorate, asked for information about rainfall and was offered "mean annual rainfall" totals for several locations between one hundred and four hundred kilometers away—despite the fact that it was the variability of the rain that had prompted his inquiry.[29] The lack of state resources dedicated to collecting and interpreting meteorological data frustrated farmers. One man wrote from British Bechuanaland, close to the border of the Bechuanaland Protectorate:

> Some time ago, the Government were prevailed upon to proclaim in the Gazette a day for Humiliation and Prayer in connection with the recent drought—good enough, perhaps, in its way; but would it not have been much more practical and to the point if they had instructed the Meteorological Department to search through the existing rainfall, etc., records, dated as far back as they can find, throughout the country, in order to see whether there exists any law to account for the occurrences of our erratic seasons.[30]

Farmers also pointed out what the experts did not measure. One wrote that he, a "non-scientific" farmer, had thought to study the rate of evaporation on his farm and was shocked at the results. Evaporation rates appeared to be so high as to call into question the assumption that most rainfall soaked into the soil or ran toward the ocean, as "we ordinary people think."[31] This early recognition of what made drylands distinct—potential evaporation rates that exceeded actual precipitation—came not from government scientists, but from farmers who had to manage the consequences.

Some white farmers proved themselves to be more meticulous record keepers than government employees. They offered up daily logs of their own rainfall records—in some cases stretching back many years. One man wrote with records from an Eastern Cape district that, he said, were more complete that those from the nearby town. His records showed declining rainfall averages.[32] A Northern Cape farmer writing under the name Jonas Klip also offered his own statistical data which, unlike the published government figures, recorded daily totals. His records highlight how apparently "average" monthly rainfall could result in agricultural failure. Essentially no rain fell on his farm from March 1 to March 19, while nearly half of the rain in April fell on the second day of the month.[33] In 1923, farmer P. J. de Wet in the Transvaal gave the Drought Commission a meteorological report based on his records going back to 1909, noting that he often recorded four or five entries a day.[34] Others produced personal records that supported the government's position, including Carl Weidner, who raised fruit using irrigation along the hyperarid banks of the lower Orange River. But farmers also understood that South Africa contained different climate zones, and that there was intense variability across space as well as time. George van Zyl of Colesburg did not reject Weidner's evidence, but simply insisted that if Weidner saw the record for *his* farm, eight hundred kilometers away, he would find clear evidence of a progressive decline.[35] Readers universally acknowledged Weidner, a frequent and opinionated contributor to *Farmer's Weekly*, to be a cantankerous contrarian. One man suggested that Weidner occasionally went "off the deep end'" because he lived in "South Africa's hottest oven, Goodhouse."[36] That Weidner supported the government's narrative of rainfall patterns was, among some farmers, further evidence of his eccentricity.

Sensing Aridity

The limitations of the available data help explain why so many farmers rejected the official position that rainfall was not declining. But dissenting views of South Africa's climate trajectory rested on more than numbers. What are we to make of a farmer who insisted in 1916 that he and his neighbors had endured "fifteen years of black drought"?[37] Such a claim smoothed out significant variations between years, creating a single unifying experience that began during the South African War. The article was written under a pseudonym, and we know nothing about the author except that he lived in the Eastern Cape and was literate in English. But his claim echoes others that sporadically appeared in letters to *Farmer's Weekly*. In 1913 a game warden in the Sabi Reserve, in the Transvaal, argued that there had been signs of desiccation since 1902, including a steady drop in the level of the Sabi River itself.[38] During the 1933 "Great Drought," a correspondent to *Farmer's Weekly* reported on a visit to his neighbor: "We started discussing the vagaries of the climate, when to my surprise the farmer informed me that it had never rained properly in the Free State after the South African War. Needless to say I was much surprised, but more surprised and amused when his wife pointed to a dish and said, 'Do you see that dish? Believe me, before the war I gathered that dish full of eggs every evening, but after the War the fowls would not lay.'"[39] In 1927, a lecturer at the Transvaal University College wrote a book aimed partly at refuting the "popular notion" that rainfall in the Transvaal had decreased since the war.[40] The men who mentioned this belief never suggested that they shared it. But the frequent references to it suggest that it was a widely held belief within sectors of the white community who did not routinely participate in literate culture. They demonstrate that "drying out" could signify different things to different people, and that it could be in multiple timelines: a long-term process driven by natural forces, a recent but gradual process unleashed by land-management practices, an abrupt change beginning with a dramatic human event, or even a timely punishment from God.[41]

Whether the assertion that rainfall had declined since the war was a critique of the postwar political order or an argument that the war itself had somehow disrupted the climate is not clear. But the idea that human conflict was implicated in climatic disorder resembles African cosmologies of climate. This similarity raises the question—which the sources cannot definitively answer—of how much Indigenous beliefs shaped white ones. Much clearer is the extent to which personal experience informed understandings of climate. Experience was the domain of "practical men"; farmers often contrasted such firsthand knowledge to the supposedly decontextualized book learning of the "theoretical" men who staffed government departments and experimental farms. Experience on the land offered much to support theories that something was amiss with the weather. Its apparent contradiction to expert knowledge cast doubt on the universality or relevance of scientific expertise.

Stories of landscapes now vanished were one source of evidence for popular ideas about climate. As discussed in chapter 1, whites had invoked the accounts of Black southern Africans as evidence of desiccation since at least the early nineteenth century. They continued to do so into the twentieth, even as Black southern Africans otherwise became almost invisible in white environmental discourses. Doubtless much African knowledge that was subsumed into white knowledge was unattributed and rendered invisible, leaving behind only "looted fragments."[42] But climate narratives were given attribution. The game warden in the Sabi Reserve warned ominously of a process of desiccation. "Old natives will point out dried-out and overgrown streambeds which in their young days they declared carried water all the year round."[43] One man noted in the early 1930s that there was no water whatsoever on his farm, aside from a twenty-foot well. But "old Natives in my employ" told him his farm—whose name, Okatambaka, referenced the presence of ducks—"used to be overrun with ducks in days gone by, as it was always full of water."[44] These "native" voices could be cited as an authority without threatening the farmer's claim to the land because their words were disembodied, removed from all historical and political context. They existed to explain the farmer's land to the farmer, obscuring other questions, such as how his laborers were more familiar with the land than he was, although he was its formal owner.

Indigenous tales of wetter times circulated across southern Africa, as they had in the nineteenth century. Whites drew on stories from increasingly distant places, reflecting the expanding reach of the South African state. South Africa wrested South West Africa from Germany in 1915, and was eventually granted the territory under a League of Nations class C mandate. The triangle of land between the Kunene and Kavango rivers, on the Angolan border, was brought under direct colonial administration for the first time. A newly appointed resident commissioner for Ovamboland found himself ruling an area in the grips of a deadly three-year famine. Schwarz's 1918 newspaper article helped Resident Commissioner Charles Manning make sense of what had just happened without indicting colonial conquest itself as the cause. Manning sent a copy to his superior in 1918 and wrote,

> I venture to endorse the opinion that the gradual drying up of the country is largely due to such important rivers as the Cunene [Kunene] flowing to the sea down deeper beds than formerly and thus draining more rapidly the accumulated water inland. . . . Although the natives do not appear to have any record of water having run in Ovamboland throughout the year, they state that at an earlier period there was more water in the Etosha area and in the watercourses leading thereto.[45]

William Charles Scully, the novelist and former magistrate introduced in chapter 1, also recounted "native tradition," including stories of African settlement and farming in areas now too dry to support agriculture.[46] Following in this vein, Schwarz invoked both African and Dutch place names as proof that the landscape in the western part of southern Africa had once been wet enough to support fauna it no longer could. (Those skeptical of desiccation, by contrast, argued that the name Karoo, with its roots in Indigenous terminology for aridity, indicated that dry conditions were normal.)[47]

Travelers and those white men who were the first to farm a given area possessed more authority than Indigenous residents as witnesses to the past, and were invariably identified as "old farmer" or "pioneer." These terms were both racialized and gendered, bestowing particular authority on the first white men to arrive in an area. One man recounted the first white farmers in the Bechuanaland Protectorate telling him that in the late nineteenth century, the water table had fallen by fourteen feet in nine years. "Many of the old residents used to speak of the good seasons and rainfall they had in former days,"[48] Scully told the 1914 committee. "You talk to any old farmers in the Orange Free State or to any old farmers in the Karoo or the Midlands and they will tell you, one and all, that the springs are drying up."[49] Those who had lived and farmed in the same area for two decades or more were very specific about the changes they had seen. One told an audience of farmers that a farm in the Eastern Cape known as Rocklands had a water mill that had been powered by the Zwart Kei river—which was "no longer a river but merely a dry sluit."[50]

This kind of personal knowledge was discounted by government officials and experts. The belief that rainfall was declining, according to the 1920 Drought Commission report, was "based on personal reminiscences which are particularly treacherous when brought to bear on meteorological data."[51] The editor of *Die Landbouweekblad*, himself an "expert" with an American PhD in agronomy, asked, "Can one say that the words of a person who was young then and is now 30 years older are trustworthy?"[52] But farmers rejected the charge that they were victims of environmental nostalgia, arguing that their authority rested in long observation of particular landscapes. A reader fired back to the *Landbouweekblad* editor, "It does not require much observational ability to see that a wetland that was a trap for oxen 40 years ago is today a place of lean cows. Earlier, a little lake; today, a crust of soil."[53] A farmer in Griqualand West wrote, "Those who speak of the good rains 'in the old days' are jeered at as being the victims of defective memory." He continued, "Many of us have recollections far too realistic to be due to imagination or faulty memory."[54]

Those recollections were grounded not just in the intellect but in the senses.[55] The parliamentarian de Jager, arguing the merits of the Schwarz scheme in 1925, argued, "We are drying up and I am convinced that anyone who is reading the history of South Africa and who gives less attention to the rain gauge than to his own head would feel that the land is drying up."[56] Scully told the 1914 committee on drought, "Your superior person will say, when you tell him that the country is drying up, that the rainfall statistics do not prove that. Well you cannot disregard the evidence of your senses and what you hear from any farmer of experience."[57] A memorable but common experience of southern Africa's weather came in the form of seeing falling rain vanish before it reached the earth—a phenomenon the Romans called *virga*, likening it to a rod or twig descending from the sky, and a sight familiar to any readers who have spent time in arid environments. Two of Schwarz's most ardent supporters recounted witnessing this sight. One was Gessert, whose experience of watching southern Africa's rain disappear led him to embrace the theory of progressive desiccation. Scully recounted a similar experience, in which he stood on a hill in the Orange Free State province and watched thunderclouds develop on all sides. "There I saw a thunderstorm beginning to break. A great blue curtain of rain came down . . . but when the rain came to 2,000 feet of the surface of the earth it did not fall further."[58]

More puzzling than virga was how seldom auspicious weather conditions yielded rain. Gessert observed the formation of rain clouds at the edge of the Namib that did not generate rain. In 1927 a man in the Karoo noted,

> Enough moisture-laden clouds have been blown up from the south-east to have flooded the country on a dozen different occasions. Each time however that such clouds bank up the wind veers round to the west, and a fierce gale blows the clouds to ribbons. It would be interesting to know, when such happens, what the cause is.[59]

In the 1930s, a government surveyor in the Northern Cape complained that the meteorology department had no answers: "Why don't we get rain when we have the right clouds a-plenty, with often thunder as well, to say nothing of wind and dust?" He then recounted his experiences of the previous day: the feel of a southeast wind, the buildup of heavy clouds, the sight of "veils of showers" around—followed by a shift in the wind and the return of the sun. "In any other part of the world where I have lived, a storm yesterday was a certainty, but not around Prieska. And no one can give the reason."[60]

The presence or absence of clouds was a near-obsession with farmers. One told a commission investigating drought, "I have noticed that, on coming out of a house, a Midland [in the eastern Karoo] man looks around for clouds. Droughts, and constantly living in expectation of droughts, seems to have got on his nerves."[61] Observers blamed the disappearance of rain clouds on the west wind, which was given almost anthropomorphic qualities.[62] Two years after arriving in South Africa, Gessert observed how the wind coming from the coastal desert seemed to prevent moisture-laden clouds in the north from reaching southern South West Africa. This "sea wind" was, for him, "the father of the desert."[63] Farmer P. J. de Wet described "the war between dry and wet conditions." The west wind was the army on the side of aridity. Writing in 1919—a year when, according to the irrigation department's figures, 100 percent of the arid and semi-arid regions of South Africa were struck by drought—an Eastern Cape man noted, "According to farmers—indeed any observer must see it—the western wind has a parching and withering effect upon land and herbage which is almost terrifying." A Karoo farmer concurred in 1927, another year of officially recorded drought: "One is apt, with parched country spread around, to associate the west wind with an evil spirit at war with the farmer. If one knew the real cause of such winds blowing it would help one the more to philosophically accept the weather as it comes." The characterization of the west wind as "evil" recurs throughout the sources. P. J. de Wet referred to the northwest wind as an "evil symptom which always makes our farmers shudder."[64]

One of the most appealing features of Schwarz's scheme was its promise to rein in the power of this westerly wind. As the Eastern Cape farmer put it: "If by a stroke of magic the wind from the Kalahari could be made moist instead of dry, the effect upon the rest of South Africa would be marvelous."[65] De Wet, noting that "when the Transvaal and Free State get good rains then the dry North-Westerly wind seems to lose its power and we get rain," was warming up the Drought Commission for Schwarz's scheme by suggesting that wet conditions begat more wet conditions. After suggesting that many of the methods of "scientific farming" would help alleviate drought—a position all the commission's members embraced—de Wet concluded his testimony with this: "Another measure, which I think would help a great deal is the flooding of the Kalahari desert. This would benefit the Midlands and tend to neutralize the effects of the hot North-Westerly wind."[66]

Scientific experts insisted that their understanding of the natural world was superior to that of people who lacked their specialized training. White farmers pushed back against these claims when they pointed out that the experts seemed unable to offer explanations for the most fundamental features of agrarian life, such as the nature of the winds and the causes of rainfall. Knowledge derived from experience and observation continued to be important to popular climate understandings, partly because the data that experts relied upon offered little insight into the forces shaping climate and weather events.[67] In the early twentieth century, even the most basic scientific understanding of climate remained elusive, in Africa as elsewhere. Most scientists believed that South Africa's rainfall was linked to Indian Ocean weather patterns. Scully disagreed; he told the 1914 committee that summer rainfall was generated by a great cloud belt across the center of the African continent that shifted with the seasons, pushing rainfall farther to the south in the South African summer, and north in the South African winter. This belt is known as the Intertropical Convergence Zone, and by highlighting its importance, Scully was promoting what would later become a dominant scientific model for understanding African rainfall patterns.[68]

Schwarz bridged this gap between experiential knowledge and an emerging but partial knowledge grounded in universal science. And he offered an explanation and a solution for rural distress. He echoed the experience of farmers when he wrote in his book, "It frequently happens at present that the winds blow up for rain; every sign is favorable, but the clouds disappear, and nothing happens. This must naturally happen, when there are no central lakes, and there is nothing but parched vegetation on the veld, from which no transpiration can be expected."[69] When he suggested, "A very little increase of moisture . . . will more than double the rainfall, and the converse is true; the more the land is droughty, the more will the rains shun it," he was simply repeating ideas shared by much of the rural white community.[70]

Explaining Drought

Schwarz's claim that rain shunned dry land reflected prevailing beliefs about the relationship between properties of the air and those of the earth's surface. Scully, reflecting on why the rain he saw falling never reached the earth, observed, "The earth could not receive it and repelled it. . . . If there is one thing my reasoning has taught me it is this that [*sic*] the incidence of rain is as much a condition of the earth beneath as the sky above."[71] Scully claimed that it was his reasoning that gave meaning to his experience, but in fact he was drawing on older concepts that were central to natural philosophy. As with Murray and Brückner's theory about the origins of rain, an idea rooted in another context took on new meaning in arid environments. A common sight like *virga* assumed particular significance in light of ideas about the forces of attraction and repulsion that governed the physical world. The belief that when the earth's surface was wet it physically drew rain down from the sky appeared prominently in a debate over the cause of declining rainfall that took place in the government-published *South African Agricultural Journal* in 1914, and it was expressed by both English and Afrikaans authors. As one letter writer put it, "It is a well-known fact that moisture attracts moisture."[72] Another wrote, "In accordance with the universal and infallible law of attraction, like attracts like, and I contend that a great deal of moisture on a large tract of country either in the soil or in the form of lakes, rivers, or marshes, unfailingly attracts moisture in the form of rain from the sky."[73]

The idea that the character of the soil surface had some bearing on rainfall was also framed in terms of emerging understandings of the water cycle. The forester James Sim told the South African Association for the Advancement of Science, "If a current of air laden with moisture—a sea breeze, for instance—meets a land surface, the question of whether or not the moisture shall be precipitated depends on the temperature of the land surface met." Recognizing that hot temperatures at the surface could cause air to rise and carry away moisture, Sim concluded that since land denuded of vegetation was hotter than forested land, the supposed deforestation of South Africa had caused rainfall to decline.[74] When "experts" were willing to consider the possibility that forests played a role in rainfall, this is the mechanism they focused on.

But readers of the *Agricultural Journal* offered lots of explanations for declining rainfall, reflecting both the quirks of individual theories and shared intellectual frameworks. In late 1913, a Durban resident named R. B. Chase suggested that the cause of reduced rainfall in the country was the elimination of vast herds of game through uncontrolled hunting. Such game, seeking cooling mud, had rolled "in certain spots of ground after rain had fallen," thereby creating shallow depressions, or *vleis*, across the landscape. "At certain times of the year during the rainy season, these pools were filled with rain, forming a constant supply of water, which by percolation formed springs and by evaporation caused a great rainfall." The loss of game and their wallows had resulted in the disappearance of springs and a subsequent decline in precipitation.[75] Chase's theory was unusual, but it was grounded in the larger idea that surface water generated rainfall. Notably, the readers who criticized him directed their critiques at the role of animals, not at his understanding of rainfall mechanics or his belief in "drying out." A man named van Zijl replied, "I have shot thousands of head of game . . . it is surprising that I have never succeeded in shooting a single head with mud on, nor have I ever seen one looking dirty or muddy." But he also admitted, "It is quite true that rainfall has decreased considerably . . . and every thinking man would like to know the reasons therefor and to improve the conditions, if within his power."[76]

Throughout 1914, a series of white men wrote the journal to offer their own explanations for declining rainfall. They pointed out the dramatic changes that had occurred in their collective memory as South Africa had industrialized and urbanized, fought a war between British and Boer, consolidated into a union, and increased the scope of state intervention into daily life. The government had, for example, begun eradication campaigns against locusts. W. Akkersdyk suggested that *this* was the cause of drought, arguing that swarms of locusts broke up pockets of hot air that forced rain clouds upward, away from the ground. "In my opinion the sooner the Government prohibits the killing of the locust and passes a law to that effect, the better for this country and its rainfall."[77] A number of city dwellers subscribed to the *Agricultural Journal*, reflecting a continued sense of connection to white agrarian society. A man living in Johannesburg speculated that the furnaces near the mines had added heat to the atmosphere, creating a vacuum that "in turn causes suction and evaporation. It also causes the wind(s) at the same time. . . . They take the major portion of the moisture with them." He suggested that a Malthusian correction was at work: "I think that an increasing population, with its attendant industries, causes a decreasing rainfall. It is nature's way of arranging matters, to prevent over-population."[78] W. W. Steers, also a resident of the city, thought that the incessant pumping of water out of the mines had disrupted the country's system of underground rivers, which reduced moisture in the atmosphere and thus the rain.[79] Farmer D. Pretorius observed that it was not only mines that drew water from underground; modern technology had reshaped many aspects of the land and waterscape. He argued that windmills and boring machinery, by extracting water for agriculture and stock raising, had caused a decline in rainfall.[80]

This debate in the pages of the *South African Agricultural Journal* played out at precisely the moment in which the Select Committee on Droughts, Rainfall, and Soil Erosion was meeting. These concurrent conversations understood cause and effect quite differently. The letter writers assumed that drought was the result of decreasing rainfall. The Senate committee saw drought as the result of bad farming. These "backward" farming practices were enumerated like a mantra in a series of government reports, in publications from agricultural colleges, and occasionally in *Farmer's Weekly* and its Afrikaans equivalent, *Die Landbouweekblad*, established in 1919. They included the concentration of livestock around water holes, overstocking, the reliance on pasture rather than stored fodder during the dry season; the indiscriminate removal of trees; the burning of the "veld," or pasture; and the practice of extensive "unscientific" agriculture. A government surveyor said that "the nomadic trekker"—the Afrikaner who moved with livestock in search of adequate pasture—was "the greatest offender" in the destruction of vegetation and subsequent soil erosion. These trekkers tended to be among the poorest livestock farmers, who did not own land. But the committee indicted landowners as well. One witness, the recently retired head forester, told the committee, "In this country you see farmers deliberately destroying vegetation and cutting down the trees on their farms. Naturally the soil gets drier."[81] Most farms had only one or two water points, and no fencing to keep out jackals. Farmers used shepherds to bring livestock onto pasture each morning and drive them back to an enclosure at night. The result, experts said, was severe overgrazing and erosion around the water points and enclosures, and the replacement of hardy and nutritious perennial grasses with more fragile annual grasses. Farmers also kept as many animals on their farms as the pasture would sustain in a good year, so that in a drought year their animals had inadequate food and overgrazed the pasture, leaving the soil bare and prone to erosion.

While government experts sought to divide knowledge into distinct categories of officially sanctioned facts and popular folklore, a careful reading of the sources reveals that this division was more imagined than real. A belief that rainfall was declining could coexist with support for experts' ideas about proper land use. As Tischler notes, vernacular versions of agricultural progressivism were "emancipatory and inclusive"—a path toward material progress—in contrast to the increasingly state-focused and didactic project of rural social engineering, aimed at lower-class whites as much as Africans.[82] At a farmers' association meeting in the Eastern Cape, a guest speaker echoed government experts and progressive farmers when he argued that burning the veld, a popular practice among both white and Black farmers, reduced soil fertility, opened the door to noxious weeds, lowered the water table, and led to severe soil erosion. But farmer S. B. de la Harpe followed this logic beyond the point where experts would go. His district had used to get twenty inches of rain per year, he told his audience, but it now averaged thirteen inches, and the rain came in more concentrated bursts. "I feel convinced that we are to blame for that," he concluded. If the water were kept in place, "as moisture attracts moisture, more rain would follow."[83] Two months later and less than one hundred kilometers away, farmer W. B. Phillips stood before another farmers' meeting and lamented the loss of reeds and pasture in his area over the preceding fifty years, a loss that he said had affected the climate. "At the same rate as the bare ground increases in dimensions will the rainfall decrease in volume. . . . the bare baked earth instead of attracting the rain repels it."[84] The reality was that many men who concurred with the senators' ideas about responsible land use also believed that rainfall was declining, while those who shared the committee's view on rainfall could nonetheless be critics of its farming prescriptions. At the Queenstown Farmer's Association meeting where farmer de la Harpe spoke, several men also expressed support for veld burning, a popular and cost-free practice that produced a flush of green grass with the first rains, but which was almost universally decried by agronomists and other government experts. The men justified their insistence that veld burning was not harmful in part by invoking the government's assertion that rainfall was not declining.[85]

The 1914 committee members could not maintain a barrier between official and unofficial knowledge even when they controlled who had the power to speak. Their own witnesses undermined them. The chief meteorologist spoke in favor of the theory of "repeated rains," and speculated that "a second ocean inside the country" could increase rainfall. (The meteorologist's testimony reveals how widely shared these ideas were years before Schwarz proposed his scheme, and again explains why people so readily embraced it.)[86] A farmer from India insisted that soil erosion was a natural occurrence that would correct itself. And Scully launched a comprehensive challenge to the experts. He gave the committee a lecture on the supposed trajectory of South Africa's climate, incorporating diagrams drawn on a blackboard, stories of his own experiences, half a dozen references to travelers' accounts and publications on climate, and even another witness: a farmer imported from South West Africa.[87] He described a land that decades ago had been covered in shallow water pans, swamps, and forests. The earth had been like a sponge: "Every drop of rain that fell was absorbed," feeding underground springs that in turn strengthened the currents of moisture rising from the earth to meet coastal winds. The loss of the bodies of water and vegetation, Scully concluded, left conditions less conducive to rain: "Before you can have rain a hand must go up from the earth and, so to speak, turn the tap."[88]

Scully's testimony interjected the main components of settler vernacular climate knowledge into an official setting. Rainfall was declining due to changes in the land created by the arrival of whites and the activities they had undertaken. But repairing the damage was not a matter of simply farming better. Earth and sky were connected: conditions on and under the surface of the land permitted or prevented the precipitation of water from above. People had the power to change the climate and thereby save South Africa but doing so required restoring the land's supply of moisture. If they did not do so, South Africa would join the ranks of once fertile lands that had become deserts.

The testimony of such outlier witnesses was simply omitted from the official report, which concluded that despite the absence of accurate rainfall records, "all available evidence goes to prove that there has been no definite diminution in the rainfall of South Africa, during historic times." Weather and climate were global, not local, phenomena, shaped by oceans and planetary air currents. In this respect, the committee agreed with an emerging global consensus that weather and climate were shaped by planetary or even astronomical forces, not local ones.[89] The mechanisms remained unclear; scientists in the early twentieth century continued to speculate about the role of volcanic eruptions, sunspot cycles, and changing levels of atmospheric carbon dioxide. But regardless of the cause, the idea that climate was relatively stable in historical timescales pointed to one conclusion: The cause of desiccation was not changing rainfall, but the behavior of farmers. "The occupiers of the land, from the earliest times to the present day, have gradually destroyed the vegetation."[90] The solution was for individual farmers to mend their ways.

Race and Responsibility

The committee completed its work and the debate in the agricultural journal petered out, but the droughts continued. Much of the country had below-average rainfall in 1916. The next two years were better; when Schwarz first proposed his scheme in early 1918, parts of eastern South Africa were experiencing historic floods. But official figures from the 1918–19 growing season showed that 100 percent of the arid and semiarid areas of South Africa were drought-stricken, while more than three-fourths of the humid and coastal regions were also struck by drought. More than 200,000 large stock and 5.4 million small stock owned by white farmers had died or been slaughtered for lack of fodder. Losses to white farmers were estimated at £16 million.[91] Schwarz claimed that his entire scheme could be built for £250,000—less than 2 percent of the amount of that loss. The following season was better, according to official tallies. But in February 1920, a Northern Cape man writing in the newly established Afrikaans magazine *Die Landbouweekblad* (Agricultural Weekly) reported that many parts of his district had not had a *waterloop reën*—a rain heavy enough to cause ephemeral rivers to flow—since the floods of early 1918. Farmers had lost thousands of sheep.[92]

Frans Geldenhuys, *Die Landbouweekblad*'s first editor, was a twenty-nine-year-old agricultural instructor who intended the newspaper to be an educational tool to inculcate progressive farming practices among the Afrikaner community. This in itself was not unique. *Farmer's Weekly*'s letterhead stated: "'The Farmer's Weekly' finds its way into the hands of every Progressive Farmer in South Africa." But Geldenhuys did not publish the wide range of opinions and ideas that the more freewheeling *Farmer's Weekly* did. Readers' contributions must therefore be read for their subtexts, and commentary on drought and dryness could indicate a belief in progressive desiccation. For example, the Northern Cape man who noted the lack of sufficient rain to fill dry riverbeds also wrote to *Farmer's Weekly*—where he argued that rainfall was declining, and advocated for Schwarz's scheme.

Most contributors to *Die Landbouweekblad* did not write simultaneously to *Farmer's Weekly*. But it is likely that many of those who highlighted the low rainfall of recent years believed that it was progressively declining, even if the editor did not. A farmer from Ladybrand in the Orange

Free State referred to the good harvests his district had gotten "when the rainfall was frequent. . . . Now that we have dry years, the grain yields are reduced by half. . . . Earlier we never had a lack of water."[93] A farmer in nearby Wepener noted that someone had described the district in an earlier edition of *Die Landbouweekblad*, "but after that many things happened." The wheat and maize harvest were complete failures, and farmers were experiencing "the worst drought that our little ones have heard of."[94]

The week after these reports were published, Geldenhuys published an editorial headlined "Dust or Rain?" Less than a year into his new publication, he expressed impatience with the continual toll that drought took on the country's farming industry. "The biggest limiting factor that restricts agricultural development, agricultural production, and agricultural prosperity in South Africa is LACK OF WATER," he wrote, asking what the government planned to do about it. Shouldn't there be more money for irrigation, for borehole drilling, and for damming springs and rivers on farms? "Will there be a standing commission to investigate and report on the issue continuously?" he asked, before adding: "What about the Kalahari scheme?"[95] Schwarz had not yet published his book. But even this self-identified expert was talking about his proposal.

The commission the *Landbouweekblad* editor called for was convened seven months later, in September 1920. The Drought Investigation Commission has been treated by historians as a straightforward study of the impact of the 1919 drought on white farmers, and a blueprint to prevent similar catastrophic losses in the future. Those were certainly among its purposes. But like its 1914 predecessor, the commission investigated drought through a particular ideological lens: one that insisted on personal responsibility for misfortune, on solutions paid for with private and not government resources, and on the centrality of bureaucratic expertise to the practice of farming. While the commission may indeed have contributed to the "rise of conservationism," as historian William Beinart argues, it also faced challenges, including those from farmers who themselves claimed to be grounded in science and progressivism.[96]

The 1920 commission was headed by Heinrich du Toit, another agronomist and agricultural educator who had lived in the United States.[97] Its members traveled across South Africa, holding publicly advertised meetings where farmers could express their views. When the *Farmer's Weekly* editor pointed out the low attendance at these meetings, du Toit defended his committee, arguing that "some farmers had an idea that the Commission was out to make rain or that some political propaganda work was at the back of it."[98] Du Toit's comments hint that what some people wanted from such a commission was not what it was offering. The committee sent a standardized letter, six typed pages in all, to experts containing questions grouped under headings like "agricultural education," "ensilage," "plant diseases," and "taxation." Taken together, they sought to reinforce the kinds of knowledge that du Toit saw as legitimate. Questions about how erosion affected underground water supplies, and about how much weight animals lost due to their freedom to roam in comparison to those confined to paddocks, presupposed certain relationships of cause and effect. The questions focused on technological solutions to agricultural challenges, and on the role of individual actions in shaping economic and environmental outcomes.

A couple of the committee's questions acknowledged vernacular knowledge. They asked whether "our country is large enough to contain its own circulatory system—for example, whether 'repeated' rains brought about by reprecipitation of re-evaporated rainwater forms an appreciable portion of our rainfall,"[99] and whether "forests, for instance, 'attract' rain in the same sense as zones of country on which rain has already fallen earlier in the season appear to do?" The questions reveal that du Toit was aware of these ideas. But they were nestled among more than two dozen other questions.

"Colonial commissions," Ann Laura Stoler writes, "reorganized knowledge, devising new ways of knowing while setting aside others."[100] The drought investigations of 1914 and 1920 were no exception. Adam Ashforth, writing about contemporaneous South African commissions that purported to investigate "native" affairs, argues that they helped "constitute the power of those who would act in the name of the state, and the subjection of those whose lives they would organize."[101] It was not only Black people who had to be disciplined and controlled in the quest for a "white South Africa"; whites, too, had to be brought into line. And so the drought commissions produced a body of knowledge about the environment, poverty, and race that ultimately held white individuals responsible both for their fates and for the larger desiccation of South Africa's lands. The commission's vision of possible futures was no less dystopian than Schwarz's; the "gloomy and ghastly future" that awaited the country if it did not change course was "'the Great South African Desert,' uninhabitable by Man."[102] But white people, not geological forces, were creating this dystopia. The final report of the 1920 commission concluded:

> Your Commissioners are convinced by the evidence submitted that, as a result of conditions created by the white civilization in South Africa, the power of the surface of the land, as a whole, to hold up and absorb water has been diminished, that the canals by which the water reaches the sea have been multiplied and enlarged, with the result that the rain falling on the sub-continent to-day has a lower economic value than in days past. Herein lies the secret of our "drought losses."[103]

Given the well-documented tendency of colonial administrations across the continent to blame African subjects for environmental degradation, this conclusion requires some explanation.

The members and witnesses of the 1914 and the 1920 commissions did not agree on everything, but they shared an assumption that coercion could be used to improve and "modernize" agriculture in the native reserves. Whites, by contrast, would have to be persuaded to change. "Unfortunately," du Toit wrote in 1920, "in a democratic country it is not possible for Governments to introduce measures too far in advance of the wishes of the average voter."[104] It went without saying to du Toit and his audience that democracy and voting, like farming, were the preserve of white men. And free white men could not be forced to do things. The 1914 report recommended the banning of pasture burning, through legislation "as stringent as circumstances will allow."[105] But in 1920 the committee concluded that there was no point in passing laws on burning the veld, because they could not be enforced. Instead, it required "the cooperation of the individual. He must be educated."[106] Education and encouragement toward "modern" agriculture were the primary approaches toward white farmers recommended by both commissions, and would remain so through the 1950s.

But with this racial privilege came responsibility. If farmers, an implicitly gendered and racialized category, were supposed to voluntarily embrace modern methods and so fend off desiccation, their compliance or refusal became a kind of litmus test of whether they deserved to possess their land. The very category of "farmer," even within its racialized and gendered parameters, was contested, as we'll see in chapter 4. As a collectivity, however, farmers' actions determined whether the white race was fit to claim southern Africa for itself. The notion that "white civilization"' had caused the desiccation of South Africa sounds, on the surface, like a critique of the settler colonial project itself. But it was a critique with many possible implications. For the men who suggested that rainfall was declining because of urbanization, mining, or other technological interventions, white culpability for desiccation was linked to modernization itself. Scully was echoing a body of ideas that extended far beyond South Africa, to the Great Plains of the United States, when he suggested that telegraph poles were also responsible for decreased rainfall, because they cause "a continual leakage of electricity into the sky."[107] For du Toit and his fellow agronomists, by contrast, "white civilization" had caused environmental ills due the absence, not the presence, of modern technology. The conquest of South Africa simply was not complete; in addition to conquering the people and the land, aridity itself had to be defeated through the efforts of racialized individuals.

Whites who identified as farmers pushed back against this discourse of individual responsibility. The Commission's findings were rejected and even ridiculed by many rural white South Africans as based on shoddy scientific evidence, insufficient rainfall data, and pre-determined conclusions that were impermeable to contrary evidence. As one man pointedly observed, du Toit had written an article for *Farmer's Weekly* insisting that rainfall was not declining before the Drought Commission had completed its investigation. "It seems a pity that he should have given his views before the Commission has finished its work, as he seems to have made up his mind already that we are wrong in thinking South Africa is drying up."[108] Du Toit's behavior therefore cast doubt on his commitment to the methods of science. A member of parliament made a similar claim in 1925: "[A]s regards the findings of the Drought Commission that in a cycle of time rain also falls in South Africa . . . this is only an opinion which is not based upon investigation."[109]

Conversations about the weather reveal that virtually no one fully embraced the emerging orthodoxy of a handful of government commissions and employees. Both the expert and the vernacular forms of environmental knowledge were grounded in moral and racial worldviews, and both argued that the desert was expanding. But the differences in how white farmers and government experts experienced, recorded, and talked about the weather set the stage for divergent views about the kind of future that whites faced and the actions that might shape that future. From an ecological standpoint, rainfall variability might be a normal feature of these lands. But for white settlers around the colonial world—from American homesteaders who believed that rain followed the plow, to South African farmers who asked whether mining or telegraph lines drove away the rain—arid lands were seen as inherently problematic and in need of transformation. Such environmental considerations did not exist in a vacuum, of course. The droughts of the early twentieth century coincided with a global shift in agriculture, which required greater investments in capital to succeed. In South Africa, Black farmers had enjoyed a period of relative opportunity and prosperity as markets expanded and they found themselves able to take advantage of these opportunities, while white farmers had struggled to find strategies that would allow financial stability amid escalating land prices and demands for greater capitalization. In the early twentieth century, legislation foreclosed Black opportunity. But whites felt more constrained by climatic instability than liberated by racist laws. Fears of the encroaching desert existed alongside fears of the Black majority and the potential for white civilization to disappear from South Africa.

4 * White Men's Fears

In 1902, as the South African War neared its conclusion, the Boer general Jan Smuts wrote to a British sympathizer with the Boer cause, describing the "peculiar position of the small white community in the midst of the very large and rapidly increasing colored races." Britain's appeal to Black South Africans for assistance "threatens this small white community and with it civilization itself in South Africa," Smuts wrote. "The war between the white races will run its course and pass away," but "the Native question will never pass away."[1] Two years later, Ferdinand Gessert observed the course of another colonial war, this time between Germany and the Herero and Nama peoples in South West Africa. He also distinguished between immediate and long-term dangers. "With the natives one will cope," Gessert wrote, suggesting that this racial war would also run its course. "Worse will be the struggle for existence when the whites are quarreling with the whites about the desiccating water bodies."[2]

For the future prime minister and international statesman, the biggest threat to white survival in southern Africa was the Black majority; for the farmer on the frontier of the colony and the edge of the desert, it was the environment. Two decades later, the editor of *Die Landbouweekblad* metaphorically connected the two in a speech delivered on "Dingane's Day," a religious holiday later known as Day of the Covenant, which celebrated the 1838 Afrikaner victory over the Zulu king and his army at the Battle of Blood River. In failing to master the natural world, farmers had not yet triumphed over the "Dingaans that are against them": drought, locusts, and plant and animal disease, as well as economic forces.[3] Here, environmental threats were explicitly equated with the threat an independent African kingdom had once posed to Afrikaner survival. More commonly, these dangers were discussed separately but in much the same language. Erosion and other environmental ills, like the parching west wind, were "evil"; so too was the problem of white impoverishment, which resulted in whites leaving the land and disrupting the racial order. Drought Commission chair Heinrich du Toit argued that failure to stop soil erosion would lead to "national suicide." The president of an agricultural union argued that a failure to increase the number of whites in the country was to "commit national suicide"; he also advocated for conserving water, afforestation, and building the Kalahari Scheme.[4]

The previous chapters explored the ecological and climatic aspects of nineteenth- and early-twentieth-century settler colonialism in drylands. But European settler colonialism has traditionally been understood primarily as a project of demographic replacement through the elimination of Indigenous peoples. This chapter considers how the fact of being a racial minority shaped the possible futures that whites imagined for themselves in southern Africa, and the centrality of land claims and agriculture to realizing that future.[5] As with popular environmental knowledge, white South Africans' diverse views about their minority status were rooted in shared concepts, including the idea that "white civilization" was a tenuous state, constantly under assault by the forces of "barbarism." Historical work on how this idea functioned in economic life is overwhelmingly urban-centered, focusing on the migration of "poor whites" and Africans to the cities, on the struggles of white industrial workers to protect their position vis-à-vis African labor competition, and on a range of segregationist policies designed to impose order on a chaotic process of urbanization and force Black South Africans into wage labor. White rural areas, when they are studied at all in this context, are largely viewed through the lens of farmers' struggle to secure African labor, or through state and church efforts to "rehabilitate" impoverished whites.

But there was general agreement among white South Africans that a robust white presence on the land was vital to the survival of "white civilization." The land was imagined as both the ultimate guarantor of the survival of that civilization and the potential instigator of its demise. The dystopian scenario, one that seemed to be emerging all too quickly, was of climate change and mass impoverishment that would force many white farmers off the land and into unemployment and racial mixing in the cities—a simultaneous loss of white rural numbers and white prestige. This was the "poor white problem," which became an increasing preoccupation of policy makers and rank-and-file whites in the early decades of the twentieth century. Its utopian counterpart was the scenario of a fertile country settled by large numbers of self-sustaining white men, supporting white wives and producing white children who would someday inherit the land. The place of Black labor in such a scenario varied. Often it was conveniently just out of sight, in the form of a docile, compliant workforce that accepted the dominance of white men without question. Others suggested that Black labor should be replaced by a productive white labor force, or simply imagined Black labor into nonexistence.

A desiccating countryside was only one of the threats to rural white population. Structural change in agriculture was another. Commercialized agriculture had spread unevenly in the half-century before the Union of South Africa was created, beginning with a merino wool boom in much of the Cape and parts of the Orange Free State. In Natal, sugar production expanded with the arrival of indentured South Asian workers after 1860. The Transvaal and the eastern Orange Free State joined the commercialized agricultural economy with the gold boom in the 1890s, which created a domestic market for agricultural products, including wheat, maize, and cattle. All of these changes exacerbated class differences between farmers. It became harder for newcomers to take up farming and for poorly capitalized farmers to continue farming. Both high land prices and the subdivision of existing farms through inheritance made it difficult to use systems of extensive farming, but intensive farming required resources for fencing, boreholes, irrigation, and other technology. The South African War exacerbated rural precarity. British scorched-earth tactics destroyed crops and infrastructure in the Orange Free State and the Transvaal. In the aftermath, the British and then successive Union governments encouraged capital-intensive commercial farming, which squeezed out poorer farmers and created high rates of farmer indebtedness.[6]

Historians have argued that the Black peasantry was better positioned to take advantage of this postwar economic context than many white farmers were. Black farmers from Basutoland poured into the Orange Free State with their families and cattle, and were welcomed by capital-poor white farmers who needed labor and plow oxen. African tenants farmed maize and other crops "on the halves," providing around half the harvest to the owner in exchange for the right to produce autonomously and keep their family and livestock on the farm. White tenants, or *bywoners*, who were reluctant to have their wives work in the fields, wanted a cash wage, and often had less experience with arable farming, were at a disadvantage and found themselves turned off farms in favor of Black tenants. In the Transvaal, a fifth of the land was owned by absentee landlords and corporations, and was worked by African tenants under a variety of financial arrangements. And in some districts Africans bought back land, either individually or collectively. The result, as Sandra Swart has argued, was a distinctly rural form of "Black peril," in which rural whites on the *platteland* (countryside) feared not sexual activity across the color line but Black economic success.[7]

Some landless whites sought pastures for their livestock in the most marginal and arid regions, which also were the most vulnerable to catastrophic drought. In many cases this simply delayed the inevitable migration to the cities, where lower-class whites lacked skills to find employment. They competed with Africans for the lowest paid jobs, and often lived in mixed-race settlements. The reality of lower-class whites, most of them Afrikaner, having a standard of living little different from that of their fellow African urbanites threatened the entire edifice of white supremacy. White poverty as a policy problem first entered public discourse in the 1880s. By the early 1900s, fears over "poor whiteism" had reached a fever pitch.

The initial response to the "poor white problem" was to try to get poor whites "back to the land" via irrigation settlements and other schemes. These were predicated on the idea that poor whites needed "rehabilitation" and closer supervision, but the ultimate goal was to make them into small-scale farmers and landowners. This reflected a shared assumption that private individual land tenure was a defining feature of the white man's land.[8] But the South African War, structural changes in agriculture, and a series of droughts threatened the idea of a white man's country built upon an agrarian base. So too did rural demographics, because Black people dominated the "white" countryside. As Beinart and Delius have noted, "There have never been exclusively 'white' farms in the sense that these were areas largely of white population." In 1911 there were more Africans on white-owned farms than there were whites in the entire country.[9]

The visions of a future in which South Africa was a white man's land were visions of unquestioned white dominance, but Afrikaner nationalists and British imperialists diverged on the means for securing that dominance: either reinforcing white prestige through a system of racially differentiated legal rights bolstered by social norms, or increasing white numbers by creating the conditions for white immigration to and prosperity in South Africa's countryside. These options were not mutually exclusive, and many people embraced both. Those who supported white immigration also worried about white prestige, while those who opposed white immigration as a distraction from efforts to ameliorate white poverty could also be drawn to the idea of a land that could support more white farmers. In the immediate aftermath of the South African War, the director of agriculture for the Orange Free State promised that agricultural modernization would make it "a white man's country in every sense of the word"—both allowing the "natural increase" of the white population and attracting white immigrants. This was largely an English-speaking vision at first, although Afrikaner nationalists like Heinrich du Toit and the editor of *Die Landbouweekblad* came to embrace it.[10]

Schwarz tapped into these shared fears and aspirations in all their variants, promising that his scheme would remove every obstacle to securing a white man's land. To those concerned with white impoverishment he promised fertile land and prosperity for all; to those worried about whites' status as a demographic minority he promised the means to attract millions of new white settlers; to those worried about the encroachment of Black South Africans into purportedly white spaces, in the form of tenant farming and urban residence alike, he promised rich new lands for Africans as well as whites, and therefore the possibility of building a segregated landscape from a blank canvas. And to the many who feared that South Africa was becoming too dependent on mineral extraction for its prosperity and that drought was ruining farming, its other major source of wealth, Schwarz promised to make it rain, to make barren lands bloom, and thereby to secure South Africa's agrarian foundation. He offered a vision of the future in which the segregationist order was sustained in the countryside, white farmers were the "backbone" of society, and whites continued to prosper when the gold and diamonds were gone.

White Numbers

South Africa's status as a settler colony was perpetually questioned by imperialists and settlers alike in the early twentieth century—and, a century later, by historians, one of whom describes the country as a "brittle, gimcrack settler state."[11] It has always been a majority-Black country. When Schwarz published his book, census returns showed whites were 22 percent of the population, a figure that changed little over the next thirty years.[12] Neighboring settler colonies had far fewer whites. In 1936, they were about 8 percent of the population in South West Africa; policies encouraging white immigration boosted the total to its peak of about 12 percent a decade later. In Southern Rhodesia, whites never constituted more than about 5 percent of the population.[13] Although there were mixed-race and Asian populations in these colonies, they were smaller than the white populations. Africans remained a decided majority in all of them, serving as a cheap labor force that undergirded colonial economies. In South Africa, that labor pool was supplemented by African men from South West Africa; the three British territories of Swaziland, Basutoland, and the Bechuanaland Protectorate; the Portuguese colony of Mozambique; and places as far afield as Northern Rhodesia (Zambia) and Nyasaland (Malawi). By the twentieth century, much of the land in South Africa, Southern Rhodesia, and South West Africa was legally owned by the white minority, but Indigenous populations continued to make territorial claims. Colonial administrations used legislation to restrict Africans' ability to compete with white farmers by limiting African land rights to "native reserves" and crafting policies designed to exclude them from agricultural markets.

This was not the outcome the colonizers had anticipated. The experience of places like North America and Australia had led many to believe that the Indigenous population would disappear before the onslaught of European settlement, leaving a society numerically dominated by white people with their crops and herds.[14] And at first, South Africa seemed to follow this trajectory. Pastoralist and hunter-gatherer populations in the Cape were amalgamated into a mixed-race "Coloured" community, pushed north across the Orange River, felled by smallpox, or exterminated by farmers who viewed them as "vermin."[15] But as both British and Dutch settlers moved eastward in the middle of the nineteenth century, the densely populated agropastoralist societies they encountered held their own, both militarily and demographically. British High Commissioner Bartle Frere acknowledged this reality in an 1878 speech to the newly formed South African Philosophical Society. Frere's war against the still independent Zulu kingdom was still six months away—a war that would begin with a Zulu victory over British forces. African military power was still a reality. Frere thought the military balance of power could change but that the demographic balance could not. To those who still believed that South Africa's Black population was "liable to die out like the Maoris and Red Indians have done," Frere responded, "My own opinion is that the race is indestructible."[16] Shortly before the creation of the Union of South Africa, a visiting British lawyer wrote that if the conquerors "had killed these black heirs to the country the inheritance would have been ours. . . . There is no chance, then, that the entire race . . . will die out. Death will not solve the problem for us."[17]

Outsider observers suggested that this "problem"—the continued presence of a Black majority—had implications for the kind of political and economic order that could emerge. The British imperial world was conventionally divided into colonies of white settlement and those where a white elite extracted resources, sometimes called plantation or planters' colonies.[18] In the first, whites created a "civilized" society of citizens modeled on the home country; in the second, a small group of whites exercised political and economic power over an Indigenous majority, but could not recreate the supposed political and social norms of Western Europe. The Australian parliamentarian Charles Pearson suggested that South Africa was destined to "deteriorate into a highly unequal, undemocratic society"—in essence, a "planters' colony"–because "colored and white labor cannot exist side by side."[19] A former secretary for native affairs told the South African Native Affairs Commission in 1905 that South Africa was not "a white man's country in the sense in which it is taken that Australia is a white man's country." The Black majority rendered this impossible.[20] The British lawyer wrote, "That South Africa is, or may be, a white man's country is a dream—a delusion."[21]

Climate was a major criterion for dividing imperial lands into categories of settlement and exploitation. At a time when climate determinism was at its peak, many Europeans believed that whites who moved to the tropics degenerated both physically and morally. Subtropical lands such as the malarial zones of northeastern South Africa could also be unhealthy. But higher elevations could mitigate the effects of latitude, as could aridity. South Africa, with its dry western half and its eastern half dominated by a high inland plateau, was lauded by its white community for its healthful, sunny climate. In this respect, it was not a typical planters' colony. An Anglican priest from London argued that "South Africa is, by climate, a white man's country. The white man, therefore, has the prior claim to possess it."[22] Demographics complicated this easy assertion. A British-born geologist and mining engineer suggested that South Africa was neither fish nor fowl, straddling the line between these two kinds of colonies. It was "a country whose climate is suited for permanent white colonization," but which, like the tropical lands to the north, contained "a vigorous and more numerous colored race."[23]

Whites across the political spectrum were acutely aware of their country's "in-between" status and its potential implications. While Alfred Milner, high commissioner for South Africa after the South African War, considered the creation of a white man's country in South Africa a "root principle," this outcome was not inevitable. Hermann Giliomee argues, "In the final decades of the nineteenth century and in the first decade of the twentieth century, it seemed possible that a substantial section of blacks would be incorporated in a way that would militate against the idea of a white man's land and white solidarity."[24] Timothy Keegan states that even as the Union of South Africa came into being in 1910, "the colonial project seemed increasingly insecure and tenuous to many white people."[25] In 1917 Smuts, then in London as part of the War Cabinet, told a British audience, "Unlike other British Dominions, our future as a white civilization is not assured."[26] As Jeremy Krikler notes, "For whites, racial supremacy in South Africa—unlike in the South of the USA—was always challenged by the facts of demography: whites were a minority in a land conquered from black people."[27]

The sense of being an embattled minority was not necessarily new. In the 1880s, *De Zuid-Afrikaan* wrote about "the struggle for existence against the natives," and a Cape politician predicted that the white population could be "utterly swamped" if Black voting rights were expanded.[28] To Afrikaners, "swamping" or *oorstroming* could refer to the threat the British posed to Afrikaner language and culture, or to the threat that a Black majority posed to the white population. Similarly, British concern with demographics was rooted in the ratio of Blacks to whites but also, in the case of the most entrenched Anglophiles, the reality that Britons were a minority among the white population, particularly in the rural areas. As a tenuous white unity was forged in the decades after the South African War, the fear of "swamping" came to refer almost exclusively to the danger posed by the Black majority.

White South Africans debated how to construct an enduring "white man's country" out of this place that had the climate of a settler colony but the demographics of a plantation colony. Some suggested that population was not destiny, and that a "white man's country" did not have to mean literally a country filled only or mostly with white men and "their" women and children. National Party leader Daniel François Malan, who later became the first prime minister under the apartheid government, argued that it could be instead a country where all whites enjoyed a higher status than Blacks.[29] On the other end of the political spectrum, the eventual Labor Party leader Frederic Creswell suggested that it "meant a country which can breed a virile white race."[30] British administrator Godfrey Lagden, head of the South African Native Affairs Commission from 1903 to 1905, agreed: "When I say 'a white man's country,' I mean a country in which the white man can live and flourish in succeeding generations."[31] In this framing, white prestige was all-important; as the editor of *Farmer's Weekly* wrote, in his first year of publication, "In a country such as South Africa with its large black and coloured population it is very necessary that the white race should at all times and under all conditions keep up their traditions and dignity."[32]

But many argued that racial demographics did matter, and their warnings run across the decades in which Schwarz's scheme was conceived and debated. The Rand Mines chairman in 1904 referred to "a small and scattered white community surrounded and hemmed in by an aboriginal race outnumbering them by about seven to one."[33] Smuts warned that in the absence of policies favoring white workers, the future would consist of "little black children playing round the graves of White people."[34] E. R. Bradfield, a regular *Farmer's Weekly* correspondent and proponent of soil conservation, warned of Africans' "greater fecundity over the white man."[35] The Transvaal Mining Industry Commission asked "whether the vast expanses of South Africa, so eminently adapted to white occupation, shall be the home of a great white people or be the habitation and breeding-place of masses of natives and other coloured people of mixed races, in all degrees of semi-barbarism and semi-civilization."[36] A correspondent to *Farmer's Weekly* argued that "the necessity for a vigorous and progressive settlement policy" was not only a "stable basis for prosperity" but also "our chief bulwark against racial extinction."[37]

The 1911 census demonstrated "a great increase of the native population," while the white population had decreased in many parts of Natal and the Cape.[38] The report of the Select Committee on Closer Land Settlement, published in the same year, wrote ominously that public indifference to the need to increase the white population "constitutes a national danger and involves risk to the very life of the nation."[39] These concerns echo through the next two decades. The Swedish engineer August Karlson, who designed Pretoria's water system, predicted in a 1919 pamphlet that by 1946 the demographic balance would be even more skewed. "Means must therefore be found to increase the European population by any means," he wrote, before proposing a rival scheme to Schwarz's.[40] National Party co-founder J. B. Hertzog, who became prime minister in 1924, reminded an audience in the Western Cape that there were "2,000,000 whites against 6,000,000 natives." This numerical difference, he suggested, was the reason why some African leaders were beginning to claim that "the white man must leave the country and the native must rule!"[41] But for Hertzog, as for most other Afrikaner nationalists, the solution was not to increase the white population but to simply solidify the racial hierarchy so that white dominance was beyond questioning.

The 1911 census and the impact of the South African War on white agriculture were among the factors that drove the passage of the notorious Natives Land Act in 1913. Debate over the act crystallized fears that Black tenants and the white landowners who partnered with them were causing white landlessness and impoverishment, thus jeopardizing the creation of a white South Africa.[42] Hertzog, arguing in favor of the Natives Land Act in 1912, observed the prevalence of African farming on white-owned farms "where no white person leads the way." He asked, rhetorically, "Do we then have the right to call the land a white man's land?" and argued that Greeks and Romans had left their slaves in charge of the farms while they lived in towns, thus speeding the collapse of their empires. Hertzog suggested that absentee landlords would bring the same fate onto South Africa: "The whites would be driven from the land." If whites did not care about and for the land, he argued, they would "disappear."[43] As one man wrote to *Farmer's Weekly*, "The white man aiding or abetting native squatters is helping to rob his own kind of their own rights."[44] This language of white against Black, coming a decade after the end of the Anglo-Boer War and in the context of what were still seen as "racial" differences between the two white communities, shows how conceptualizing the land as the preserve of white men could help foster a shared white identity and reorient racial categories toward Black and white.

White landowners faced pressure from their peers to rein in the independence of their Black tenants in order to level the playing field for white workers. One farmer who insisted that his Black tenants were more efficient than his white *bywoners* was physically threatened by his white peers.[45] But he was hardly unique. Many landowners lacked the money to pay white laborers cash wages, and most preferred Black tenants to white ones. Black tenants were cheaper to compensate, brought their entire households into the labor pool and their plow oxen to the farm, and were generally recognized as more skilled and industrious workers. In purportedly white parts of the countryside, Africans were the primary cultivators and effectively ran many farms; some owners, in rare but revealing moments of honesty, referred to them as "partners," and acknowledged whites' widespread dependence on Black agricultural skills. As one wrote, "Most of the farmers here are [recent] settlers, and could not get a quarter of the crops they do now were it not for having [African] squatters who have oxen."[46] In the Transvaal, where Black land purchases were legal prior to 1913, some Black farmers pooled profits derived from livestock accumulation to buy farms.[47] Even if such successes were statistically rare, they were evidence of a possible future in which Black landowners would replace white ones. In his biography of the sharecropper Kas Maine, van Onselen suggests that such fears were always in the background. After a conflict between the sharecropper and his landlord, Maine departed with the threat that one day he would own a farm and hire the former white boss as a laborer—a threat the landlord remembered.[48]

Historians have noted that the sponsors of the Natives Land Act were MPs from the districts where Black land purchases and Black sharecropping were most prominent. In the Orange Free State, wealthy farmers created a *Boerenbond* to lobby for limits to Black sharecropping. One man argued that the goal of this farmers' union was "to ensure 'the natural rights' of white men and to make South Africa a 'white man's country.'"[49] Implicit in this language was an assumption that destroying the sharecropping economy was a prerequisite for solving the "poor white problem." Hertzog suggested that poor whites could only succeed if they did not have to compete with Africans, and that the choice was existential: "We must choose whether we keep the land or the native."[50] But such social policies sat uneasily with the incentives that individual white farmers had to rely on Black tenants.

The Natives Land Act sought to regulate and limit Black tenantry and land purchases, and it has been understood as a watershed in South African history. Its formal terms limited African land ownership outside native reserves, except in the Cape; limited sharecropping in the Orange Free State; and delineated native reserves that constituted only 7 percent of the country (expanded, in 1936, to 13 percent). But it was more a "statement of intent about segregation on the land" than a catalyst for dramatic change. Its provisions did not apply to all of South Africa and, at any rate, were largely unenforced. While some wealthy white farmers used the act as an excuse to expel their Black tenants and replace them with wage labor, or to renegotiate the terms of tenantry agreements in their favor, Black sharecropping continued for decades. It was only with widespread mechanization in the 1940s that tenants were replaced with tractors. In the years after the act was passed, both the extent of sharecropping and the number of cattle kept by Africans on white farms actually increased; so, purportedly, did the number of "poor whites."[51]

The US South offered another approach to maintaining white supremacy. National Party politicians and theologians speaking on the "poor white question" at a 1916 church conference praised white American southerners for how they had "managed" a large Black population through the enforcement of segregation. (They did not acknowledge the racial terror that white Southerners had unleashed upon African Americans.)[52] British observers, too, suggested that the US South had lessons for South Africa. The British social reformer Violet Markham, who wrote extensively about her travels in southern Africa, made ominous predictions of the "swamping of the white vote by the black and colored vote" if nonracial suffrage expanded. "The experience of the Southern States shows that no white race will tolerate such a situation. . . . The white race will by fair means or foul keep the power in its own hands and thus safeguard its political institutions."[53] But others drew different lessons, arguing that the experience of the United States demonstrated the importance of white immigration. The engineer Karlson asked, "Without immigration, where would the United States be today? Instead of 8.41 whites to one coloured there would be only 4.23 white [*sic*] to one coloured, and in some Southern States there would be a coloured majority."[54]

The distribution of the white population was also a problem. Malan spoke of a "black tidal wave" engulfing the rural areas.[55] Whites across the political spectrum expressed alarm at the changes revealed in the 1911 and 1921 censuses, and historian Philip Bonner argues that the data had "major repercussions on public perceptions, public debate and public policy. Here, graphically, the widening trajectories of white and black population plotted the route to Armageddon."[56] Drought Commission report author and National Party supporter Heinrich du Toit, writing for *Farmer's Weekly*, used the census returns to justify his claim that farmers should buy land sufficient to bequeath to their sons, and that even those living in towns or on rented land should seek to buy their own farms. "This drift [to the towns] can only be stopped effectively by keeping people on the land."[57]

White rural depopulation led to calls for government policies designed to keep whites economically secure. D. R. H. Featherstone, who had farmed in the eastern Karoo for more than forty years, told the Drought

Commission, "This is a white man's country and we must help and save every white man in it."[58] Dr. Stals, a National Party MP who supported Schwarz's scheme, echoed Featherstone: "In this country, there is a small number of whites against the natives, a few civilized people against uncivilized hordes, and for that reason it is so important that not a single white person should be allowed to go under. . . . T]he existence of the European civilization in this country hinges on it."[59] A year later he tied this to agriculture, which was "the backbone of the country" and needed to be "guided in a direction which makes the future of South Africa safe." It was "our duty," said Stals, to insure the "possibility of existence" for South Africa's white population.[60]

This alarm over the decrease in the white farming population across a wide swath of the Cape and the Orange Free State, particularly in districts that were relatively arid, suggests that historical writing on white impoverishment has overlooked a crucial aspect of white fears at the time. As others have noted, scholarly writing on "poor whites" has tended to focus on the gold-producing area in the Transvaal, thereby situating the causes of white impoverishment in mining and the subsequent capitalization of agriculture, and thus the proletarianization of newly landless whites whose only other option was to compete with cheaper Black labor.[61] Fears of Black competition affected those who worked on and owned farms as well. Malan argued that white population losses were greatest in regions adjacent to large "native reserves"—proof, he argued, that it was Black competition, not drought, that was most threatening rural white livelihoods.

But others who singled out the native reserves focused on both labor and the climate. "You cannot segregate a native and lure him away from the reserves at one and the same time," wrote Eric MacDonald, imagining a world in which all Blacks worked on white farms. Abolishing independent Black agriculture would have the additional benefit of preventing a decline in rainfall, MacDonald argued. Native reserves were "situated in that part of the country that has the greatest bearing on rain precipitation," he wrote, adding that they were "fast deteriorating into desert." Given prevailing beliefs that the condition of the land itself determined rainfall, many thought that the degradation of African lands threatened to decrease precipitation across South Africa.[62]

MacDonald was rare in his penchant for holding Africans responsible for declining rainfall. As we've seen, white farmers were more commonly blamed for South Africa's "drying out." But he was not unique in connecting white precarity to climate change. In 1919 William Miller MacMillan, a lecturer in history and economics at Rhodes University and Schwarz's colleague, compared the 1911 and 1918 censuses and observed "an almost solid stretch of country, right across the center of the Union, in which there has been an absolute decrease in [white] population," while other areas had seen an increase. The most affected area, MacMillan wrote, mapped neatly onto the Karoo and adjacent areas; there was therefore an ecological component to declining white population.[63] Schwarz borrowed MacMillan's ideas, citing him in his book and emphasizing that drought was the ultimate threat to white numbers.

The sense of existential threat was rooted in whites' minority status and their economic struggles. As urban historians have noted, it was also rooted in competition with nonwhite workers. Whites demanded higher wages than their African and Colored counterparts—the result, as it was popularly expressed, of the greater material demands that a "civilized" life required. And yet employers saw little benefit in employing more expensive white workers. As a result, Colored artisans dominated many of the trades in the Cape.[64] An Orange Free State farmer stated frankly that Black artisans were as good as their white counterparts—and, since they would work for less, there was no market for white labor no matter how skilled. "This country is never going to be a white man's country," he concluded.[65]

The 1922 "Rand Revolt," when white workers demanded the reinstatement of a color bar, exemplifies these long-standing fears about white survival in South Africa. The rhetoric of the strikers and their leaders was existential: the union warned that defeat meant that "the Kaffir in future will take up the place of the white man and then we are doomed to national annihilation."[66] White women petitioned the prime minister to "uphold the right of existence of the White Population of South Africa."[67] The 1922 revolt was more than a labor strike; it was a race riot. Groups of white men attacked Africans on the street, while others armed themselves and sought to seize control of the city. Krikler argues that in the face of competition from Black laborers who would work for lower wages, white mineworkers believed "that not only was their very existence at stake, but even their ability to leave their trace upon the future." Their imagery, Krikler writes, was "of mortality, of lines snuffed out."[68]

The harsh government response to the 1922 riot—which included aerial bombardment of the strikers and the hanging of some strike leaders for treason—prompted a political backlash that vaulted the Labor Party into the governing coalition with Hertzog's National Party. The years after the strike marked the moment at which, as Giliomee states, "the state increasingly defined the relationship between white prosperity and white supremacy." From mine owners to workers to farmers, whites developed "a symbiotic relationship with the state."[69] The very people who were so critical of state experts insisted, in most cases, that the state create the infrastructure for their success, whether through economic policies or through ambitious engineering schemes.

White Precarity

Afrikaner nationalists were skeptical of solutions that rested on increasing white numbers. They argued that immigrants would likely come primarily from the British Isles, thereby entrenching British dominance within the white community. This was a charge that many of the immigration boosters denied. But the presence of some of the most jingoistic British South Africans in the pro-immigration camp did little to assuage Afrikaner concerns. For this group, then, the greatest threat to the survival of the white race in South Africa was not its minority status, but the apparent increase in white economic precarity—a situation that threatened whites' claim to superior social, economic, and political status.

This was not purely a concern of Afrikaner nationalists. As Africans acquired more education and prospered as farmers, artisans, and professionals (however small their number), they demanded more rights. A missionary in the Cape suggested that "if this is to be a black man's country," Africans should be trained in skilled work, but that "if this is to be a white man's country," Africans should not receive such training, because "otherwise you will displace the white man."[70] In a rare example of testimony from a farmer and an Afrikaner to the South African Native Affairs Commission, Dirk Cornelius Uys stated, "If we are going to make this a white man's country, we must not give them [Africans] the franchise."[71] Hertzog argued for stripping Africans of their limited voting rights lest South Africa eventually experience "the transfer of civilization to barbarism." And Violet Markham—daughter of the man who eventually helped fund the first investigation of Schwarz's scheme—observed "the fear which haunts many South Africans of a swamping of the white vote by the black and coloured vote," an outcome which would "run unnecessary risks with the foundations of society."[72]

But white economic impoverishment was seen by most whites as the greatest threat to white prestige. The "drift" of poor whites to the cities, in the language of the time, was part of a larger process of urbanization that was also reshaping Black societies. In 1890, only about 2 percent of Afrikaners lived in cities. By 1926, the figure was 41 percent, and by 1936 it was 50 percent.[73] The official number of poor whites increased as well, from an estimated 106,000 in 1921 to 300,000 in 1933. White poverty disproportionately affected the Afrikaner population: When the Drought Commission published its report in 1923, perhaps 10 percent of the white population had an economic status that classified them as "poor white," but in the Afrikaner community it was as high as 25 percent.[74] The "poor white problem" contributed to the growth of Afrikaner nationalism and, ultimately, to the formation of the National Party itself.

The alarm over white poverty was partly due to the fact that urban poverty was more visible than its rural counterpart. But visibility was a problem in part because "poor whiteism," as it was sometimes called, jeopardized the racial order. The idea that white South Africans might live a material existence indistinguishable from that of "natives" threatened the entire edifice of white supremacy, casting doubt on the inherent superiority of white people. A whole series of early twentieth-century government commissions, whether they dealt with drought, unemployment, land settlement policies, or "native" affairs, was ultimately about "the poor white problem." "The native question"—of which policies ought to govern the political, economic, and land rights of the African majority—was simultaneously "the poor white question." Giliomee observes, "The more concerned the state became with alleviating white poverty, the more 'racialized' it became."[75]

In order to be a "white man's country," South Africa, with its Black majority, needed to abolish white poverty. This was a point on which British and Afrikaners agreed. Alfred Milner, the British high commissioner and architect of Britain's postwar reconstruction policy, said as much at the end of the South African War. "The position of the whites among the vastly more numerous black population requires that even their lowest ranks should be able to maintain a standard of living far above that of the poorest section of the population of a purely white country."[76] J. J. Naudé, the superintendent of white labor for the South African Railways, agreed, arguing that poor whites who lacked ambition or were "utterly lazy" were "a danger to the State"—a "peril" akin to the *swartgevaar*, or Black peril.[77] This was a moral and existential issue. The "poor white question," former Cape Colony Prime Minister and current MP John X. Merriman told Smuts in 1915, "constitutes a great and growing evil" and threatened "the very foundation of our national existence." Whom the "nation" included is revealed in Merriman's argument that the problem of "indigent Europeans" was "bound up with the whole future of our race in this country."[78] In the same year, Malan asked, "Do we Afrikaners have the right to exist or would it perhaps not be better to commit collective suicide?"[79]

A series of letters from the 1920s illustrates what downward economic mobility among rural whites looked like, and underscores the point that landlessness was just one component of rural white precarity. On the same day in 1924 two different men, both English, wrote the governor general of South Africa, asking for his help. H. F. Wirsing had owned an "estate" of 22,000 acres in northern South Africa for at least twenty years. He and his brother had a long history in the region. In the 1890s they had been general dealers with stores in Rhodesia, the Bechuanaland Protectorate, and northern South Africa, and were contracted to deliver the mail across the region. Wirsing reminded the governor general—who was cousin to the British king—that he had once hosted him on a hunting party. Now sixty-five years old, Wirsing had fallen on hard times and owed the bank £8,500. His letter was from "a drowning man, clutching at straws to try and keep himself from sinking into ruin." Wirsing insisted that if he just had "a couple of good seasons," he could repay the entire loan. But if the bank foreclosed, "about 20 native families and myself will have to walk out ruined by drought, locusts, and interest"—a statement designed to appeal to liberal sensibilities about "native" welfare, but also one that reveals that Wirsing relied on Black tenants to work his farm.[80]

The other letter writer, Fred Milsom, had been granted a farm on a settlement for former soldiers in 1920. Milsom could not claim a personal connection to the governor general; instead, he highlighted his service to Britain during World War I. He then related a tale of the last four years, during which drought had ruined his crops and left him with no income. "My only fault is the unfortunate circumstance that I branched out during the year of the severe drought," he wrote. "I have no wish to be turned into a poor white."[81] Behind on his debt payments, he left his farm in charge of a neighbor and took up wage labor elsewhere. The Lands Department stated that by absenting himself, Milsom had breached his agreement and lost the right to his farm lease. An outraged Milsom detailed the tenancy arrangements of many of his absentee neighbors, all with Afrikaans surnames, and argued that he was being singled out as an Englishman: "If the Department only requires Dutch on the settlement, why don't they say so."[82]

More letters followed.[83] A. W. Godwin, like Milsom, suggested that he was the victim of prejudice. He accused the Lands Department of misleading him as to the potential of the farm he leased, and claimed that no one had made a success of farming in his district. He demanded that his correspondence with the Land Board be forwarded to King George, so he could see "how an Englishman is being treated in this country." Listing his dates of service and his war medals, he concluded, "I think after the above I am entitled to make an honest living in this country."[84] "Mrs. A. E. Roberts" asked the Prince of Wales for assistance. She had been born in Britain, as had been her husband, who was "well-known in Johannesburg as an honorable and right dealing man." But for the five years they had held their farm, they "were unable to grow even food, owing to terrible drought and locusts." The government was threatening to expel them for falling behind on their debt payments, and "we are threatened with starvation and homelessness through no fault of our own."[85] F. G. Staples-Cooke, another former soldier faced with losing his farm in the Transvaal, wrote, "This is my home. I have no other to go to. I have my two old parents to look after. . . . I have been farming nearly all my life."[86] His brother noted, "My brother served his King and Country all through the Great War. . . . My Dad who is English and home born fought the natives during the two rebellions."[87] C. H. Rotherforth, a returned soldier trying to support a wife and eight children on the same settlement as Milsom, wrote, "I am ruined in health by War Service. . . . And now through a run of bad luck I am being turned off my Plot into the veldt."[88]

The letters demonstrate that a consolidated white identity was far from reality. The writers claimed a right to state assistance because of their war service to Britain, their good character, and their families' history of helping to secure white dominance. And they questioned whether the South African government practiced ethnic favoritism in its rural land policies. But the correspondence also demonstrates how white identity was fractured along class lines. The 1913 Natives Land Act was motivated in part by a desire to curtail competition from independent Black farmers, and it led to the scaling up of organized Black resistance as rights and opportunities were constrained. But it did not lead to universal white prosperity. Farmers at the time recognized in letters to *Farmer's Weekly* that the law would benefit wealthier farmers who could afford to pay cash wages to their laborers, at the expense of poorer landowners who compensated tenants by allowing them to farm partly for themselves.

These tales of white desperation, through a quirk of the archives, are clustered over just two years, 1924–26, as leases for farms granted to former British soldiers terminated and farmers were required to purchase the farms outright. The writers are unusual in that they were English-speaking whites, a minority within the "poor white" community. Most, but not all, were beneficiaries of schemes to put English-speakers on the land to counter Afrikaner dominance and increase the overall white population. The years they wrote were not uniquely bad, at least not according to official records: 1925 was in fact a year of record rainfall in southern Africa. But their farm leases had begun during drought years, putting them behind from the start. These men and women, many of them newcomers to farming in South Africa, did not argue that the climate was changing—a contrast to the ideas embraced by many more experienced white farmers. Instead, they understood their experience as the result of bad luck in the form of an exceptional string of poor seasons. Most insisted that just one good year would turn things around.

None of these appeals to the governor general's sympathies worked. Their requests were forwarded to the Lands Department, whose employees tallied the money the farmers had been loaned and how little they had paid back to emphasize that in fact, one good year would *not* turn things around. Officials at the Lands Department had little sympathy for farmers' struggles, and blamed the letter writers rather than larger circumstances: "I do not think there is a more hopeless case than this. . . . [Staples-Cooke] always has some new and impossible scheme on."[89] "We need waste no sympathy on Milsom. He is a very unsatisfactory settler."[90] "Rotherforth was quite hopeless as a settler. . . . He would never be able to make a success of his farming operations."[91] Rotherforth ended up working at a labor camp for five shillings a day: "I am sorry to have to say that in a few weeks time I shall not have any clothing left for myself and family to be able to take up any respectable position again except it be on a farm as you can hide a lot of poverty on the farm."[92]

These desperate farmers then fade into archival oblivion. But the correspondence reveals divergent understandings of the causes of white poverty and the limits of government programs designed to subsidize white agriculture no matter which party was in power. Many in the government bureaucracy thought that poverty was the result of personal shortcomings, and that charity was a moral hazard to be avoided. This instinct among the political and bureaucratic elite superseded party politics. These men asked for help during the early years of the Pact government—a coalition of the ruling National Party, newly elected to power by rural Afrikaners, and the urban, worker-focused Labor Party. (This may have formed the context of the letter writers' complaints about favoritism toward "Dutch" farmers.) The National Party, under the leadership of J. B. Hertzog, criticized Smuts's South Africa Party for favoring the "big man" over the "little man," and rode to power on the strength of support from white farmers. As an opposition politician, Hertzog had expressed sympathy for poor whites, and argued that they were the victim of external forces. But, as Giliomee notes, after becoming prime minister in 1924, "he was soon exasperated by their incessant demands." Certainly, the bureaucrats working under him seemed to be.[93]

The most famous example of expert bureaucratic discourse about poor whites is the Carnegie Commission of Investigation on the Poor White Problem, which was funded by the American Carnegie Corporation but staffed almost entirely with Afrikaner social scientists. Like the commissions on drought and rainfall, this commission selectively appropriated popular knowledge about white poverty while ignoring aspects of public evidence that did not fit the agenda of the chair, University of Cape Town lecturer E. Malherbe. Census evidence for rural depopulation, in the view of the commission, was due less to drought or Black competition than to increased demand for workers in the cities, and to the capitalization of agriculture. When it came to the origins of "poor whiteism," its conclusions largely repeated descriptions by three decades' worth of government commissions of the supposed moral and physical deficiencies of individual poor whites, including malnutrition, laziness, and poor parenting. Drought losses were due to land mismanagement. But Malherbe argued that changing the environment in which poor whites lived could change all of these factors. This message was often distorted by popular media to better fit with existing eugenicist narratives, such as that of the Drought Commission, which concluded in its report, "Indigency is due, not only to economic reasons, but also to a psychological and physical state, brought about by such causes as unfit parents, inbreeding, underfeeding, disease, and climatic conditions."[94]

Schwarz, as usual, found a way to represent both perspectives in his plan. "I well appreciate the difficulties of the redemption of the Poor Whites; they are sunk into such a hopeless state of degradation," he wrote, adding, "It is no use doling out charity to them; I have had personal experience of this myself." But Schwarz also suggested that interventions had done little to offer poor whites an attractive future. The Kalahari Scheme would succeed in the redemption of poor whites as well as that of the Thirstland, providing land sufficient to allow settlers "to save money to buy for themselves some of the comforts and amenities of civilized existence" while ensuring that they stayed on the land, "for which they are by nature and inheritance best fitted."[95]

Whites outside the bureaucratic ranks often took a more sympathetic view of poor whites than did experts and the commissions they led. Reverend Marchand, the treasurer of the Dutch Reformed Church's Kakamas irrigation scheme for poor whites, told the Select Committee on Drought Distress Relief, "I do not think the number of absolutely lazy people is as large as some of us think it is."[96] People who saw the struggles of their rural neighbors often had a more expansive view of the causes of white poverty, and their sympathy for poor whites was often embedded in larger critiques of government policy and the operation of the capitalist economy. They argued that a lack of markets and transportation infrastructure, the high cost of land, subsidies for large absentee landowners, and unfavorable taxation policy all made farming an economically risky proposition, and put farmers in a constant state of precarity.

This more expansive view contained within it a sense that any farmer, through an unfortunate twist of fate, could end up as a poor white. Transvaal farmer Temple Nourse wrote that he understood the forces that generated white poverty due to landlessness because "I narrowly escaped it myself."[97] Witnesses for the Drought Distress Relief Committee recounted cases where "well-to-do farmers" were forced to take up heavy manual labor building roads for minimal pay—work intended to assist destitute whites—in order to tide themselves over during droughts.[98] Members of Parliament, summarizing the views of their constituents, reflected this generalized anxiety. The pervasiveness of the belief that anyone could become a poor white was acknowledged by Merriman when he wrote in 1915, "There is a general conviction that this misery, and degradation, is widespread, indeed universal, and that it is on the increase."[99]

In testimony to the Select Committee on Drought Distress Relief, Johannes Hendrik Schoeman, the MP for Oudtshoorn, an ostrich-farming region of the Karoo, suggested that the farmers who were in financial trouble were in many cases "the largest asset to this country." But they would require government assistance to survive the current drought. Schoeman argued that it was partly the government's fault that they were in this situation:

> I can assure the Committee that the position of the unfortunate people, even in the agricultural parts of the country, is not due to carelessness or to laziness, but it is owing solely to the unprecedented drought that prevails in the country. If we had had proper irrigation works I am certain that even in the parts of the country that are now worse affected we should not have had this poverty at all."[100]

Government inaction sent a clear message: "'We will not do anything, and we will drive you off the land into the towns where you will become a class of poor whites.'"[101]

The dominance of Afrikaners in the ranks of "poor whites" does not seem to have prevented English-speaking whites from imagining that they could one day join their ranks. One wrote to *Farmer's Weekly*: "Our Government, instead of solving the poor white question, will make us all poor whites, if we follow their advice."[102] He denounced, among other things, "expert" advice that farmers should borrow from the National Land and Agricultural Bank, formed in 1912, to purchase farm equipment, noting that a few bad years would cause them to fall behind on debt payments and lose their land.[103] His critique echoed the observation of a Karoo farmer who on an abandoned farm had seen "windmills, a large irrigation furrow from the river, cement tanks," and whose friend remarked, "I think we farmers spend too much on improvements on our farms"—implicitly linking the farm's failure to the precarity imposed by debt.[104] Farmers also argued that tax policies prevented them from maintaining the white man's country. One wrote that small farmers in his district had offered a good living to white tenants until a tobacco tax was levied: "Down went the small farmer; in came the moneyed man, and off went the poor white to the town, and also a good few of the small farmers. In a few more years, if the tobacco tax is kept up, all the small farmers will be off to join the poor whites, and a few rich farmers will own the district."[105]

This rhetoric to some extent matched that of the National Party. MP Wilcocks, one of the founding members, spoke in 1917 "of the man who at one time had been a farmer, but had now become a poor white."[106] A few years later, M. L. Malan, a National Party MP from the northern Free State, argued that "even the man who owned a good farm had a difficulty in making ends meet under present conditions."[107] The shared specter of economic failure and land loss, like that of "drying out," was one element of a collective rural white identity, which included a shared sense of white superiority that coexisted with a sense of Afrikaner/British difference. In 1912, a correspondent writing under the name Unity urged *Farmer's Weekly* readers to "unite" and "demand what is right and good for the whole country. . . . Be white men. There can only be two classes, white and black, and white must dominate and cooperate."[108] Another man insisted on the equality of all *white* men: "The native is intended by God as a servant to the white man. . . . He is a savage by nature and an inferior by birth, no matter what his education and qualifications, and the 'poor white' is a brother to us."[109] And a defender of poor whites argued that they wanted education and social mobility, and appreciated assistance when it was given: "They are not prehistoric beings." He asked readers to "put their hands out and help their poor white brothers."[110] The paternalistic vision of segregation as a system that would help Black as well as white people had little purchase among these men.

There were differences of opinion about what help should be given to poor whites, however. Government commissions and their witnesses repeatedly expressed their opposition to what Marchand called "pure charity." And they created typologies of poor whites. The "honest" and deserving poor were the victims of bad luck, while those who were lazy and "degraded" required "rehabilitation" via paternalistic, authoritarian interventions. Marchand drew a distinction between a poor *man*, who was a victim of circumstance, and a poor *white*, who was a victim of his own moral shortcomings; some white poor people made a similar distinction.[111] In this view of the world, self-imposed poverty stripped white men of their manhood as well as their racial identity. By contrast, Black poverty was largely understood as appropriate to Africans' supposed level of "civilization," while Black economic success was a threat to racial hierarchies and white futures. Despite its disparate causes, white poverty was seen as a pathology, a sign of racial ill health, and poor whites were treated as the patients of the state.[112] For the honest poor, "preventative" measures were required, whereas for the degraded poor, "curative" measures would be needed.[113] Marchand, the treasurer of the Kakamas irrigation scheme for poor whites, referred to a man who had been a "first-class farmer" and "a big landowner": "He came down, but he is now recovering."[114] Malherbe, later chair of the Carnegie Commission, argued for social science research into the causes of "this malady."[115]

Not surprisingly, few white men saw themselves in the image of a poor white who had "degenerated" and lost his work ethic. They resisted the harsh discipline of labor colonies that were aimed at their "rehabilitation." A man working on the Hartbeespoort Dam irrigation scheme complained that workers were treated like "slaves" and asked, "Why must I be treated like a white Kaffir? . . . I am an old burger of the country and have been shot to bits in its service."[116] His language reveals not only a rejection of state ideology around white poverty, but a robust sense of racial privilege. It was the state, not his own economic precarity, that undermined the racial hierarchy. His complaint echoes those of poor white farmers who, unable to compete with rich farmers for African labor, argued that they had to "use their children as Kaffirs."[117] White identity here, as Duncan Money and Danelle van Zyl-Hermann point out, was fractured by class tensions and resentments.[118] Yet whites were united in their sense of racial privilege, their level of material privilege notwithstanding.

For poor white men, that racial privilege was imperiled by the kind of labor they performed and the terms under which they performed it. Those who were compelled to engage in manual labor for other white men felt the precarity of their racial status as well as their economic status. This was especially true of the poorest and most destitute whites. When J. J. Naudé, the superintendent of white labor, was interviewed by the Drought Distress Relief Committee, he emphasized the need to give the children of poor whites a "practical" education rather than focusing on academic subjects. This prompted Merriman, who was questioning Naudé, to ask him if he had read "a book written by a negro, Booker Washington, about the value of working with the hands."[119] The idea that Black people in the US South and poor white people in South Africa required the same intervention underscores the extent to which white poverty blurred racial hierarchies and categories not just in a theoretical sense, but also through the policies designed to combat it.

This exchange helps explain why some destitute whites saw certain kinds of labor as compromising their claim to whiteness. People commonly observed that white South Africans refused to perform the kinds of manual labor that their counterparts routinely did in North America, Australia, and New Zealand. By the early twentieth century, the idea that white men in the majority-Black country would not perform manual and unskilled labor, including on farms, was repeated in the pages of newspapers, in testimony to government commissions, and in public speeches. Their fellow whites were divided on what poor whites' resistance to performing manual labor signified. Certainly, many saw it as a sign of laziness and moral failing. But others argued that it was a positive sign that poor whites remained invested in the racial hierarchy. E. B. Watermeyer, an MP from the arid district of Clanwilliam, when questioning a witness on the Select Committee on European Employment and Labor Conditions, suggested that the "false pride" that prevented a white laborer from willingly working alongside a Coloured laborer was a good thing, because the white laborer was aware that "the coloured man would have no respect for him." Racial pride "is one of the redeeming features of the poor white question, which gives us the hope that when we try to do something for this people they will respond."[120] One man wrote to *Farmer's Weekly* that this racial pride meant that total territorial segregation would be required to solve the poor white problem, even to the point of removing Black labor from white land. A white man "will not or cannot labor himself while the black is there."[121] This advocacy for "pure" partition was never the dominant vision of segregation. But some agrarian-minded English speakers and Afrikaner nationalists in the first half of the twentieth century saw it as the only feasible path to a country of white farmers.

The Agrarian Mystique

Given the growth of South Africa's cities and its industrial sector in the early twentieth century, one might assume that attempts to solve the problems of white poverty and a small white population focused on its cities. But in fact, the countryside loomed large in efforts to create a white man's land. From highly regimented labor colonies where destitute whites could work toward land ownership to schemes to lure well-resourced British farmers to South Africa, and the Natives Land Act of 1913, which attempted to legislate white-Black relations in the countryside, rural spaces were central to the quest to shore up the racial hierarchy. The twin dangers of white poverty and small white population numbers were met with an expansive "back to the land" campaign that drew on a variety of intellectual traditions, crossing political lines and rural/urban divides. Its reach helps explain why Schwarz's vision of creating new lands for white settlement initially drew support across the political spectrum.

The idea of a white man's country was both spatial and racial: it meant a white population distributed across the territory's landmass, rather than simply concentrated in a few urban centers and prime farming districts. Whites' security and their claim to the country lay in their physical presence on and ownership of South Africa's agricultural lands. The necessity of increasing the number of rural whites united pro-immigration English speakers with anti-immigration Afrikaner nationalists focused on the "poor white problem." As Fedorowich writes, "Politicians from all bands of the political spectrum agreed on the urgent necessity to secure a large white population on the land," but differed on whom and from where.[122] What they shared was a sense that the white population, whatever its size and wealth, could not be secure if it was garrisoned in South Africa's towns and cities.

Like many ideas at the heart of Schwarz's scheme, "back to the land" was a concept both transnational and racial. Southern African whites were part of a larger cultural world that romanticized white agrarian life. This agrarianism drew from multiple sources. Aristocratic fantasies animated white settlers from Kenya to South West Africa. As *Die Allgemeine Zeitung* in South West Africa opined, "We did not come to Africa just to share the blessings of culture with our 'black brothers.' . . . The invader, especially when he stands on a higher level of culture, becomes the ruler. . . . The ruler wants to shift as much of his work as possible on the ruled and live as comfortably as possible."[123] But those seeking to build a white man's land imagined a more egalitarian rural order—one that was dominated by self-sufficient, independent white men. Some British officials, including Milner, held romantic views of English peasant life and culture, albeit paired with a focus on commercial, "scientific" farming—views that Schwarz appeared to share. But if that romanticized English past potentially contained within it a dose of benign hierarchy, most South African whites expressed more affinity with its American variant: the Jeffersonian ideal of enlightened, independent, and white yeoman farmers who would form the foundation of a democratic and egalitarian white man's country. Early Afrikaner nationalism, which saw independent farming as central to Afrikaner identity, drew on some of the same beliefs about the importance of agricultural life to the political order. In the late nineteenth and early twentieth centuries, capitalizing agriculture and rapid urbanization generated anxieties about rural depopulation in Europe and North America, and led people to repurpose older agrarian ideals. In South Africa, too, these changes seemed to threaten agrarian life. It was in this context, then, that literature and government policies in all of these places valorized the idea of a return to the land as the key to individual self-fulfillment and national vigor.[124]

This agrarian mystique was inseparable from transnational white supremacy: it was white men whose vigor and self-fulfillment were prioritized. American agrarians were not encouraging Black migrants in Northern cities to go "back to the land" as independent landholders. Germany's *Blut und Boden* policy was explicitly white supremacist. In South Africa, the "back to the land" movement was embedded in a shared conception of the central place of agriculture in a white man's country. And in all cases, the vision was gendered: the patriarchal household was seen as the proper unit of independent farming. It not only maintained order within families but maintained a social order in which white men were all theoretically equals. As one farmer wrote to *Die Landbouweekblad,* it was "fatal" for an Afrikaner to sell his farm and move to the city. "On his farm, he feels like a man even though he doesn't own as much as his richer neighbor. There he has a sense of independence and self-worth, because he is the boss of his farm, and as far as his social status is concerned, he is on an equal footing with his fellow farmer and they treat each other as equals."[125]

This agrarian ideal could exist alongside the individual disillusionment of farm families, whose ambition for their children might be secure jobs in the towns and cities. Farming was a national, not just an individual, virtue; agrarians argued that it was the only truly sustainable industry and the bedrock of South Africa's character as a white man's land. *Farmer's Weekly,* on its first anniversary of publication, wrote, "Agriculture is the backbone, the greatest industry of South Africa."[126] This language was embraced by urban professionals as well as (and perhaps even more than) by farmers. Peter MacDonald, a civil engineer proposing ways to increase the rural water supply, wrote that "the farmer and food-producer is the real foundation of the country."[127] Claude Lowe, who called himself a "professional business organizer and something of an engineer," urged the public to support "climate redemption" and argued:

> When a drought is on and the farmer is making nothing, business comes to a standstill in every line. If he cannot pay, as he cannot if he is making nothing, the doctor, grocer, tool merchant and storekeeper generally, insurance companies, and in fact everyone down to the parson, goes short. Why? Simply because all wealth in an agricultural country comes from the land.[128]

Lowe wrote this in 1933, during the Great Depression and the final year of the "Great Drought" to be discussed in chapter 7. Agriculture in that year earned just over half of what mining brought to the economy, and even less than the country's nascent manufacturing sector. But Lowe, living in Queenstown, still saw South Africa as "an agricultural country."

Support for Schwarz's scheme reflected this agrarian mystique. When his supporters sounded the alarm over desiccation, they understood the danger in terms not just of individual farmers but of survival of white society as a whole. The Schwarz supporter and National Party MP Albert Jacobus Stals argued that agriculture was "the backbone of the country upon which our existence depends."[129] In an article for the *Cape Times*, Karoo farmer and Schwarz supporter W. G. Collins wrote, "We farmers are so often told that we are the backbone of the country. Well, unless something really big is speedily done to assist us, I am afraid we will become all backbone and no country, for it is as surely being taken away from us (in that it is capable of supporting less each succeeding year) as if by invasion of a foreign race."[130] The equation of an expanding desert with invasion by a "foreign race"—is there any question that this hypothetical race was imagined as not-white?—illustrates how desiccation was understood to threaten whites' existence.

Agriculture served the cause of white dominance in several ways. In the early twentieth century, Afrikaner nationalists naturalized the link between the *volk* and the land—a unifying idea at a time when a growing percentage of Afrikaners were living in urban areas, removed from their rural origins. This agrarian mythology operated at an individual level: Coetzee suggests that in the farm novels of prominent Afrikaner authors, people recovered their true selves by returning to the earth.[131] It also operated at a collective level, legitimating white possession of the land and reinforcing the idea that the land was "the exclusive preserve of the white man and his family."[132] But the nostalgia and racial politics inherent in "back to the land" were not the exclusive preserve of Afrikaner nationalists. Those seeking to recruit British immigrants to the agricultural economy used similar language but drew on romantic visions of English village life, revealing agrarianism's transnational reach. When pro-imperial Union Party Senator J. J. Byron published a pamphlet aimed at luring British farmers to South Africa, he titled it *Back to the Land* despite targeting people who had never been to the subcontinent, much less its farming regions. The return in this case was to an agrarian way of life, not a particular place.[133]

"Back to the land" therefore consolidated multiple agendas. One was the return of poor whites to independent farming. This idea was not limited to Afrikaner nationalists; many reformers, Afrikaner and British alike, believed that placing poor whites "back on the land" was vital for their redemption because cities were places of corruption, particularly for those whites who "drifted" there without workplace skills. National Party Senator Andries D. W. Wolmarans, former secretary of state for the South African Republic, advocated setting aside money for irrigation schemes "in order to persuade poor whites now in the Union to leave their urban environment, where many of them are morally ruined."[134] Rural living was virtuous for everyone. Wolmarans's political opponent, Union Party Senator Byron, argued that "the sound settlement of a greater white population on the land" would "make the country great both morally and materially."[135]

The campaign to expand white farming served political and economic purposes as well as moral ones. Afrikaner nationalists saw their political power threatened by the economic precarity of their ethnic compatriots. British imperialists hoped to counter Afrikaner dominance in the countryside by settling English speakers on farms. They also wanted to increase South Africa's low agricultural productivity, which was blamed on "backward" farmers—generally understood to be Boers. Supporters of immigration argued that bringing progressive, scientifically minded Englishmen to the land would modernize agriculture. They published pamphlets, set up recruiters in Britain, and established programs to settle ex-soldiers, most of them British (and some of whom we've already met), on farms.[136] The goal was the creation of a yeoman-style racial democracy in Africa, one founded on modern farming practices.

But few wealthy Britons took the bait, and the settlement of poor whites and ex-soldiers could not alone ensure a white countryside. Other measures were required. South West Africa, seized from Germany by South African forces in 1915 and then awarded to South Africa under a class C mandate in 1919, was imagined as another source of farms for landless whites. There were 1,138 white-owned farms in the territory, Gessert's among them. Over the next decade, 880 farms were allotted to new settlers, and the colony's white population doubled.[137] But South West Africa offered limited potential as a country of yeoman farmers, not least because its aridity meant that farms had to be enormous to be economically viable. Instead, South West Africa looked very much like the version of a white man's land that had a small group of white men sitting firmly atop a subservient class of Black laborers. Agronomists, colonial boosters, and supporters of "progressive" farming argued that "closer settlement"—placing whites on smaller farms—was the safest route to a white South Africa. The ecological arguments for this idea will be discussed in the next chapter. But the call for "closer settlement" also contained social arguments. Expansive farms were a recipe for isolation and a tenuous white presence on the land. Schwarz expressed this idea in 1919, when he wrote about an advertisement for a 32,000-acre (50 square mile) farm in the Karoo. "How can we in South Africa expect to make our country a habitable one under these conditions?" Schwarz asked. "The population must remain extremely scanty, and with the enormous distances between the farms there can be no proper communication, no cooperation, and none of the amenities of civilized existence." People like the Orange River irrigator Carl Weidner were happy in their desert solitude, Schwarz wrote, but men who thrived in such conditions "are few and far between; I have seen too many cases of men having gone to pieces in their lonely situations to wish to perpetuate this state of affairs." While irrigation schemes allowed for "closer settlement" in limited areas, Schwarz wrote, "my scheme aims at making the country habitable all over."[138]

What would be the racial makeup of such a country? The campaign to get people "back to the land" and Schwarz's scheme to "make the country habitable all over" were rooted in a vision of South Africa's future that required a white countryside. Some who had advocated for the Natives Land Act had hoped that it would create opportunities for white laborers by foreclosing the possibility of relying on Black sharecroppers. But the plan to achieve universal white prosperity by limiting Black prosperity foundered on white class antagonisms. Farmers were not eager to employ white labor, and many landless whites were not eager to do the kind of work farmers required. A man who wrote, "The poor whites on the whole are a useless lot" expressed an opinion widely shared by farmers defending their use of Black tenant labor.[139] Another insisted that willing white labor did not exist. "The so-called poor whites . . . are out of the question, having been tried and found wanting. We want real workers."[140]

Where would those workers come from? Some proponents of white labor insisted that there were abundant and willing white workers—if not in South Africa, then overseas—waiting to be recruited. An anonymous *Farmer's Weekly* correspondent suggested that the widespread importation of white labor "would act as a safeguard against the black question, which is greatly required."[141] But reality did not bear this out. Many African farm workers suffered brutal labor and living conditions, with few protections against violence, malnutrition, and withholding of their wages. Given white farmers' reliance on this hyperexploitable, ultracheap labor, it's hard to imagine that a campaign to switch to white farm workers could gain any traction. Through the 1910s and 1920s, people asked how to manage the differences in the supposed material requirements of African and white laborers without burdening employers with high labor costs. In a memo to government departments, Hertzog referred to "civilized labor" as a "standard of living generally recognized as tolerable from the usual European standpoint." By some definitions, this included the ability to hire a servant.[142] Others suggested that whites simply had to learn to do without Black labor. As Temple Nourse, the farmer who had "narrowly escaped" becoming a poor white, wrote, "Every able-bodied poor white who came from the land should be taken in hand, brought back to land where a living can be made . . . and taught how to make this living with his own hands and not depend on the black's."[143]

While proposals to manage without Black labor were usually aimed at lower-class whites—few wealthy men seemed inclined to give up their servants or Black farmworkers—the idea that there was a choice to be made between "the land" and "the native" led some whites to contemplate a future without Black labor. One man sent *Farmer's Weekly* a clipping from a Baltimore newspaper about a supposedly successful experiment using only white labor on a former Virginia plantation.[144] Another reader arguing for white immigration suggested that "if white labor is not encouraged and the blacks driven out, there will be no land for the white children when they do grow up. This is going to be a white man's land or a black's; it can't belong to both."[145]

An Invisible Majority

This idea—that "driving out" the black majority was somehow a possibility, one that would yield a white majority country—is striking both because it foreshadows the ideas around the Grand Apartheid project in the late 1950s and early 1960s and because it indicates a popular commitment to radical segregation or some other form of population removal—even if only in the abstract—at time when, historians have argued, those ideas were still only vaguely conceptualized by political leaders.[146] The first use of the word "segregation" in South Africa is conventionally dated to 1903, and its meaning remained fluid into the 1920s.[147] Hertzog, the politician most associated with segregation, was vague about its meaning. "The elusive quality with which he invested segregation was its very strength, for it drew differing groups into its discourse, always promising, never quite revealing."[148] Deborah Posel has argued that scholars have too quickly dismissed Afrikaner nationalist voices that argued that the preservation of white supremacy required "total segregation," including the replacement of Black labor with white labor in the "white" areas. While this is not the policy that came to dominate in the 1950s, it was a significant strain of Afrikaner thinking.[149] But it was not confined to Afrikaner nationalists, and it was discussed well before the 1940s.

In the early twentieth century, faced with the reality of an African majority, some whites suggested that a different outcome was still imaginable. Hulett, the former secretary for native affairs, told the South African Native Affairs Commission, "If you can get the black races out of the country, then we can have a clean slate, as other countries do."[150] A man wrote *Farmer's Weekly* noting that he had seen enough letters over how to treat and employ Africans on farms: "Now, I should like to read a few as to how to get rid of them altogether."[151] While historian John Cell argues that "the monopoly of the dominant group over the political institutions of the state" lies at the center of segregation, many whites saw conversations about the place of Black South Africans as lying *outside* of politics. The man who asked how to "get rid of" Africans opened his diatribe with, "This is not a political letter, only a racial one." *Farmer's*

Weekly, which published many such letters, stated in its first anniversary issue that it was a "non-political" newspaper that stood for "agriculture and the advancement of the agriculture of South Africa . . . an honorable cause, a cause which should command the allegiance of all true South Africans."[152] Those true South Africans were by default white and male, and they already controlled the state.[153]

Like "South African," "farmer" was a term reserved for white men. Those white men debated its specific meaning quite fiercely; there were entire months where the letters pages of *Farmer's Weekly* ran debates over the question "What is a farmer?" Progressive farmers argued that real farmers used modern methods and conserved their land. Middling farmers accused progressive farmers of being "checkbook" farmers who had professional paychecks to subsidize all those fancy improvements. Real farmers, for them, were those who made their living off the land, not those who could afford the stud animals that won prizes at the fair. But certain groups were unquestionably excluded from the category of farmer. White women were called farm wives, even if they farmed. Whites called Black workers "natives" or "squatters," regardless of their role on the farm and despite the fact that, in the cases of absentee owners, many ran farms. Even those farmers who insisted that Africans were key to white agricultural success, including a few who dared publicly refer to sharecropping as a "partnership," never glorified the Black partner with the label of farmer. By contrast, white men who were described as lazy and destructive in their land use practices still qualified as farmers. A British immigrant and former sheep farmer who accused trekboers of breaking gates and fences and of grazing livestock, cutting trees, and burning pasture on other people's land without permission complained that they "walk all over one's farm as if it was their own." This letter encapsulates every prejudice that self-identified "progressive" farmers had against mobile pastoralists. Yet the offending livestock owners were still bequeathed the title "farmer," a title never given to a Black cultivator or a white woman in the sources I have read.

The taboo on referring to any white woman or African working the land as a "farmer" indicates the ideological importance of this word to the constitution of a white man's land, and hints at why experts, as well as government officials (especially in Hertzog's Nationalist government), "hailed back-to-the-land policies even when they—in an age of mechanization, scientific farming, and (potential) overproduction—meant prioritizing social engineering aims over economic rationality."[154] While many white men relied on Black farming partners to keep and profit from their land—and while we may assume that their wives were heavily involved in the farms—this was commonly denied in public discourse. Black farming ability was minimized: Proper white farmers plowed the soil deeply, while "natives . . . only scratch at the surface." African farmers used "small, insignificant little toy-plows."[155] South Africa's history prior to the arrival of white settlers was not portrayed as agrarian, even though agriculture had been practiced there for nearly two millennia. A man who manufactured agricultural machinery told the Select Committee on Closer Land Settlement that "100 years ago South Africa was a huge big game preserve, and was only peopled in a few parts."[156] The acting secretary for agriculture told the same committee, "That the pastoral possibilities are great one knows from the amount of large and small game it originally carried"—rendering African cattle as invisible as their owners.[157] There were contradictions in these narratives. While some argued that Africans scarcely worked the soil at all and thus had minimal impact on it, others accused Black farmers of "soil annihilation."[158] But claims of African destruction were relatively rare in the early twentieth century.

The invisibility of Black farmers reflected a broader Black invisibility that characterized everyday white discourse. At the most basic level, whites did not include Black South Africans in the category of people. The magistrate Scully wrote to his wife from Griquatown, stating, "There is no one else in the place—except law agents, police officers, etc." In another letter, from Caledon, he wrote, "I have not seen a soul outside the house and the office since you left."[159] Scully was a magistrate; his job was to oversee the local population. But they were invisible when he accounted for who populated his days. The language of the committee on closer land settlement also provides multiple examples. In the report, "unoccupied" land was land with no resident white farmer but only Black tenant farmers.[160] When a Cape ostrich farmer and member of the House of Assembly spoke of the urgency of "having more people in this country," he did not include Black South Africans in his category of "people." The agronomist William MacDonald told the committee, "We must never forget that we possess millions of acres of some of the richest land in all the world lying absolutely unpeopled."[161] Senator Southey stated, "We want more people in this country. This is the great want of the country."[162] Here, and throughout writing about the place of agriculture in South Africa, "farmers," "South Africans," and "people" were white by default. This fact was so obvious to the entire white community that it did not require stating; it was simply linguistic convention to extinguish Black existence. When Schwarz asked one of his audiences in 1924 how many people were in South Africa, he answered his own question by stating "one and a half millions," thereby erasing the entire nonwhite population from the category of "people."[163]

Africans were acknowledged in discussions of farm labor, segregation, and, more rarely, soil erosion. But they are almost entirely absent from conversations about the role of farming and farmers in South Africa, and indeed about the "drying out" of the country. Their views were not published in *Farmer's Weekly* or in daily newspapers, at least not in any letter where the writer identified as Black, though whites sometimes legitimated their own authority by citing "native informants" or using names like "moKalahari" (person of the Kalahari). Historical mythologies of empty landscapes that justified colonization are a well-known feature of Afrikaner identity and, later, nationalism. But this sleight of hand by which a world full of people was rendered empty was by no means limited to Afrikaners or to historical narratives. Long after the land was claimed and occupied, the Black majority continued to fade in and out of the white minority's field of vision. In everyday conversation, Black South Africans had *already* been "driven out" of the country that whites imagined they inhabited. A few scholars have noticed this relative lack of white attention to Black people. Ivan Evans observes that the "Afrikaner civil religion" that emerged after the South African War was notable for "the marginal attention it devoted to Africans."[164] J. M. Coetzee writes, "Blindness to the color black is built into the South African pastoral."[165]

This blindness formed the content of white imaginaries about a "white man's land." In much of day-to-day life, white South Africans inhabited a world in which a Black majority went unacknowledged. Most whites in South Africa thought about their Black neighbors in terms of labor; much of the rest of the time, they thought very little about them at all. Such erasures were enacted through a variety of practices, including everyday discourse, policymaking, and idealizations of the future. At first glance, the concern with white numbers sits oddly alongside the erasure of Africans in conversations about the future whites desired for their country. But this conceptual erasure, to use Lorenzo Veracini's term, is a key aspect of settler colonialism around the world; it only seems remarkable in the context of South Africa if we accept racial demographics as the determinant of settler attitudes. A focus on population numbers discounts the importance of settler imaginaries, those shared visions of social order that reflect a community's deepest aspirations. Those imaginaries, revealed in casual linguistic conventions, help us understand why few whites even considered the fate of the African majority as they weighed solutions to the drought problem.

Another reason to not think about Black South Africans was that by the early twentieth century there was already, as Jakob Zollman writes, a growing sense of the "impossibility of continually upholding white dominance over the disenfranchised majority."[166] Cohabitation, much less equality, was considered out of the question by the vast majority of whites. But the version of the "white man's land" in which a white minority formed an aristocracy unquestioningly obeyed by the Black laboring masses became less and less imaginable over the long term. This lack of political options compounded whites' existential fears: the perception that there was no path to a future that would be secure and would maintain white dominance over the long term. In promising to overcome environmental and territorial limits, Schwarz's scheme offered something unattainable within current realities.

In that scenario, water was the key to a secure future. As a Transvaal man wrote in 1925, a lack of water on many arid farms was "a very real cause of poverty which sooner or later leads to 'poor whiteism.'" South Africa's arid lands were "enormous and, given water, are capable of carrying a large European population." A little state intervention could have a large social impact, he argued. "Will it be better for the State to have these areas owned and occupied by large numbers of men, each responsible for a home sheltering a wife and family, or to have it owned by a land-grabbing company with one object—dividends?"[167] An MP from Oudtshoorn, in the Karoo, insisted that the "distress" in his district predated the most recent drought and was due primarily to a collapse in the price of ostrich feathers—a product that had led to an economic boom in the area, followed by a crash when global fashions changed. Drought was "the whole thing," Schoeman insisted. "People were making money right through the depression, but when the drought came it was impossible for them to keep on doing so. In some parts of the district there has been no crop for the last 18 months or 2 years." He urged the government to invest in irrigation schemes in the district, stating, "The only thing we want is the water."[168]

5 * Watering the White Man's Land

In March 1918, Schwarz's supporters in Parliament called for a government investigation of the Kalahari Scheme. Their colleagues responded that the government also ought to study irrigation possibilities within South Africa itself, but Schwarz's supporters argued that domestic irrigation was an entirely separate issue and did not belong in discussions of the Kalahari Scheme. In fact, however, these were simply different approaches to a problem that everybody acknowledged: South Africa did not have enough water, and it allowed the water it had to "run to waste" to the sea. Experts asserted that only about 4 percent of the country's rain became overland runoff. But the shared belief that vast quantities of water were lost to human use united people who held a variety of views about drying out.

This chapter explores how white southern Africans understood water—where it was, where it belonged, what futures it might enable. Water was necessary for South Africa's industrial and urban expansion, but in popular conversations and in government commissions, water was perceived primarily as a means to consolidate a white agrarian world. Placing more whites on the land meant dealing with ecological constraints as well as labor shortages, and water was the biggest ecological constraint. Water was therefore crucial to the quest to create a white South Africa, and it helped construct white identity. The refrain of water "running to waste" reflected the idea that controlling water and using it to its full potential was synonymous with civilization. White South Africans lamented that their water rushed away unproductively to the sea. It was their capacity to change this, and the extent to which they succeeded, that distinguished them from Black South Africans who supposedly did nothing to manage water other than appealing to charlatans who claimed to make rain. Water management therefore became another marker of civilizational status and racial identity.

Hydrology was a well-developed field in the British Empire, and many of the men who debated the proper use of water in South Africa had experience in India and Australia. Others had visited or worked in the United States, widely praised for its efforts to "reclaim" its arid lands. Within the British imperial context, engineers' efforts to "conjure water in places where they deemed it to be lacking" were described as "improvement."[1] Hydroengineering was partly informed by a view of nature as something to be conquered and dominated—an attitude summed up in an oft-quoted line from the commissioner of the East African Protectorate, in 1905: "Marshes must be drained, forests skillfully thinned, rivers taught to run in ordered courses and not to afflict the land with drought or flood at their caprice."[2] Such hubris derived partly from Britain's technological and scientific achievements. But colonial attitudes toward water were also shaped by the more distant past. The glory of hydraulic works in the dryland empires of antiquity inspired many men who sought to dominate water in the early twentieth century. Ferdinand Gessert reportedly went to Egypt to study its irrigation schemes before he began building dams on his farm. The irrigation engineer William Willcocks, who designed the original Aswan Dam and worked in Iraq and South Africa, saw himself and other colonial water engineers as following in the footsteps of ancient irrigation societies such as Egypt and Mesopotamia.[3] So did Schwarz and several of his competitors, each of whom claimed they were going to create a new Egypt in southern Africa.

The idea of water running to waste fired the imaginations of both Schwarz supporters and his opponents. Orange River irrigation farmer Carl Weidner asked, "Why go hunting in the desert for theoretical water schemes, when we stumble over hundreds of practicable ones at our very door? When we have at our feet an enormous volume of water rushing daily to waste right across the Union . . . ?"[4] Large-scale irrigation projects proposed in the early 1900s—including that of a Swedish engineer named August Karlson, who designed the dam at Hartbeespoort, outside Johannesburg—generated public enthusiasm because they promised to exploit South Africa's supposedly vast and unused water resources. Yet establishing southern Africa's hydrological potential was difficult. The Drought Commission's report noted that falling rain could follow many paths: it could feed watercourses, evaporate, transpire from plants, or sink underground. "The relative proportion of the rainfall which follows the various alternatives is of the utmost importance," the chairman wrote.[5] But those proportions were a mystery. Even the percentage that became runoff was an educated guess and, as some frustrated farmers noted, there had been virtually no attempt by government scientists to calculate rates of evaporation.

Surface water is rare in southern Africa. There are no natural lakes of any size, and relatively few perennial rivers. Springs occur infrequently, and white settlers usually found such prized sites already occupied. Securing access to them required negotiating with existing communities or engaging in violent conquest. White settlers sought to conjure water on, under, and above the land's surface, and relied on indigenous knowledge to do so. They dug wells in places where the water table was shallow. They used African techniques of harnessing ephemeral rivers by building dams in seasonal riverbeds in order to augment groundwater supplies that could be tapped once the rivers dried up, or by diverting floodwater onto farm fields via *saaidams* or sowing dams.[6] They also relied on technologies used by white settlers around the world. They constructed small-scale masonry dams and weirs on their farms, and those who could afford it drilled boreholes where hand digging failed. Both were beset by technical difficulties. Expensive boreholes often failed to reach water, and establishing maximum river flow proved impossible. Many farmers watched their dam walls disappear in the raging torrents that followed a downpour. This was not a problem limited to amateurs. Floods destroyed the Hartbeespoort dam project in the 1910s, delaying its completion. As late as the 1950s, consecutive years of historically high floods on the Zambezi damaged the massive Kariba Dam—twice—as it was under construction.

Even as these torrents destroyed the infrastructure meant to contain them, the spectacle reinforced a collective belief in water abundance. It followed that a lack of water was due to the failure to capture those torrents and hold them on the land. "Water running to waste" offered an alternative explanation for the "drying out" of southern Africa. The chairman of the Tarka Farmers' Association, responding to members who stated that the rains were failing, asked whether "the rainfall had decreased in this district, or whether the water only ran to waste."[7] Waste was a moral issue, the result of improvidence that threatened civilization itself. In language reminiscent of that used in discussions of white population, the Drought Commission's report suggested that "the waste of [South Africa's] natural resources . . . can but lead to national suicide."[8] In the year Schwarz wrote his book, "W. W." from the Transvaal lamented that "billions of tons of water run to waste every season" along the Vaal river, and that it was allowed to pass by "lazy" farmers who cared only about raising livestock and had no interest in the hard work of putting water to productive use.[9] There were repeated public calls to dam up every river and "spruit," or seasonal stream. In this battle for the future of South Africa, the Black majority was once again invisible. Both the destroyers and the redeemers were white men.

That redemption rested on white farming, which helps explain why the state was so involved in propping up the white agricultural sector from the time the Union was formed, in 1910, through the 1960s. During that half century, South Africa was governed by leaders and political parties with visions very different, at least in some respects, from Milner's British imperialism or the Afrikaner nationalism of the National Party. Despite some differences in approach, state involvement in agriculture has been a constant—reflected in a historian's claim in 1941 that the country "was farmed from the two capitals" (Cape Town and Johannesburg). Throughout the early twentieth century, the state took an "interventionist approach" to white farming, subsidizing the movement of agricultural products via favorable railway tariffs and paying for research into livestock breeding, plant and animal diseases, and crop improvements. Tariffs put price floors on food exports and taxed food imports.[10] In 1912, the National Land and Agricultural Bank, known as the Land Bank, was formed to offer loans to farmers.

State intervention in agriculture intensified between 1924 and 1937 under Hertzog's leadership, but it built on the actions of earlier governments. Many of Hertzog's policies sought to solve the "poor white problem" rather than seeking to maximize agricultural productivity. Those policies included work-relief schemes, irrigation settlements, the extension of roads and railways, land settlement schemes, and even direct subsidies to producers. Over time, price control and tariff policies artificially kept rural farm incomes commensurate with urban incomes, passing the costs onto consumers, who paid far more for food than the global average.

Historians have noted that such policies, especially those that offered easy access to loans and kept domestic food prices high, allowed small and inefficient farmers to survive—a reflection of the importance of rural votes as well as a populated white countryside. Despite alarm over the supposed depopulation of the countryside, in reality the number of white-owned farms actually increased by 23 percent between 1918 and 1928.[11] And despite a consistent refrain about how "backward" South African agriculture was, food production rose markedly during the early decades of the twentieth century as the country shifted from importing food to exporting it. As Jeeves and Crush note, agricultural transformation usually fuels industrialization; but in South Africa it was the reverse, thanks to the late-nineteenth-century mineral discoveries. Commercial agriculture became more, not less, important to the economy over the first half of the twentieth century.[12]

Despite all the state intervention, however, many white farmers were deeply in debt, and thus were in a poor position to weather crises like the Great Depression. Government policy ultimately did little to alter the process of commercialization in agriculture and thus the widening gap between prosperous and precarious farmers. Over the course of more than three decades, correspondents to *Farmer's Weekly* consistently accused government programs of harming, rather than helping, struggling farmers. Meanwhile, even as it handed out financial benefits to farmers, the state lectured them for their poor behavior and insisted that their problems were their own fault. This was the social engineering of a white man's country: Using revenues from the mines and the pocketbooks of consumers to subsidize a white presence in the countryside alongside the moralizing language of individual responsibility and white fitness to rule.

Moralizing Farming

In 1912 a sheep farmer told a joke, in Afrikaans, to an assembled crowd of his peers in the eastern Karoo. A man asked what an expert was. The response: "You know what *een paard* (a horse) is? Well, an *expaard* (an ex-horse) is a donkey."[13] After the laughter subsided, the teller insisted to his audience that, all joking aside, "experts had done much good for the country." But many attendees, like white farmers around the country, were more ambivalent. They resented the condescension that the country's experts, many of them foreign, directed toward "practical" men who had decades of farming experience. One called the government agriculture department's point man on wool, who had been brought from Australia, "a so-called expert." It was a conversation that could have taken place in almost any part of South Africa.[14] A man, "born in South Africa, who has never been across the water and never reads farming literature," voiced the view of many farmers when he claimed that "so-called experts" would "starve if they had to work on a farm."[15]

The experts that rural whites encountered were overwhelmingly agricultural experts, whose mission was to implant "progressive" or "scientific" farming in the country and so fit white farmers for their role as the master race and custodians of the land. But, as these accounts make clear, the relationship between white farmers and this burgeoning bureaucracy of agricultural departments, colleges, and experimental farms was as fraught as that between farmers and the experts who told them rainfall was not declining. Institutions promoting scientific farming and agricultural expertise came with British overrule in the wake of the South African War.[16] Milner arranged for Boer POWs to tour Canada, New Zealand, and Australia so that they could witness "modern" agriculture grounded in scientific fields such as chemistry and botany.[17] In the two decades after the South African War, the Department of Agriculture created an educational and extension service that included agricultural colleges and demonstration farms in every province. Drought Commission chair Heinrich du Toit was a major figure in this process. He ran the government's experimental farm at Lichtenburg, and later became the head of the country's agricultural education and extension services. His views helped to define the relationship between farmers and this emerging world of expertise.[18]

Scientific agriculture was rooted in the same settler revolution that pushed people to take up farming in dry lands. The United States, which established a network of state-funded agricultural institutions in the late nineteenth century, served as a source of global knowledge and had a special place in the South Africa agrarian imagination.[19] South Africans compared American success in drawing white immigrants, settling arid lands, harnessing rivers, and applying science to agriculture to their own dismal track record. Many of the first employees of the new institutions designed to spread scientific farming in South Africa had studied and traveled in the United States, including du Toit and his agronomist colleague at Lichtenburg, William MacDonald. Their experimental farm drew inspiration from the 1862 US Homestead Act: The farm units were 160 acres, and they employed only white labor.[20]

South Africa's privately published agricultural newspapers worked alongside government institutions to spread the gospel of progressive farming. Columnists described their modern methods and decried the tenacity of "ignorant" ideas among some farmers. Articles lamented the "evil" of soil erosion. Photos of farm ponds and prize-winning livestock adorned the pages of *Farmer's Weekly*. By the 1930s, such images shared space with photos of haystacks, silos, and windmills, and in the 1940s tractors and other machinery were added to the mix. *Die Landbouweekblad* was even more wedded to the project of modernizing agriculture than *Farmer's Weekly*. Its founding editor, Frans Geldenhuys, had a PhD in agronomy from Cornell University and was explicitly didactic.[21] In contrast to *Farmer's Weekly*, with its freewheeling correspondence section, the letters that Geldenhuys published asked for help rather than expressing opinions. *Die Landbouweekblad*'s reader correspondence section was titled "The Doctor and Pharmacy," and it dispensed advice to farmers on all manner of topics. Geldenhuys, far from challenging the stereotype of "Boer farming" as wasteful, unprogressive, and uncivilized, embraced it and made it his mission to bring his fellow Afrikaners into the modern era.

"Backveld" farmers were the antithesis of "progressive" farmers, and were characterized in many of the same terms as African farmers. They did not plow deeply, but only "scratched" the soil.[22] They were irrationally attached to their animals, and did not farm with an economic mindset. They were nomadic and had no respect for private property. They did not understand the role of financial institutions.[23] A surveyor named Benedictus Watermeyer told the 1914 Senate Select Committee, "There are so many people who live for to-day and the sheep farmer especially will destroy the whole country for the sake of his sheep."[24] Johannes

Hendrik Schoeman, an MP from Oudtshoorn testifying before yet another drought committee, distinguished between "the small farmer who treks about" and "the real farmer who has a farm."[25] Francis Kanthack blasted the "abhorrence of hard work, and . . . utter want of ambition" of the "average backveld pastoral farmer."[26] As Susanna Maria Elizabeth van der Watt notes, Afrikaner farmers who "failed in their duty to extract from the land in the way the 'civilized' nations do" were "in danger of losing their identity as a 'civilized' people."[27]

Not everyone rejecting the practices of scientific farming was an isolated and uneducated Afrikaner. Milner lamented that South Africans, "British just as much as Dutch" (Black farmers were again invisible), "snort at . . . scientific agriculturalists" offering them advice.[28] A 1912 article reported on a meeting of more than fifty farmers in Uitenhage, in the Eastern Cape, in which the assembled crowd rejected scientific consensus—in this case, medical knowledge about livestock disease. The chairman of the meeting claimed "that rinderpest had been brought and not stopped by inoculation." Others suggested that their attempts to use chemicals to kill ticks had been ineffective or had even harmed their animals. They argued that East Coast Fever and other diseases were not tick-borne at all. The plea from a farmer who supported dipping that "they must not expect to see any miraculous results" fell on deaf ears. Five voted in favor of compulsory dipping, and forty-five voted against.[29] A month later, a newspaper columnist opined, "Had this meeting taken place away up in the backveld . . . it might still have been sad, but not so surprising." But this community was "near the coast, surrounded by civilization . . . how can a country expect to prosper when its main productions are guided by such men?"[30]

Farmers selectively adopted the language of progressivism and backwardness. One, writing *Farmer's Weekly* to ask about the correct sizing for windmill cylinders and pipes, signed his letter "ignorant farmer." The next, asking about machines to help him locate groundwater, called himself a "would-be progressive farmer."[31] They pushed back against the idea that there was a clear division between progressive and backward farmers—a critique reminiscent of those aimed at scientists who claimed to understand the workings of the climate. Studies of the creation of a class of scientific experts in South Africa have not acknowledged the extent to which expertise was contested terrain. They have instead tacitly equated the creation of a state infrastructure of expertise with its wholesale embrace—a conflation that was criticized even at the time. The Bloemfontein paper *The Friend* stated that the agriculture department

> is completely out of touch with the farmers of the Union for whose benefit it primarily exists. . . . A Department concentrated in Pretoria issuing

> bulletins, which not one farmer in a thousand ever reads . . . is just what the country does not require. . . . It is altogether a wrong assumption that the farmers exist in order that the country may have an Agriculture Department.[32]

Historians have tended to adopt the arguments of government officials and experts themselves when they state that most white farmers failed to "modernize" and adopt the methods of industrial farming.[33] They also have defended experts against farmers' accusations that they were detached from the realities of farm life, noting that in fact experts often had deep knowledge of the places they worked.[34]

But the most persistent and incisive critiques were not about experts' scientific knowledge or even their local roots. They were about their limited field of vision—about the disembodiment of their knowledge from the political-economic realities of South Africa. As one man wrote in 1912, "We have wool experts, fruit experts, dry land experts, agricultural experts, and experts galore, to teach us farming. Of what use is all this teaching, since we are not able to get our produce to market owing to bad roads and bridgeless rivers?"[35] Men who identified as farmers critiqued agricultural experts and ideas about agricultural modernization from multiple angles. They challenged the idea that modernity was an unmitigated good by pointing out, repeatedly, the soil erosion caused by poorly constructed roads and railways—thereby blaming the infrastructure of the modern economy and the state that built it for environmental ills often pinned on traditional farming practices.[36] They argued that most farmers understood banking and business principles, but that lenders arbitrarily denied loans even to those with good collateral.[37] And they sought to challenge expert definitions of who comprised the country's *real* farmers, insisting that many of the men who were praised as "progressive" for their adoption of new methods and animal breeds were in fact "check-book farmers" who subsidized their experimentation with profits from jobs in the mines or in the urban professions.[38] Such people, readers argued, were not true farmers because they did not have to make their living from their agricultural activities. Less well-resourced farmers were unable to embrace the practices recommended by experts without taking on risky amounts of debt.[39]

These critiques of progressive farming prescriptions argued that scientific knowledge could not solve political and economic problems that were also rooted in land speculation, labor shortages, and a lack of water, transportation, and marketing infrastructures. In 1912, a *Farmer's Weekly* reader scoffed at government-sponsored farmer tours of other parts of the world, stating that the value to South Africa was "exactly nil." The lessons learned on such tours, as he summarized it, was "'Be progressive, keep pedigree stock, fence, etc.' In other words: 'Have a fat bank account before you start farming in South Africa.'"[40] Timothy Keegan, writing of the highveld area across the Transvaal and Orange Free State, argues that the survival of many farms required "access to fairly substantial funds from outside agricultural production, at least initially, and continued reliance on the state's resources as a cushion against climatic variations and market fluctuations."[41] In other words, there was no model for a progressive farm that left farmers solvent if the farm was their sole source of income. Even those highly critical of "backward" farmers recognized these realties. The surveyor who told the 1914 commission that a backveld farmer would "destroy the whole country for the sake of his sheep" admitted that people could not be coerced into fencing their land into paddocks for rotational grazing because "it would ruin more than one man."[42]

White farmers questioned whether embracing scientific farming, with its requirements for expensive infrastructure, would secure white futures in the economic context of a country with limited markets. A report on economic distress among white farmers in South West Africa noted, "Nearly all the Farmers belong to the Debtor class many of them having taken up loans at very high rates of Interest." When farm incomes were good, they expanded their herds, bought cars and farm equipment, and increased their property holdings, so that a decrease in prices left them unable to pay on their loans.[43] A farmer writing in *Farmer's Weekly* in March 1916 observed the abandoned farms in the Aberdeen district of the Eastern Cape and surmised that they had succumbed to the same fate. He described approaching the homestead of a "prosperous" farmer: "There were proofs that money had been spent liberally on improvements." The writer and his colleagues had observed "gardens, lucerne lands, windmills, a large irrigation furrow from the river, cement tanks"—the infrastructure of multiple water points, irrigation, and fodder cultivation that the experts recommended. Yet the farm was abandoned. "Being practical farmers, we took in all that the scene conveyed. Our own position—our own prospects—what are they with this wicked drought, which has us entangled, as it were, in its fangs of misery?" Drought threatened to make everyone a poor white, and investing in progressive agriculture could speed the process rather than prevent it.[44]

In South Africa, global changes in agriculture toward consolidation and capitalization coincided with several dry decades, and highlighted the disconnect between those who worked the land and those who set policy. The minister of agriculture visited the Cape Midlands in late 1919, a year *Farmer's Weekly* described as "one of the blackest periods in the annals of farming pursuits."[45] The Midlands were one of the hardest-hit areas. But after acknowledging enormous stock losses from drought and disease, the minister urged farmers to invest in irrigation. A report of his visit paraphrased his speech:

> He was inclined to follow expert opinion, which told him that this would yield a better result than the proposed flooding of Lake Ngami at enormous expense. Many farmers had to learn how to use their ground to the best advantage. . . . He concluded by telling his hearers that districts like Graaff-Reinet had a great future.[46]

The same week, a writer for the *Eastern Province Herald* observed "this dreary waste of waterless and treeless desert" marked by "the decaying remains of ostriches, and . . . grim mounds of skin and bones" where horses and cattle had died of thirst. "Those whose lot in life it is to wrest a living from the soil stare ruin blankly in the face—ruin that stalks nearer every pitiless, unchanging day."[47] Two weeks earlier, a farmer in the Orange Free State had reported that the throats of lambs were being cut as they were born in order to save the lives of the mothers, who were being fed costly grain in the absence of pasture.

Amid these depictions of farmer despair, government experts and officials argued that rainfall variability was normal and was no impediment to a bright agricultural future for those who followed the principles of scientific farming.[48] The agronomist William MacDonald argued that for "intelligent" farmers, drought "is no more than a passing storm to the skillful mariner at sea."[49] The Australian author of an article on the "drought-proof farm," reprinted in *Farmer's Weekly*, argued that "much, if not all, of the losses arising from dry seasons are preventable . . . attributable more to the neglect of preventive measures than to the falling off in the rainfall."[50] The idea that agriculture could be independent of the climate, combined with concerns about white respectability and fitness to rule, made farming a moral project in the first decades of the twentieth century. Farming practices were discussed using the same Manichean language that white farmers used to describe drought and the dry west wind. In the words of the Drought Commission, increasingly large floods due to soil erosion were evidence of "man's evil-doing."[51] References to the "evils" of soil erosion recur in *Farmer's Weekly*'s columns and letters, while a white farmer in the Eastern Cape referred to modern farming methods designed to check erosion as a "work of salvation."[52] Another reader described numerous rivers whose waters might be impounded, arguing that currently they were "wickedly allowed to run to waste."[53] This moralizing language spanned political regimes and their different policies toward farmers. An individual had responsibility for his fate, and his success or failure indicated his fitness to be a farmer. In this view farmer actions, not drought, explained white rural poverty, farm failure, and land losses.

The ideal of "progressive" or "modern" farming that such experts and their supporters promoted was bound up with conversations about Schwarz's scheme. Progressive farming and a belief in declining rainfall addressed questions of responsibility and revealed the stakes in debates over climate change. By denying that rainfall had decreased, in the absence of definitive evidence one way or the other, government experts were denying that farmers' losses were due to circumstances beyond their control. They were also staking out their territory, claiming that they possessed the solution to the problem of farmer precarity. As the Drought Commission opined, "Farmers as a general rule do not make provision for droughts, even though they quite realize that droughts must be expected."[54] Some farmers were "fully cognizant of the evil" of their actions when they went about "wringing out the life blood of the farm in order to make a quick profit." Others were simply "backward" farmers, ignorant of the consequences of their actions. But in all cases, "The individual has brought about the damage and without his cooperation the damage cannot be repaired."[55]

Heinrich du Toit had an outsized role in creating this image of individual farmers who were responsible for their fate. He had little sympathy for poor whites, whom he described in eugenicist terms as the product of "inbreeding" and "unfit parents." In the Drought Commission report, he wrote of the "poor white, rejected by the land"—a vivid image that endowed the land itself with agency, using rainfall deficits as a test of farmer fitness. B. P. J. Marchand, the minister who was treasurer of the Kakamas irrigation settlement, told the Drought Distress Relief Committee in 1916 that the poor whites living at Kakamas were hunters who had benefited from the easy access to land that marked the frontier, and who had never learned to farm properly. He suggested that drought served a constructive purpose in forcing people to change their ways and adopt more modern methods of farming, or to give up farming altogether. "When you speak about the distress, I do not think perhaps that a bad drought like this is entirely an unmixed evil for the country," he told the committee.[56] Du Toit, in a 1919 pamphlet, wrote,

> When someone who cultivates his lands shallowly, badly, and in an untimely manner, who sows the wrong and bad seed and then expects a good harvest that he does not earn and is not entitled to, gets no harvest, he is then often inclined to blame the cause of the failed crop on Heaven or God, while his own bad brain and handiwork are actually the cause.[57]

If a farmer suffered from drought losses, he (women were never considered in such calculations) was to blame for not working harder and smarter.

Du Toit was by no means a solitary voice. Francis Kanthack, irrigation director of the Cape Colony and then the Union of South Africa, was if anything even harsher in his judgment. He lambasted the attitude of "backward" farmers toward irrigation and other "progressive" farming practices. In 1909, he wrote, "We have been told that God made the rivers so that the water would run in them, and hence [it] should not be taken out by artificial means, and also many other quaint objections of a similar nature." Claiming to represent these farmers' thoughts, Kanthack wrote, bitingly,

> Why work? Why irrigate? God sends the rain and makes the veld sweet. He also sends locusts, diseases, and drought to humble us. Taken as a whole, I have never met with, nor heard of, any white race who have less ambition to improve their material conditions for the benefit of themselves or future generations than the average backveld farmer of South Africa.[58]

In their supposed willingness to live with no thought of the future—also a common stereotype of Africans—the whiteness of these "backveld" farmers was called into question. If they were ruined by drought, this would benefit the country, because it would separate the fit from the unfit. Some *Farmer's Weekly* readers agreed. As one wrote, "I am inclined to think that it is not so much the drought, but the working of the soil that is wanting."[59]

John X. Merriman, a member of the 1916 Select Committee on Drought Distress Relief, echoed such sentiments. "There is no body of men in any district of this country who, if they have a good year, dream of putting something away against a bad one." But an MP from the Ladysmith district challenged him. "You are not acquainted with their conditions," he responded. "Men who, two years ago, were making their fifteen hundred to two thousand a year are now looking for charity."[60] It was often members of Parliament who argued against expert pronouncements, in part because farmers were their constituents, and they knew them outside times of financial crisis. They also saw farmers who had followed expert prescriptions, yet were on the brink of insolvency. The *Farmer's Weekly* editor, too, was more sympathetic. He wrote in 1912, "Probably we shall have plenty of Job's comforters telling us to study the lessons of the drought in the last three years so as to obviate the losses in the future. The bulk of the farming population are probably of the opinion, and rightly so, that they have had more than a sufficiency of such lessons since 1911."[61]

Ideas about the potential for "good" farmers to avoid financial hardship due to drought had real-life repercussions when they were shared by those running the institutions designed to assist white farmers. For example, T. B. Herold, the general manager of the Land Bank, suggested, "The combination of cooperation, water, conservation, winter feeding, dipping and fencing should remove distress permanently under ordinary conditions and enable [fodder] reserves to be accumulated to tide over for a period of unfavorable conditions."[62] It followed that farmers who suffered drought losses and appealed to the Land Bank for relief on their loans could be denied on the grounds that they had brought on their own misfortune through bad farming.

Creating the Drought-Proof Farm

What made a farmer immune to drought? Experts suggested that livestock farms designed around modern practices such as fencing, rotational grazing, storing fodder, and managing the vegetation cover would be less vulnerable when drought struck. They argued that on mismanaged farms, rain rushed off the surface of the bare and compacted ground, taking water and topsoil with it. *Farmer's Weekly* profiled livestock farmers who emerged unscathed from drought thanks to their full silos.[63] A reader suggested that this amounted to denying the realities of agriculture in South Africa. "You mention only successes. . . . What about the numerous farmers whose names appear amongst the insolvents week after week in the Government Gazette?"[64]

A "drought-proof" crop farm required other methods, including dry farming and irrigation. "Dry farming" was a term coined in the US West to describe techniques for growing wheat on lands that received very limited rainfall. Much of South Africa's white-controlled land, including nearly all of the Cape, received less than the twenty inches of annual rainfall deemed necessary for traditional non-irrigated farming. In books on dry farming published in English and Afrikaans, du Toit gave examples of South African farmers who had successfully raised wheat and oats despite little or no rain during the growing season.[65] In theory, dry farming allowed the conversion of pastureland to cropland. In the 1910s and 1920s, experts viewed this as a desirable shift, because it increased South Africa's self-sufficiency in food, reduced farm sizes, and allowed for denser or "closer" white settlement. One man wrote that the results at the Lichtenburg experimental farm proved "that the whole unirrigable area of plain in the Union can be made to produce crops successfully. . . . South Africa can, and I sincerely hope will be, carrying a vast white population before long. . . . Salvation lies with the white man, not with the aboriginals." As with meteorology, however, the experts did not control how their ideas were used, and some of their most enthusiastic supporters combined expert prescriptions with ideas that experts rejected. In this case, the letter writer elaborated on his conclusion, arguing that salvation "lies with the white man" because erosion in the native reserves was changing the climate: "As that rainfall deteriorates, or the land surface deteriorates in suitability to receive it, the tell-tale Karroo advances east to register the fact. . . . We know that when rain is repelled it eventually falls in devastating torrents."[66]

Dry farming was labor intensive and land intensive. It was therefore a difficult prescription in a country where white farmers clamored for more cheap labor and where land prices were skyrocketing. In the absence of tractors—rare until the 1940s—it also required extensive use of plow oxen and thus theoretically could reinforce the African sharecropping that wealthy white farmers abhorred. In his book *Conquest of the Desert*, the agronomist MacDonald included a chapter titled "A Rainless Wheat." MacDonald told readers that preparing fields to produce such "rainless wheat" required an entire year of frequent plowing followed by harrowing after every rain to restore the "soil blanket" of pulverized earth—steps that locked in whatever moisture fell from the sky. This meant, in essence, cultivating twice as much land as one actually planted. And because constant cultivation exhausted the soil, additional land had to be rested in true fallow, covered with a green mulch in order to rebuild the "humus bank." All this excess labor and land was worth it, MacDonald stated, because "it also means a certain crop in seasons of drought."[67]

The agricultural press, some progressive farmers, and politicians embraced dry farming. A senator began giving cash prizes to dry farmers in 1911. *Farmer's Weekly* repeated MacDonald's claims that "a rainless wheat" had been grown at Lichtenburg.[68] Yet just six months later, the editor of *Farmer's Weekly* suggested that the experts were overstating their case. Dry farming was a way of storing moisture in the soil, but it still required water. "Crops will not grow without rain and it is only drawing ridicule on an otherwise sound doctrine to advocate that they will." Others agreed that dry farming was overpromoted by the experts and led to the mistaken impression that it was actually possible to farm without rain.[69] Further, in places with irregular rainfall—which includes most of South Africa's semiarid and arid lands—dry farming could increase wind erosion. In a description reminiscent of the US Dust Bowl, an experimental farm official in the Northern Cape reported that high winds blew a foot of topsoil off the fields, piling it three feet high against the fences.[70]

Irrigation was the other component of drought-proof farms and, unlike dry farming, it could theoretically be done on truly rainless land. Carl Weidner's farm at Goodhouse, on the south bank of the lower Orange River, received almost no rain; he irrigated fruit trees and lucerne with water pumped from the river itself. Irrigation gripped the imaginations

Figure 5.1 A man surveys his farm dam. Standing water on white-owned farms was a frequent motif for photographs and postcards in German South West Africa. Source: National Archives of Namibia.

of experts, policy makers, and empire builders as well as rank-and-file farmers and urban residents. It was aesthetically more pleasing than dry farming, since it created wet, green spaces out of formerly brown and dry ones. The power of this aesthetic can be seen in the innumerable photos of farm dams and filled "vleis" that featured in agricultural publications, often sent in by readers themselves. In archives, too, photographs of farms feature water even in the most arid environments. (figures 5.1 and 5.2). It is rare to find a photograph that depicts drought or unusually dry conditions.

Western US farmers proclaimed dry farming to be radically new, even though its techniques had long been practiced elsewhere. By contrast, irrigation was understood as the foundation of many of the world's ancient civilizations, a time-tested template for making the world's arid lands productive. British magistrates arriving in the Transvaal at the end of the South African War were shocked that there was no irrigation there, confirming the already suspect civilizational status of Afrikaners.[71] As Donald

Figure 5.2 A postcard from a farm called "Paulinenhof" offers an idealized scene of German settlement in South West Africa, and illustrates how water infrastructure was imagined as a place of leisure as well as production. Source: National Archives of Namibia.

Worster noted in 1992, the extension of the new imperialism to arid lands meant that irrigation became an important part of securing economic dominance.[72] In South Africa it would also help secure white dominance, saving the country from aridity and a Black majority. A correspondent to the *Agricultural Journal of the Union of South Africa* wrote in 1913, "Irrigation, and irrigation alone, can redeem South Africa, home of the happy idle native races, content to live a precarious existence, from becoming a barren wilderness." It was crucial to whitening the South African countryside: "Irrigation must and will convert it into perhaps the most fertile and productive country in the world, bringing in its train a large industrious white population."[73] Modern technology applied to irrigation would increase its potential far beyond what ancient civilizations had accomplished. Edward Bradfield, a regular contributor to *Farmer's Weekly* and an ardent conservationist and Americophile, extolled an account by the engineer and future irrigation director Alfred Lewis, which described the work of the US Bureau of Reclamation. The bureau's projects had succeeded in "overcoming the greatest physical difficulties in bringing people on to the land."[74] "People" here were, as usual, implicitly white and male.

In the early twentieth century, the promise of a more densely populated white rural world was captured under the phrase "closer settlement." Closer settlement required creating the ecological conditions for an elusive class of white independent farmers. But closer settlement was also seen as *generative* of those ecological conditions—a virtuous cycle not unlike that Schwarz proposed for rain itself. The agronomist MacDonald wrote, "Plant more people on your desolate lands, and then you will cease to fear drought."[75] This hyperbolic language was rooted in the expert belief that land degradation, not a lack of rain, accounted for farmer losses, and that closer settlement would reduce soil erosion. The notion that *more* intensive farming would lead to *better* ecological outcomes appears paradoxical. But it was an article of faith among the proponents of smaller farms and closer settlement, based on the idea that farmers could not apply best farming practices to large tracts of property.[76] When it came to land degradation, both the problem and the solution involved human intervention.

Even as some experts promoted "closer" agricultural settlement, others tried to dampen enthusiasm for the irrigation such population densities would require. Kanthack was disdainful of farmers who refused to irrigate, but he was equally dismissive of those who were overly optimistic about irrigation's potential. "The ordinary individual"—Kanthack clearly did not put himself in that category!—

> looks out the window of the railway carriage, gazes upon thousands of square miles of seemingly flat Karroo, and immediately proceeds to moralize about our backwardness in not turning it into waving cornfields by means of water brought from the distant Orange River . . . intervening mountain ranges of great height being considered no obstacle, nor the inconvenient fact that water runs down, and not uphill.[77]

The 1920 Drought Commission report noted that while many people believed rainfall was declining, almost as many believed "that it is possible to turn the whole of the Union into a flourishing garden by irrigation."[78] Since only 1 percent of the South African land was suitable for irrigation, "the roseate dreams of extensive irrigated areas in South Africa can only remain dreams."[79]

Experts were at least partly responsible for these roseate dreams, however. In the same article in which he belittled those who overestimated the country's irrigation potential, Kantack also moralized about the consequences of letting water run to waste:

> We have no rivers in this land now, only torrents. . . . By restoring to the country its natural clothing of vegetation, which the white man, by his

> thoughtless and improvident actions, has largely destroyed, leaving the face of the country naked, we will control the run-off in such a manner as to make it all, or nearly all, available for irrigation.[80]

The scientists' insistence that white settlers were responsible for the draining away of the subcontinent's water was contradicted by their own statistics. The meteorologist John Richards Sutton noted that South Africa had few large rivers; this fact alone "should be sufficient to show that the popular idea that most of our rainfall runs off at once into the spruits, and away down the rivers to the sea, must be wrong." The rate of runoff was "very small nearly everywhere."[81] The Drought Commission reported that the focus on water rushing away to the sea had obscured the extent to which water was lost to evaporation. "The lay mind, estimating the run-off would probably place the figure in the neighborhood of 90 percent of the rainfall. It is with incredulity one hears for the first time that average run-off for South Africa is only about 4 per cent."[82] But experts were partly to blame for this misperception, since they insisted that white farmers were "drying out" the land by reducing the capacity for water to soak into the soil.

In the face of such contradictory messages, expert pronouncements on the country's limited irrigation could not compete with the experience of seeing of once-dry rivers suddenly swelling dramatically in flood and then dropping to a trickle. Such images, combined with the moralizing language of scientists about human contributions to runoff, reinforced the perception that vast quantities of water were rushing away to the sea, with catastrophic results. It made grand projects to trap that water seem conceivable and wise.

And because letting water "run to waste" had taken on moral overtones, it raised the question of who was being immoral. If experts used the idea of water running to waste to critique farmers' land management practices, farmers in turn used it to critique experts and the government for their failure to manage the water supply. Government agents and committees placed responsibility for farmer misfortune on the individual, thereby minimizing the state's responsibility for securing white rural prosperity. But many people insisted that the problems facing white farmers were due primarily to a dearth of water, and that it was the duty of the state to provide it. Many white farmers therefore imagined a different kind of state intervention: one in which the state did not police the fitness of white farmers, but provided them with the means for success and prosperity. They demanded draconian policies to secure Black labor that would be captive in all but name. They also demanded radical interventions to secure the water supply via large-scale engineering schemes that would be underwritten by the state. Without such state intervention, farmers argued, they could not be judged for their failures.

Engineering the Future

Public enthusiasm for the Kalahari Scheme reflected not just vernacular environmental knowledge but also a pervasive faith in the power of science and technology to solve any problem if the scale were large enough and if the state assumed the risk. The techno-optimism at the foundation of Schwarz's scheme was widely shared by white South African men, whether struggling farmers, government engineers, urban professionals, or eccentric tinkerers. Indeed, the relatively uncritical embrace of technology and state planning united Schwarz's critics and his supporters. As they debated the feasibility of the Kalahari Scheme and its necessity, neither group asked a question that we would ask today: What might be the unintended consequences of such a drastic reordering of the natural environment? The *Times of Natal* summarized this optimism:

> Enlightened by the wonderful results obtained by the application of pure science to industrial purposes during the last half-century, it is with growing confidence that the people look to the patient scientist in the realm of research work to solve the problems which go hand-in-hand with our complex civilization.

One of those problems, the newspaper argued, was South Africa's drying up—and one of those solving it with "pure science" was Schwarz.[83]

Such comments reveal a shared commitment to science, modernity, and environmental mastery that was becoming a cornerstone of white identity. But they also reflected a willingness to place environmental and social engineering in the hands of the government—evidence of the "bureaucratic paternalism," to use Ivan Evans's phrase, that came to define the relationship between the South African state and its white citizens.[84] When Marchand spoke to the committee on drought distress, he insisted that it was the state, not church or private enterprise, that needed to solve the problem of white poverty. "You cannot throw on the Churches or on the people work that belongs to the state," he said. "The poor people belong to the State."[85]

Newspapers and government archives are littered with the plans of tinkerers, citizen scientists, and engineers who sought to recruit the state's assistance in surmounting South Africa's climatic limitations. Some schemes were local. Two men suggested covering more than 165 acres outside Grahamstown with concrete or steel sheeting in order to collect rainwater for the city.[86] Other plans were as territorially expansive as Schwarz's, or even more so. They promised canals across the subcontinent, railways and electrical lines from Cape Town to London, the transformation of the "Karoo Desert into a Garden of Paradise," and the creation of new citrus and coal industries. They were technologically elaborate, even fantastical: one man suggested building a condensing plant at a high elevation to create rain clouds or, as an alternative, to "have a very large lens made and erected near a sheet of water."[87] They often involved distant places that were abstractions to the schemers and their audiences alike. A Transvaal man named G. P. Robarts proposed an elaborate scheme to water the Karoo, while Ernest Long, who lived near the Karoo, proposed a plan to divert rivers in the Transvaal and Swaziland.[88]

The Kalahari featured prominently in the most ambitious schemes, including those proposed by some of Schwarz's most vocal critics. A surveyor named Johann Leipoldt offered readers of the *Star* an extensive critique of Schwarz's scheme based on his travels along the Kunene, and then proceeded to describe his own plan for the Kalahari's redemption: using wireless radio waves to create rainfall in the desert.[89] Carl Weidner published a 1925 pamphlet that proposed the construction of a combined waterway and railway across the breadth of southern Africa, from Walvis Bay on South West Africa's Atlantic coast to the Indian Ocean port town of Beira, in Mozambique. This would, he wrote, "open up for settlement the biggest and most suitable cotton and wheat country in the world with an abundant labor supply at hand."[90] August Karlson, the Swedish civil engineer who had come to South Africa in 1892, proposed diverting water from Angola into the Kalahari. Karlson had some credibility as the man who had designed Pretoria's water system and who had first proposed the Hartbeesport Dam, which was under construction. He was wary of Schwarz's claims for increased rainfall and also insisted—correctly—that Schwarz's weirs would not work, based on what was known about the area's topography and hydrology. But Karlson embraced the idea of remaking these same river systems in order to "reclaim as much as possible of the Kalahari" for white settlement—language that reflected the influence of the US Bureau of Reclamation on technocratic discussions of arid lands. He proposed deeper reservoirs in Angola, which would reduce evaporation losses, and a massive irrigation project in the Kalahari. He also suggested that large-scale irrigation and a forest belt along the length of South West Africa could increase rainfall.[91]

The authors of all these projects assured their audiences that they were technologically feasible. Any flaws that might later be discovered were simply design problems to be resolved through additional engineering and the application of state funds. After critiquing Schwarz's proposal, Weidner assured his readers that he was offering them "a practical scheme." Karlson said he had studied the Kalahari problem from "an engineering point of view"—subtly highlighting the fact that Schwarz was not an engineer—and called Schwarz's scheme "entirely chimerical" while characterizing his own as a "practical possibility."[92] Long wrote, "I do not wish to waste your time by indulging in fanciful dreams which cannot be realized," before reassuring the governor general that his scheme—to create a cement industry, electrify the railway system, and grow citrus in Bechuanaland—was a paying proposition.[93] Robarts confessed that he lacked the funds to test his scheme on a large scale. But his small test had yielded good results, "so trust your Engineers will get to work early."[94] These white men had great confidence that the state would take them seriously. The "robust sense of self and agency" that van Sittert identified among the white "organic intellectuals" who wrote letters for publication around the turn of the century remained a feature throughout the twentieth.[95]

Speaking to the Senate, Prime Minister Jan Smuts admitted—to much laughter—that he dreamed of Schwarz's theories at night. But he suggested that "it might be that the problem was too huge for human solution."[96] Such caution was rare in popular discussions. Two days after reporting Smuts's comments, the *Cape Times* wrote of the Kalahari Scheme,

> Merely because it is vast in its conception and might involve very large expenditure it is not to be idly dismissed. On the Nile immense irrigation schemes have been carried out at the cost of millions, with results which have astonished the world. In Mesopotamia there is very good reason to believe that a country which has been a desert for thousands of years, though once the home of great civilizations, will be brought once more within a very few years, thanks to British engineering enterprise, into the very forefront of the grain-producing territories of the world.[97]

People defended the scale of Schwarz's scheme by pointing out that many successful large-scale engineering schemes had initially been dismissed by skeptics. One recalled that the Australian man who had designed a system for bringing water from Perth to the Kalgoorlie goldfields had been attacked in the press so viciously that he had died by suicide before the construction ended.[98] A mine engineer named George Henry Blenkinsop reminded those who claimed that Schwarz's scheme was unrealistic that "the same thing was said of the Suez Canal and the Panama Canal, and, mind you, there were experts among those who said it."[99] Another wrote, "Surely the work would not be such a big thing as the Suez Canal or the big dam at Assouan [*sic*], in Egypt, both accomplished facts."[100] This defense continued to be important to Schwarz's supporters. A decade later, the superintendent of roads and public works in the Orange Free State told an audience, "It should be borne in mind that the vast projects such as the Panama Canal and the Colorado [River] reclamation schemes had at first been considered impracticable."[101]

The men who proposed and supported these various techno-optimist schemes insisted that they and their kindred spirits had higher ideals than a base quest for wealth. As one wrote, "The majority of such men do not put 'lucre' first, but feel intense pride in the fact of being able to do what no one has thought capable of accomplishment."[102] Marc Reisner, writing about the engineers who flocked to work at the Reclamation Bureau in the United States—the department dedicated to turning arid lands into productive agricultural lands—argues that they operated in "a fog of idealism, ready to take on the most intractable foe of mankind: the desert." They saw themselves as "a godlike class performing hydrologic miracles."[103] The men who proposed engineering a white man's land from southern Africa's rivers had similar qualities.

Mobilizing technology on a grand scale was the path toward an independent, prosperous, and populated white South Africa—and, ideally, toward a world in which southern Africa's Indigenous populations not only were subordinated once and for all, but ceased to matter. Like the plans for "closer settlement," ambitious hydroengineering schemes promised a vastly increased white population that could be settled in rural areas as autonomous, independent farmers—in contrast to Black farmers who worked for white landholders or under chiefly authority in the reserves. Schwarz, as we have seen, planned for millions of acres of irrigated land for white settlement, as well as abundant and predictable rainfall in regions that were currently losing their white population to drought. Ernest Long fretted, "At present our whole economic system is founded on the native and with our methods there will never be enough of him."[104] He dreamed of a South Africa "freed from the necessity of pandering to the native element."[105] Karlson argued that irrigation and increased rainfall would "allow a total increase of 2 million white inhabitants in South Africa"—by which he meant the Union of South Africa and the surrounding territories of South West Africa, Bechuanaland Protectorate, and Rhodesia. He even proposed pumping water uphill from Etosha into the karst region of central South West Africa, in order to open up land to settle two hundred thousand whites: "This is a 'heroic' means, but when there is a question of getting white men into the country any means must be employed."[106] This tension, between modern farming that turned a profit and massive state investment in engineering a white agrarian world no matter the cost, continued to shape the debate over how to create a white man's country out of spaces that had too little water and too few white farmers.

Rain and Yeoman Farmers

Of the many schemes that promised to harness technology to solve the twin threats of drought and white precarity, Schwarz's was the only one that developed a popular movement behind it. Here it is worth pausing to ask why.

One distinguishing feature of Schwarz's vision was that it was remarkably low-tech. He suggested that a simple "weir built of piles and filled in with rubble and branches" would be sufficient to divert the Chobe River, and that a similar technique might also be enough to turn the Kunene River inland.[107] The image is one of a glorified beaver dam; it stands in stark contrast to the elaborate blueprints Karlson submitted to government officials, and to the futuristic networks of pipes, metal sheeting, glass lenses, and underground tunnels suggested by other techno-dreamers. But more important is what Schwarz's scheme did *not* emphasize: the conservation of water through individual effort and expense.

In 1923 the civil engineer Reenen van Reenen joined Schwarz and others at a "drought symposium" organized by the South African Association for the Advancement of Science, in the wake of the release of the Drought Commission's final report. Van Reenen, who had been educated in the United States and had worked on irrigation projects in Nebraska, was a presence on multiple government commissions related to drought, irrigation, and South West Africa's northern border. He argued that there were three ways, other than appeals to the divine, to solve the problem of insufficient rainfall. The first was "artificial rainmaking—the extraction of water from the atmosphere by the direct action of man." The second was "the induced increase of rainfall brought about indirectly by man"—a direct reference to Schwarz's scheme. The third option was "the more efficient use of the rainfall which we obtain naturally." Van Reenen insisted that more attention had to be paid to the third option: the conservation of rainfall, including through irrigation and dry farming. *Farmer's Weekly* had made the same point three years earlier. But its editor recognized that people found plans to increase rainfall more attractive than strategies to optimize aridity. Experts' pedestrian solution to water scarcity, *Farmer's Weekly* stated, left farmers "indignant and disappointed."[108]

One reason rainmaking appealed to the public was that van Reenen's third option was not proving a success. By the 1920s, many white farmers and some politicians were skeptical that a "drought-proof farm" could exist. This was not necessarily because they accepted the environmental arguments of Kanthack and others that the irrigation potential of South Africa was very limited. Instead, they were wary of piecemeal irrigation projects that put most of the risk on the irrigators themselves. Farmer after farmer told stories of failed irrigation schemes that were distant from any market, that had poor soils, and that burdened farmers with debt and water rates that were impossible to pay. Some settlements recruited overseas, bringing in men who knew nothing of the country or how to farm it. A senator who had visited an irrigation settlement in the Transvaal called it "an absolute failure." Upon their arrival, the immigrant settlers discovered that the soil was infertile, and could not obtain fertilizers. Their rent exceeded their profits. "They were dumped down there, and I think it was cruel the way these people were misled."[109]

Although Schwarz included irrigation in his plan, in his public writings and speeches he emphasized the anticipated climate benefits of the scheme. Irrigation, Schwarz claimed, benefited the rich. The world's only profitable irrigation schemes were in India and Egypt—places whose large indigenous populations could be coerced into hard labor by a ruling elite. Irrigation could not form the basis of country of white men who were equal. "In America, all are financial failures; California, after 50 years of a struggle, has become more or less solvent, but only through the checkbooks of the Eastern millionaires who have settled there to enjoy their old age amid rural surroundings."[110] The reference to "checkbooks" echoed popular critiques of "checkbook farmers" in South Africa. The rents and water rates required for investors to break even caused the farmers to go broke and default on their obligations. Irrigation schemes were also plagued with ecological problems such as high evaporation rates, salinization, the rapid siltation of reservoirs, and shortfalls in water delivery.[111] Meanwhile, an anonymous "Australian correspondent" had told *Farmer's Weekly* readers in 1915 to look elsewhere for a model of successful irrigation. "The irrigation business is costing us huge sums of money, and is producing very little in the way of tangible results." Schemes were heavily indebted, and people assumed that the government would ultimately cover the shortfalls. The correspondent praised landowners who had created small irrigation schemes: "These are the men who are really solving the irrigation problems."[112] And yet some South Africans believed that small schemes had little collective impact given the scale of the problem. MP de Jager lamented, "Our irrigation works are just as good as sending a child with a cup to empty the sea."[113]

Kanthack responded that Schwarz was too pessimistic, and that the schemes he cited as failures were older schemes that "have for years past been examples of how not to do things."[114] He insisted that proposals were now carefully vetted for financial soundness, and offered the Sundays

River irrigation scheme as an example. When it was begun in 1917, Kanthack wrote that the scheme was "one of the soundest and most promising ones I have ever been associated with in South Africa."[115] This was no poor white labor colony. The scheme's founder, Percy FitzPatrick—politician, Anglophile, and author of *Jock of the Bushveld*—had traveled to the United States to study models for the citrus industry he hoped to create. The Sundays River scheme was intended as a close settlement project populated by "the best class of settlers" from Britain. One investor effused that while mining could produce rapid wealth, "gold schemes could be worked out, while a land scheme with irrigation could not be worked out." True and lasting wealth lay in white men's production on the land.[116]

Yet again, expert optimism was out of step with reality. Construction of a storage dam along the Sundays River began in 1917, but the site's remoteness made transporting supplies difficult, and its aridity meant that huge quantities of fodder had to be brought in to feed the transport animals. World War I created a shortage of building supplies and labor, and the 1918 influenza pandemic compounded problems. In 1921, floods washed away a pumping plant. Alfred Lewis, who had succeeded Kanthack as director of irrigation, noted that 30 percent of the money advanced to farmers was in arrears, due to "the fall in the price of farm produce, to drought, and to the greatly-increased cost of the works."[117] The dam was completed in 1922, but it took six years to fill the reservoir, far longer than expected. The government assumed control of the scheme in 1925, and in 1934 it wrote off £2.3 million in farmers' debt.[118]

The Sundays River scheme was not unique. In Graaff-Reinet, where the agriculture minister had told an audience in the drought year of 1919 that the area had a great future if only farmers would invest in irrigation, an irrigation scheme begun in 1925 was declared unviable just three years later.[119] In 1924 the minister of lands told the House of Assembly that irrigation schemes were so expensive that the "ordinary farmer" could not turn a profit. The minister suggested allowing farmers forty years to repay their loans.[120] A year later, the anthropologist C. Daryll Forde suggested that very few parts of South Africa were worth irrigating.[121]

The Sundays River community, comprised of handpicked modern farmers, decided to stop looking to the river for their water. Instead, they looked to the skies. In the early twentieth century, a number of experiments were aimed at making "artificial" rain in the US West, Australia, and South Africa—the arid spaces that had been incorporated into the late nineteenth-century "settler revolution." In South Africa, the production of "artificial rain" was part of the general fascination with American water management technology. But it was the famed California rainmaker Charles Hatfield, rather than any government agency, whom farmers identified as their most promising American option. In the early twentieth century, multiple farmers' associations sought to hire Hatfield to come to South Africa to make rain.[122] The Sundays River farmers were among them. They had done their homework, writing to a University of California, Berkeley professor to check out Hatfield's credentials. The professor said Hatfield was a fraud, but testimony from an Alberta man whose farm had received rain after Hatfield's visit trumped the expert's advice. The Sundays River farmers proceeded with their efforts to recruit the controversial rainmaker. The conversations on the *Farmer's Weekly* letters page over whether Hatfield's experiments had been successful reveal the extent to which white farmers were collecting information on meteorology and rainmaking from North America. After a period during which the subject dominated the correspondence columns, the editor of *Farmer's Weekly* finally declared a moratorium on further debate.

Local techno-enthusiasts also proposed their own rainmaking experiments. Some sent their ideas to Schwarz, who told the *Sunday Times* of London:

> A Johannesburg gentleman asked me recently why I did not advocate obtaining more rain by catalytic action; a Bloemfontein gentleman said, "Why not boil the sea by electricity"; and a gentleman from Windhoek explained that as rain was formed by the combination of oxygen and hydrogen in the atmosphere, which were united by the action of the electric spark, otherwise the lightning flash, would I use my influence with the Government to induce it to take up his patent for blasting off electricity into the air, producing more sparks and consequently more rain?[123]

A reader told *Farmer's Weekly* that he had been urging the Australian press for forty years to support the creation of "condensing plants" to create clouds and then send up airplanes to release bombs to "explode the moisture."[124] In the 1910s, a Karoo farmer named Charles Hall began his own experiments, constructing "rainmaking boilers" on his farm.

Trained as a civil engineer, Hall had returned home to run his family's farm and was a regular contributor to *Farmer's Weekly*. He illustrates the entanglement of expert and popular knowledge. Hall had developed a reputation in the Eastern Cape, where desperate farmers in the Sundays River Valley—who were still waiting for irrigation water five years after the dam had been completed upriver—turned their attention from Hatfield to Hall, and lobbied the governor general to support Hall's experiments. Hall managed to gain the support from Kanthack, as well as from Charles Stewart, the country's first meteorologist; and in 1929 South African officials agreed to loan him a military plane to conduct rainmaking experiments over Pretoria.

An aide in the governor general's office wrote that artificial rainmaking "is a rather problematical question and may be held to savor somewhat of witch-doctor methods!"[125] Another man suggested that Hall's technique of seeding clouds with dust particles compromised his racial identity: "Personally, I have as much faith in a Kaffir rain doctor's powers, as in the dust theory."[126] A *Farmer's Weekly* article referred to Hatfield as "the California *Inyanga*"—a reference to the word many African communities used for a ritual specialist.[127] Other white South Africans rejected these comparisons, however, seeing little similarity between rainmaking that used "scientific" principles and that supposedly based on African "superstition."

Hall's experiments yielded few results. But his public profile indicates that promises of increased rain appealed to some white South Africans. Schwarz's focus on increasing rainfall, which made him unique among his competitors, may therefore best account for his popularity. The geologist Alex du Toit, who led an investigation of the Kalahari Scheme in 1925 (to be explored in the next chapter), recognized this fact: "The farming community in South Africa, because of this propaganda, have come to have great expectations from the scheme and have hailed it as the solution of one of their greatest handicaps, a normal deficiency in the rainfall."[128]

In contrast to the dependence that irrigation imposed on farmers, rain fostered independence. It was a leveling mechanism that benefited everyone, rather than pitting them against each other in a battle over scarce water resources. It was therefore an appropriate means to build a prosperous society of equal white men. Increased rainfall meant more land opened to farming without relying on the state or on capitalist investors to build dams and create irrigation settlements. The promise of more and dependable rainfall, and with it more land that could be farmed by autonomous white men, offered a means to make reality what was currently aspirational and mythical: a citizenry comprised of self-sufficient, landowning white farmers who could succeed without a lot of capital.

The aesthetics of Schwarz's imagined future underscored this promise of an agrarian utopia. With more rainfall, he wrote, the harvest "will not be followed by the clean-sweep of every sign of vegetation." Irrigation schemes, he said, transformed only "narrow strips of land along the rivers" into green oases; the rest was left brown, parched, and unusable.[129] Schwarz's photos, too, made an aesthetic argument. They counterposed images of the bones of 1915–16 famine victims littering the landscape with those of the Kavango River and the oshanas in flood. A land with more surface water symbolized the desired future, as was so often the case in the pages of *Farmer's Weekly* itself.

Another reason for the visibility of and public support for Schwarz's scheme was his vigorous publicity campaign. Most techno-dreamers

Figure 5.3 A doodle of Ernest Schwarz, likely self-drawn, from his personal papers. This was drawn on the back of a flier announcing an event for the South African Philosophical Society in the late 1890s.

shared their plans with government officials and newspapers; a few, like Weidner and Karlson, self-published pamphlets. Weidner's ornery reputation and his isolation in the remote northwestern Cape made his scheme easy to ignore. Karlson, meanwhile, overwhelmed his readers with engineering details. He sent the prime minister's office hundreds of pages of technical specifications, much of it written in miniscule handwriting, and diagrams on blue drafting paper. He also came across as paranoid. He told officials that the Hartbeespoort Dam, which he had proposed but not designed, was doomed to catastrophic failure thanks to design flaws he had pointed out years ago; and he hinted at a conspiracy to silence him.[130]

Schwarz was a more congenial presence: a bespectacled professor full of enthusiasm for his subject (figure 5.3). And he presented his scheme in a full-length book, with prose that was highly accessible. In May 1920, with drought still buffeting the country, *The Kalahari; or, Thirstland Redemption* introduced the English-reading public to his scheme, as well as to a geological history of the African continent that placed folk hydrology within a scientific framework. Situating southern Africa within a continental context of hydrological and climatic change, Schwarz described a process of "river capture" that encompassed the Congo, the Nile, and the Niger. He included a map that showed Africa's ancient river systems draining into inland basins. He presented his original argument concerning southern African desiccation, but offered far more detail, reviewing the genealogy of the concept among travelers and scientists. And he repeated his call to divert the Chobe and Kunene Rivers inland. "The country is going backwards," he wrote, "and the process of drying up will go on till the country, as a whole, will become uninhabitable."[131] In soaring language, he described a transition from dystopia to utopia. "Everyone in South Africa, whether he wants it or no, will receive the additional rain, will see his land rendered more fertile, and all his difficulties from drought, famine, and pestilence disappear." In daily life this would mean "steady rains instead of sudden bursts," "clearer water," "no veld erosion," "protection of surface by continuous vegetation," "no brak [salinization]," and "no need to irrigate, except for special crops."[132] This was heady stuff for South African farmers reeling from two decades of drought, locusts, livestock disease, war, and economic recession.

Language was not the only thing that distinguished Schwarz's 1920 book from his 1918 presentation to fellow scientists. His book also tackled issues that were absent in his earlier writings, including an explanation of how solving environmental problems would solve social ones. The Kalahari Scheme would provide enough land to offer every "poor white" a farm—and, unlike the current irrigation settlements of South Africa which offered only a "bare subsistence," these would be large and fertile enough to guarantee a good living. Further, Schwarz wrote, the increased rainfall and absence of drought would remove a psychological barrier to the success of poor whites, who had come to feel that "no effort on their part is of any use."[133] His scheme would also shift the balance of racial demographics in favor of whites. Not only would an improved climate dramatically increase the amount of land that could support rain-fed agriculture, but the diverted rivers would open up 13,250 square miles of the Kalahari to irrigation. Almost all of this land would be reserved for white settlers. But Black South Africans would, like their white counterparts, also benefit from improved rainfall, thereby stabilizing whites' labor supply.

The imaginative power of Schwarz's vision is reflected in the fact that popular writers—not just William Charles Scully, but also Ernest Glanville and Lawrence Green—found themselves drawn to his scheme.[134] It also was almost infinitely flexible in the futures it could accommodate. Schwarz was a populist, appealing to people's senses and emotions, empathizing with their frustrations, encouraging their suspicion of experts, and positioning himself as a scientist who was standing up to an unresponsive and hidebound club of government experts. His claims shifted with the popular mood, and over time they became increasingly grandiose. A few months after the publication of his book, he wrote that his scheme would cause all of the ephemeral rivers in east-central South West Africa to become perennial.[135] As criticism of the purported rainfall benefits mounted, Schwarz and his supporters changed gears and emphasized the scheme's irrigation potential.[136] By 1925, Schwarz was claiming that three million whites could be settled in the Kalahari if his scheme were built—double the number of whites then in South Africa.[137]

And all this could be achieved with two simple barriers of posts, mud, and sticks, placed across two distant rivers. Together, they would transform southern Africa into a "white man's land" once and for all.

6 * "The Kalahari Dream"

Newspapers and journals across southern Africa and the English-speaking world announced the publication of Schwarz's book in May 1920.[1] While reactions ranged from intrigued to enthusiastic, virtually no one dismissed his ideas outright. The former Africa explorer and colonial administrator Harry Johnston published an effusive review in *Nature*. The *Times* of London stated as fact that the Ovambo were threatened by "extermination from famine" as a result of the Kunene River's recent change in course. A reviewer for the American Geographical Society's journal expressed reservations about Schwarz's meteorological claims but concluded, "Only theories oppose the consummation of it and not laws. In other words there is a chance that Professor Schwarz is right." Media coverage was generally favorable in South Africa as well. The *Times of Natal* wrote that Schwarz's scheme would "secure the prosperity of South Africa for all time." The *Diamond Fields Advertiser* suggested that Schwarz could be considered either "a visionary enthusiast" or "a practical reformer in advance of his day and generation." *The Queenstown Daily Representative and Free Press* called the scheme "an ingenious proposal." *Zuid-Afrikaan* reported that the scheme would "completely solve our poor white problem." *Grocott's Daily Mail* highlighted the possibilities for large-scale white settlement, and the *Cape Times* ran an article aptly headlined "The Kalahari Dream," which observed that the "reclamation" of the Kalahari "would solve our land problem, and perhaps our native problem."[2]

Government archives also reveal broad public support for the scheme, whose flexibility allowed it to fit many agendas. Pro-British parties imagined that it would support the creation of a rural yeoman class through white immigration and opportunities for the white urban unemployed to go back to the land. Nationalist politicians focused on its potential to solve the poor white problem. And outside of South Africa, the *Bulawayo Chronicle* envisioned a future of dense white settlement in the interior: "Apart from the hope of an improvement in our climate, there is an attraction for Rhodesians in the picture of vast areas of natural irrigation to our westward, with a closely settled agricultural country on the direct railway route between us and the Atlantic seaboard." South West Africa's *Mitteilungen der Farmwirtschafts-Gesellschaft*, the newspaper of the white farmers' organization, published a speech given by the German veterinary scientist Rolf Hartig to farmers in South Africa. Hartig noted the "indulgent smile" that experts often gave those who spoke of the drying up of the region, then proceeded to argue that *lamsiekte*, or botulism, in livestock was an immediate result of increasing aridity. At the time, scientists were uncertain of the cause of the disease, which was devastating South African cattle herds. Hartig told his listeners that reports of the disease had only surfaced after 1820, around the time that Makgadikgadi ceased to be a lake—an implicit endorsement of the Kalahari Scheme, although Hartig did not mention Schwarz by name. The National Party newspaper *Die Weste* was an outlier: it called Schwarz an "English Imperialist" and criticized the idea of planting "a large English colony" in the Kalahari.[3] But even its criticisms assumed the scheme would work as Schwarz described.

Amid this general faith in the scheme's efficacy, *Farmer's Weekly* was a rare skeptic. Its editor pointed out that the only direct benefit South Africa could possibly obtain was increased rainfall—and that many scientists doubted this would happen.[4] Its readers, meanwhile, were divided. During the summer of 1920–21, *Farmer's Weekly* published articles and letters on the origin of rain, the trajectory of the climate, and the viability of the Kalahari Scheme almost every week. Over the next few years, such letters formed conversations that extended over multiple issues. Daily newspapers in South Africa's towns and cities also published letters, if less frequently. Some expressed doubt that Schwarz's scheme would work, but even skeptics often suggested that it deserved investigation.

In these early years, ecological affiliation determined enthusiasm for Schwarz's scheme to a greater extent than party or ethnic affiliation. Letters to *Farmer's Weekly* did not always identify where the writer lived, but the most dedicated supporters came from the Karoo, the Northern Cape, and the most drought-prone districts of the Orange Free State and the Transvaal. Within South West Africa, which was more arid than South Africa, there was strong support among some German farmers as well as South African colonial officials. The extent of the National Party's support waned after it came to power in 1924, suggesting that its initial embrace of the Kalahari Scheme had been rooted in a desire to bludgeon Smuts's pro-business government for being unsupportive of distressed farmers. And broad-based enthusiasm waned over the years, as scientific critiques of the scheme were publicized. The most enduring support came from the most arid areas. National Party representatives from the Cape's arid regions, in the Karoo and Northern Cape, remained steadfast in their support even as the National Party in general pulled away from Schwarz. Those MPs reported that their constituents were pressuring them. In 1925, a National Party MP elected as part of the new Pact government (in which the National Party and the Labour Party became unlikely partners) told his colleagues, "There is great interest today in the country in Professor Schwarz's scheme, perhaps more than the members of the Government think. Almost all our country representatives are from time to time asked how the scheme is getting on. I have only today received two letters [about it]."[5]

Schwarz actively cultivated public support, writing articles for scientific journals, South African newspapers, and popular British colonial publications such as *Tropical Life* and *United Empire*. Immediately after his book was published, he left for England and spent six months promoting his scheme there. At the Africa Society, colonial boosters and members of the British Parliament offered him their support. His private papers contain records of his lectures at the Royal Colonial Institute and at the Ipswich Scientific Society, the latter indicating that his tour included small, local scientific societies.[6]

Schwarz's time in England was not just spent speaking to gentleman scientists. He also sought private financial support for his project, and government approval for a concession in the Bechuanaland Protectorate. "The Kalahari Development Company" would be responsible for surveying the area and for building the infrastructure for "irrigation and settlement on a large scale." It would also engage in a wholesale reshuffling of the local population. Schwarz repeatedly promised "full protection of native rights," but this did not mean that Africans would keep the land they currently inhabited. Irrigated farms would be prepared for them in places "determined by committees composed of the native chiefs concerned, officials of the Government and of the Company." In turn, the chiefs would renounce their historic land rights to clear the way for farms with irrigation canals, houses, outbuildings, livestock, crops, and implements—all made ready for the whites who would be settled there. Smuts indicated support for a survey of the area and, if the plans were viable, for Schwarz's concession. His office sent the governor general an extended description of the US Carey Act, arguing that it was an effective way to develop desert lands through a partnership of the state and private investors.[7]

The correspondence leads naturally to the suspicion that Schwarz's goals were financial all along. Less than two months after his book was published, and despite his assertion that irrigation was largely a failure, he was seeking to create a private company that would invest in an irrigation settlement! Yet despite the fact that the concession was rejected in 1921, Schwarz spent the rest of his life advocating for his scheme.

This chapter looks at the debate over the Kalahari Scheme, and at attempts to assess its viability in the 1920s. Within white society, fantasies about the Kalahari and its hinterland, popular beliefs about arid environments and climate change, fear of the "swamping" and impoverishment of the white minority, and a utopian—and largely masculine—faith in technology came together to generate deep and lasting support. White farmers, members of the South African Parliament, native commissioners, popular writers, academics, urban professionals, and the country's top munitions manufacturer all wrote or spoke in support of Kalahari redemption. The opposition by almost all local scientists—many of them Schwarz's professional colleagues—simply hardened public support, revealing the depth of contempt for scientific elites and "experts." This suspicion of "theoretical men" coexisted with enthusiasm for wildly ambitious technological solutions—an enthusiasm that challenges scholarly portrayals of "high modernist" schemes as initiatives of state bureaucrats who foist them onto an unwilling or resistant public. The fact that the government relented and sent an expedition into the Kalahari underscores the racial state's vulnerability to populist science despite the criticisms of its own scientific experts.

"A Fiery Cross"

It is not easy to quantify support for Schwarz, since the evidence is largely confined to print culture—and a primarily Anglophone print culture at that. The historian Bill Schwarz, suggesting that ideas about race in the colonial context became part of popular culture in the metropole, argues that historians must rely on hints and traces rather than on an empirical assessment of sources.[8] Similarly, there are hints that the Kalahari Scheme was debated beyond the world of those literate in English. *Die Landbouweekblad* did not offer a public forum comparable to *Farmer's Weekly*, but Afrikaans surnames feature prominently in the letters to the editor section of *Farmer's Weekly*. And the letter writers were not necessarily part of a literate elite. They sometimes apologized for their lack of education, or admitted they had never written to a newspaper before. A Schwarz supporter from Gordonia, in the Northern Cape, asked *Farmer's Weekly* readers "not to expect too much of me, as I am not much of a scholar, and have had but a very meager education."[9] The published letters also hint at broader community conversations. Several people living on the margins of the Kalahari wrote to Schwarz and offered to subscribe toward printing his book in Dutch.[10] Another asked *Farmer's Weekly* to publish a sketch map of the region Schwarz proposed to engineer, because he had been asked by several people where Etosha Pan was located.[11] Although none of Schwarz's private correspondence appears to survive, he reported that people sent him anecdotal evidence that supported his theory of lakes and rainfall. One example was a man who had lived in South West Africa in the late 1880s and who stated that for several years after water levels were high in Etosha Pan, rainfall improved as far away as the coastal Namib Desert.[12]

The words of government scientists and leaders also reveal the reach of public discussion. Kanthack, an outspoken opponent of Schwarz's scheme, told the Royal Geographical Society that it was "most attractive to the people of a country chronically suffering from drought. . . . He has certainly got the ear of farmers and the public generally."[13] In 1925, Jan Smuts delivered the presidential address to the South African Association for the Advancement of Science, which was meeting at Oudtshoorn, in the Karoo. He credited Schwarz with focusing "popular attention" on the question of climate change, and lamented that scientists had not paid more attention to the issue.[14]

The scheme also animated conversations in town halls and at agricultural society meetings. A few months after Schwarz's book appeared, the acting irrigation director—Kanthack had just retired—told an audience at the Congress of the Cape Province Agricultural Association that "the scheme was a fascinating one and was being given serious consideration by his Department. The great difficulty, however, was that the Department had not the men available and present to carry it out." The farmers were having none of it. One responded that he was "fed up" with the government's delays. Why not hire private engineers to do the work? "It was a simple thing" they were asking, he insisted. The minister of agriculture stepped in to enumerate other difficulties: scientists had serious criticisms, the rivers lay outside South Africa, and the location of the border between Angola and South West Africa was still disputed. The farmers applauded the minister "for several minutes," according to a newspaper account. But after he left to catch his train, they approved a motion asking the government to provide estimates "for the preliminary survey and investigation of the Kalahari problems in connection with irrigation and also the schemes suggested by Professor Schwarz for restoring the inland lakes."[15] The farmers' response established the pattern for the next two decades. Neither deflection nor the listing of scientific critiques and political obstacles quelled public enthusiasm for the redemption of the Kalahari.

Additional evidence of the public impact of Schwarz's ideas comes in the form of two commissions created in 1920: the Drought Commission, discussed in earlier chapters, and the Angola Boundary Commission. Each was shaped by the specter of Schwarz's proposal.

The northern border of South West Africa had been contested almost from the moment that Germany and Portugal signed the original border agreement, in 1886. Until 1926, a seven-mile-wide no man's land extended across much of South West Africa's northern boundary, reflecting territory claimed by both it and Angola.[16] Aside from complicating colonial administration, the disputed border meant that it was not clear what part of the Kunene River lay wholly in Angola and what part could legally be accessed by South West Africa. In Angola, a Portuguese military engineer and surveyor named Carlos Roma Machado alerted the Portuguese government to Schwarz's proposal to divert the river. He argued that Angola's southern neighbors had designs on Portuguese water and planned to impose conditions that the Portuguese "cannot and must not accept."[17]

When the Boundary Commission was formed two months after Schwarz's book was published, Smuts explained the importance of access to the Kunene River in terms of climate change. "In view of the drying up of Ovamboland and the Kalahari"—Smuts stated this as fact despite the contrary opinion of his own government's scientists—"this right [to the river] may prove valuable, provided it is feasible on engineering and financial grounds."[18] Kanthack did not agree that lands untouched by white farmers were drying up. But he was tasked with surveying the Kunene and establishing whether Schwarz's proposed weir would work.

The Drought Commission also cannot be understood outside the context of public enthusiasm for the redemption of the Kalahari. Established in September 1920, it was a response not only to the 1919 drought but to Schwarz and his supporters. The agronomist and dryland farming enthusiast Heinrich du Toit, the head of the commission, initially sought common ground with Schwarz, insisting that they had similar goals. "I admire your scheme and your courage," he wrote Schwarz in early 1921. "You are, on the one hand concerned with the establishing of an inland lake system, and I am supplementing your work by endeavoring to stop our rain water running off to the sea." Du Toit suggested to Schwarz that they meet when the commission came to Grahamstown, and asked him to provide a paper that could be included in the committee's report.[19] Somewhere along the way, however, things changed. There is no record of such a meeting, and Schwarz contributed nothing to the final report, which does not even mention his name. Indeed, some people argued that the commission had quietly declared war on Schwarz and his followers. Scully accused du Toit of deliberately excluding Schwarz, and reported that a farmer hosting the commissioners was told that they would not hear comments on Schwarz's scheme.[20] In fact, Schwarz makes an indirect appearance in the report's appendices, in reprinted testimony where multiple witnesses cited his arguments about climate, irrigation, and the problem of white agricultural precarity, and spoke favorably about his scheme. But the commission's central conclusion—that rainfall was not declining and that drought losses were due to human behavior—left little space for common ground.

Schwarz's meteorological claims were among the most contentious because they were among the least certain. The feasibility of the engineering aspects could be tested, given enough time, money, and personnel. But few would unequivocally opine on whether lakes could produce rainfall. The American *Geographical Review* noted that Schwarz's claims for reprecipitation were "exceedingly weak," but added that this was not Schwarz's fault, "as no case of complete experimentation and research along these lines is available."[21] The geologist John Gregory, who a decade earlier had argued against planetary desiccation, wrote that Schwarz's estimates of his scheme's climate effects were "highly speculative," but he did not completely dismiss them.[22]

Schwarz's supporters and those sitting on the fence acknowledged that his climate claims were controversial. In South West Africa the German farmer Paulsmeier, who had proposed a similar scheme in 1914, suggested that popular opinion was on the side of reprecipitation:

> There are, I believe, men who do not agree in the view that the presence of a large fresh water lake, or for that matter, of a large body of any water, would have the anticipated beneficial influence on the whole of South Africa, but I can say, practically all old South Africans, to whom I spoke about the matter, are enthusiastic on the point.[23]

Schwarz skeptics pointed out that many bodies of water in the world's arid areas had no apparent impact on local precipitation, among them Utah's Great Salt Lake, California's Salton Sea, and the Caspian, Aral, and Red Seas, not to mention all of the oceans bordering the world's coastal deserts. Schwarz claimed that some combination of topography and wind patterns distinguished all of these places from the Kalahari.[24]

While Schwarz was in London, the Royal Geographical Society convened a panel on this issue that revealed a divide between South African and foreign scientists. While the British panelists were generally supportive, Kanthack, the only South African to comment on Schwarz's paper, was uncompromising in his criticism. Kanthack claimed the status of both a scientist and a "practical man" whose knowledge derived from experience. Confining his comments to the western half of the scheme, he told the assembled guests that Schwarz had never even seen the Kunene River, and that his own on-the-ground surveys proved that a lake at Etosha could never assume the dimensions Schwarz imagined. There were no logical sites for a dam that would not require "a work of colossal dimensions." And once diverted, the water would have to travel over a "dead flat plain" consisting of highly permeable soils. Kanthack concluded by telling the audience that Schwarz had few supporters "on the scientific side in South Africa."[25]

The country's most prominent scientists had been cautiously skeptical in 1918, but two years later their opposition had hardened. Their case rested on partial data from topographical surveys, doubts about the mechanism of reprecipitation, and botanical and other evidence that southern Africa's arid environments were ancient. But these criticisms did not provide a watertight argument for or against the scheme; they only created uncertainty. And uncertainty, it turned out, was not enough to end popular support. This is partly because the stakes were different for supporters of the Kalahari Scheme and for scientific experts. Even people who were agnostic about Schwarz's scheme argued that scientists could not be 100 percent certain it would not work. And for those who believed rainfall was declining, it was worth taking a chance on it, given the threat they perceived desiccation to pose to their futures.

Experts and "practical men" actually asked many of the same questions about Schwarz's scheme, and both groups focused primarily on its feasibility. But those outside the expert community rejected scientists' idea of where the burden of proof lay. Was it necessary to prove that rainfall was changing, and that lakes in the Kalahari could produce rain—or to prove that rainfall was not declining, and that lakes could not produce rain? One man wrote, "Who is bold enough to say that Professor Schwarz is in error when he states that the filling up of the lakes will cause moisture to rise, and from it the rains will come?" Another reader responded: "Will the rains come? The crux of the whole scheme rests in the answer to that question."[26]

These men were on opposite sides of the Schwarz debate, but they agreed that declining rainfall constituted a pressing problem. The skeptic, who had lived in South Africa for fifty years, wrote, "I quite agree . . . about the regularity and frequency of the rains in the early days.'"[27] Popular publications also accepted as fact that southern Africa was "drying up." *De Zuid-Afrikaan* stated that the reality of a drying climate "is widely recognized."[28] The *Cape Argus* warned that "the country is rapidly drying up, and is in imminent danger of becoming a desert."[29] The *Allgemeine Zeitung* of South West Africa wrote, "One sees in the Union that something must happen to save South Africa from desiccation."[30] Outside of southern Africa, many geographical and meteorological journals accepted this desiccation as fact, reflecting the transnational conversations rooted in encounters with the globe's aridity. A reviewer for the *Journal of the Royal African Society* wrote, "One of the great problems of the future is to arrest the steady drying up of the African Continent."[31] The *Quarterly Journal of the Royal Meteorological Society* concluded that rainfall statistics "do indicate a decreasing rainfall since 1874. . . . On the whole the evidence is in favor of desiccation"—a conclusion opposite scientific consensus in South Africa itself.[32]

An article by "a South African scientist of the first rank" made the case for climate stability, but the writer acknowledged, "The belief that the rainfall has deteriorated is very general among farmers, in some districts."[33] (Notably, the newspaper did not identify the author, perhaps a tacit acknowledgement of the unpopularity of his views.) South Africa's scientists continued to insist that Schwarz was operating from a false premise, and thereby lost credibility with much of the public, whose experiences of aridity were now reinforced by favorable media coverage domestically and abroad.

Kanthack and his colleagues assumed that ending support for the Kalahari Scheme simply required collecting and publicizing the evidence against it. In this they were disappointed. Not only did it remain popular; some supporters used the findings of Schwarz's critics to make their case. Government scientists thus found themselves losing control of their narratives. Someone writing to the *Allgemeine Zeitung* under the name "Outis" cited an unnamed English researcher who had traveled to Ovamboland in 1924 and found that the Kunene half of Schwarz's scheme was feasible. "Outis" used Kanthack's own estimates of the Kunene's flow as evidence in favor, rather than against, a diversion scheme.[34] Heinrich du Toit, in his report for the Drought Commission, had warned of the impending "Great South African Desert, uninhabitable by man" if farmers did not change their land-use practices. In the hands of Scully, a Schwarz supporter, the phrase was weaponized as additional evidence for declining rainfall.

Given the extent of agreement over rainfall decline, experts' continued insistence that rainfall was not declining led many people to discredit other aspects of their critiques and to insist that they were exaggerating the scheme's flaws. As one farmer wrote, "To the plain man, the situation . . . does not exactly bristle with the difficulties that were attributed to it."[35] National Party MP Conradie, who represented the Kalahari region of Gordonia, suggested that an investigation need not be expensive; if the government sent a couple of engineers who knew the land, they could figure out within a year where the water would run, and make a new estimate of costs.[36]

Experts found their inability to dampen this popular enthusiasm perplexing and, on occasion, infuriating. John Sutton, a Cambridge-educated meteorologist, publicly accused Schwarz of acting more like a demagogue than like a scientist. Calling Schwarz "a personal friend of mine of many years' standing," Sutton justified his decision to publicize his concerns:

> If his arguments be sound, his hypothesis would become a theory; if they be not, it would die a natural death. But he has elected to make his scheme a fiery cross, and to start a crusade, with the support of a lot of freelance politicians quite incapable of weighing scientific evidence of any kind. He asserts and keeps on asserting and waiting and lecturing that South Africa is drying up, and that he alone knows how to convert it into a Garden of Eden.[37]

Such passionate critiques were no more successful at dislodging public support for Schwarz than a recitation of scientific data had been. The chasm between the two sides remained the question of progressive desiccation. Scully replied, "From the general tenor of Dr. Sutton's article, I take it that he does not accept the idea that the rainfall of South Africa is decreasing." Scully then offered a synopsis of fifty years of rainfall returns from the farm of the noted progressive farmer Walter Rubidge.[38]

Scientists continued to describe "lay opinion" as uninformed and, in Sutton's words, "incapable of weighing scientific evidence of any kind." Divergence from scientific consensus carried with it some of the same connotations that divergence from "civilized" material standards and modes of living did. If poor whites had a shaky claim to whiteness based on their mode of living, those who subscribed to environmental narratives that diverged from scientific orthodoxy had aspersions cast on their identity as whites, based on their supposed resistance to science and rationality. Weidner, perhaps the most prominent Schwarz opponent outside the government, not only portrayed Schwarz as operating beyond the bounds of science; he implicitly questioned his status as a white man. Weidner called Schwarz a "lake-charmer," and described his scheme as "rain-making magic." For most whites, these conjured images of an African charlatan—an image refined and popularized through countless missionary and traveler accounts that described rainmakers preying on the gullibility and desperation of others. Weidner's racially charged language implicitly compared Schwarz's supporters to supposedly superstitious and gullible "natives" who put their faith in their rainmakers.

Such accusations further eroded the legitimacy of "expert" opinion because Schwarz's supporters understood their own knowledge and worldview as embedded in rationality and science, not superstition. Many identified as progressive farmers, itself a gendered and racialized category. Many, though not all, presented themselves as educated men aware of the larger world and conversant in the basic principles of science. As in broader conversations about weather and climate, a scientific elite controlled the categories of "expert" and "lay person," but many of the latter rejected this binary and, in their actions, blurred the lines between them. Schwarz supporter Townshend wrote *Farmer's Weekly* encouraging farmers to plant trees and offering recommendations on specific species.[39] Transvaal farmer P. J. de Wet told the Drought Commission in 1923, "In my opinion the desert of Central Africa is the source of all our droughts. It is noticeable that the desert conditions of the Kalahari are encroaching on a large slice of South Africa." His solution was to expand the use of jackal-proof fencing, a favorite prescription of government experts, as well as "the flooding of the Kalahari desert."[40] A man who wrote under the pen name "A Sufferer from Drought" suggested that the solution to the drying up of South Africa was to "preserve moisture by not burning the grass, by damming the spruits, and filling up the pans and lake Ngami."[41] One of the men who suggested that rainfall had changed since the Anglo-Boer war was a regular user of agricultural extension services. A Schwarz supporter and farmer wrote government experts asking for advice on controlling grain moths and seeking a supply of beneficial insects to help with aphid control.[42] These men saw no contradiction between their support for Schwarz's ideas and their engagement with government scientists.

As with weather predictions, government experts could offer useful advice to these men only some of the time. For example, they could not always identify an insect, describe its life cycle and food source, or provide the requested means to eradicate it. Farmers' continued outreach to government agricultural services suggests that they understood that expert knowledge had limitations, something the experts themselves were often reluctant to admit. It therefore did not trouble Schwarz's supporters unduly that he did not yet have all the details worked out. For these men, Schwarz's scheme embodied the goals of progressive farming: the application of scientific knowledge to increase the productivity and sustainability of the natural world's resources and the viability of "white civilization," all done with the support of the state. It was in that spirit that many members of the public called for an investigation and argued that the state was abdicating its responsibility by not conducting one.

A few men suggested that the Kalahari Scheme could be investigated and even built by mobilizing community support.[43] But most insisted that it was the responsibility of the state, whose primary purpose was to ensure the stability of a white South Africa. They also argued that an investigation could easily be done if the government felt so inclined. William Jacobs suggested that the scheme was feasible "if the Government would contemplate and tackle the job," and that it would offer "a source of wealth and prosperity . . . to South Africa in general."[44] MP Munnik, an engineer representing a National Party stronghold in the Transvaal, insisted that "from an engineering point of view, there are no great difficulties."[45] This idea that the scheme was a simple one, despite its intended transformation the land and waterscape, was reinforced by Schwarz's insistence that it would be inexpensive to build: £250,000 or, as a supportive senator put it in 1918, "just about the amount that the Union Government spends in a single year on printing and stationery."[46]

The demand for an investigation provided an opportunity to air grievances about government priorities and expenditures—grievances that could be expressed because white men assumed that they encountered the state on an equal footing, and that their opinions deserved serious consideration. Despite recurring assertions by government commissions that it was not the job of the government to give handouts and that state resources were limited, white men—in a pattern familiar to people in other parts of the world—expected the state to meet their needs rather than allocate money to causes they found unimportant.[47] This was not merely a matter of exploiting political divisions. While the pro-British Smuts was prime minister, the pro-British *Star* opined, "The Government appoints Commissions with or without excuse to enquire into all matters of issues of various shades of importance dwindling to insignificance." So why not a commission to investigate the Kalahari Scheme?[48] Five years later, when the National Party was in power, a Nationalist MP from the Karoo town of Beaufort West said, "We spend £25,000 or £30,000 annually for the League of Nations. This is also only a theoretical matter to avoid war, so also the scheme of Professor Schwarz is a theoretic [*sic*] matter to prevent drought."[49] A man in Gordonia, where the National Party dominated, detailed the cost of several dams—"all failures, and nothing to show for it."[50] Someone in Queenstown, where the South Africa Party consistently prevailed, scoffed: "For the benefit of humanity Governments can find no gold—but for the waging of wars for their destruction millions per day are forthcoming as easily as drawing water from the town reservoir."[51] Several years later, a fellow resident of Queenstown echoed these ideas:

> The Government of the day, whichever party may be in power, is quite prepared apparently to gamble with the people's money on steel factories and other wild-cat schemes, which may or may not show a profit; to spend millions on war which, of course, is necessary at times, but when it comes to a question of the actual life of the country, the matter is referred to "experts."

He concluded, "From a purely South African point of view, no war was ever one-tenth so important as the drying up of the country."[52]

Some called for making support for Schwarz's scheme a litmus test for political office. The scheme should "be the first plank in every political party's platform," wrote a man from the semiarid and drought-prone western Transvaal. "Dry Farmer" on the margins of the Kalahari agreed: "What good are politics to a farmer who gets no rain on his farm?" he asked. An Eastern Cape man described a political platform of "moisture-laden wind from the west—showers in place of dust—green grass in the lambing season—no more parching Octobers—lakes in place of desert."[53]

White Men and African Land

For Schwarz and his supporters, hydroengineering and social engineering were inseparable. Much of southern Africa's land had been seized by white settlers over the course of 250 years. But the Kalahari Scheme, like the other techno-optimist schemes described in the last chapter, claimed rights to transform land that was still occupied by Africans, much of it closed to white settlement and not under the political control of South Africa. These men justified the expropriation of African land and water by the supposed failure of Africans to fully utilize their natural resources. When Ernest Long proposed diverting rivers that "now run idly through Swaziland" and through "the native territories of the Transvaal," their idleness justified his claim over them.[54] When the mine engineer Blenkinsop contrasted the "present miles and miles and miles of arid unproductive wastes" with the "literal earthly paradise" that South Africa could become, Schwarz's scheme became a moral imperative.[55] Schwarz asked, "have we the moral right to undertake irrigation works in native territory?" and answered with a definitive yes.[56]

Schwarz and Carl Weidner disagreed on nearly everything, but they shared the belief that whites had a legitimate right to the lower Kavango River, located in the Bechuanaland Protectorate. When Weider recounted the story of the chief who unwittingly destroyed his country's ecosystem, he concluded that it was "unquestionably the 'White' man's duty to repair the harm done by Chief Morimi."[57] The language of duty implied that white men had the responsibility to seize African resources for their own good. The Swedish engineer Karlson wrote, "Nature has given us these equatorial Rivers and it is our duty to see and examine in which way we can make the best use of them."[58] He promised that in his scheme, "The natives would be fully provided with irrigated land and would greatly participate in the improvement of the country"—assuming, of course, that they settled down as farmers on whatever land they were given.[59] Schwarz also emphasized that Africans would not be left landless: "The natives have first right to the ground, and any scheme for irrigation must first of all make ample provision for them. We do not want to see repeated what has happened in other parts of South Africa, that is, small reserves for the natives surrounded by white farms"—a subtle criticism, perhaps, of the Natives Land Act.[60]

But in his promise to protect African land rights, Schwarz was also protecting "white innocence"—the comfortable conviction that whites were not responsible for the environmental degradation around them and that "drying out," not white expansion, was the real threat to African land rights. If the Kalahari continued to spread, Schwarz told a town hall in Queenstown, "they would have to find land in the Union for 150,000 natives."[61] Karlson noted that Basutoland was already overcrowded, without acknowledging that huge tracts of its most fertile land had been seized by white settlers. His plan to render arable sufficient land to provide a living for "1,600,000 natives" in the northern Kalahari could therefore sound magnanimous.[62] Schwarz claimed that Ovamboland's chiefs "ardently desired" his scheme, while in the Okavango Delta those who would be most affected would reap substantial benefits.[63] African opposition, meanwhile, was couched in racialized tropes of superstition and irrationality. Schwarz said that the Mbukushu community, living upstream of the Kavango Delta, would be "annoyed" when his scheme increased rainfall, because their chief, a noted rainmaker and "comptroller of the floods," would lose the basis of his power and wealth.[64]

Concern over African land shortages was rooted primarily in concern about the well-being of the white community. Schwarz focused on the availability of Black labor. "It is wise policy to support an industrious, virile nation like the Ovambos, if only from the fact that they constitute a valuable supply of native labor," he wrote—a theme taken up by *Allgemeine Zeitung*.[65] One of his supporters argued, "This very point of ground suitable for native occupation is becoming an increasingly serious problem. We cannot do without the native for his labor; thus it is up to us to see that he has room to exist in, so that the next generation will also have sufficient labor."[66] This vision of a white man's country—of Black labor that showed up when needed and then disappeared when it was not—depended on African access to land. It is a problem that historians of this period have recognized only fleetingly, but it was understood at the time. Karlson argued that for segregation to work, Africans needed adequate land, and "this cannot under the present circumstances be provided."[67] Hydrological engineering would create a more perfectly segregated society by solving the problem of land shortages that forced whites and Blacks to live side by side, threatened the supply of exploitable labor, consigned whites to the status of a racial minority, and pushed some whites into poverty, thereby undermining claims to racial superiority. Schwarz argued that so much land would be available that it should "completely dispel the fear expressed by some people, that this irrigation scheme will cause trouble with the natives; the country is so vast that they need not come into contact with the white settlements at all."[68] Karlson's imagined Kalahari settlement would be segregated by the Botletle River, which was the downstream outlet of the Okavango Delta, with land to the north reserved for Africans and that to the south reserved for whites. This effectively limited Black land occupancy to the northern quarter of the country, where the Delta and associated pans were located, and gave whites the bottom three-quarters. "Keeping whites and natives absolutely separated" would not only prevent conflict, but would prevent the problem of malaria among whites. Segregation would protect white bodies as well as white supremacy.[69]

Such utopian thinking revealed a great deal about the existing world, and not only because this white utopia constituted a Black dystopia. Proposed farm sizes would be racialized on the grounds that Africans had fewer material requirements than whites—the same logic that governed already existing differential wage scales and meal rations in the labor force. Karlson allocated seven acres for each of his proposed 1,170,000 white settlers, while 1,600,000 Africans would get an acre each.[70] As in existing irrigation schemes, an irrigated Kalahari would also draw distinctions among whites. Schwarz suggested twenty-acre plots for poor white families, allowing them to "buy for themselves some of the comforts and amenities of civilized existence." These were larger than the plots offered to them in current irrigation settlements. But those with the means to purchase their own land would receive one-hundred-acre farms.[71] Nowhere, of course, was there a suggestion that women of any race would be landholders. Engineering the Kalahari's rivers would simultaneously engineer a society in which distinctions of race, gender, and class were preserved and fortified, with white settlement organized around the nuclear family and white class distinctions mapped onto the landscape through farm size. By segregating the settlement of this supposed *terra nullius* from the start, it would avoid the messiness that characterized South Africa itself, where white and Black land claims and farm occupancy remained intermingled.

In these plans for their fundamental transformation, local populations could be rendered as abstract or invisible as the Black majority within South Africa's borders. Karlson incorrectly stated that Ovamboland was inhabited by Herero cultivators. He offered no further information about the region's economy, politics, or even agriculture, though the engineering aspects of the scheme—the cubic feet of water available, the height to which water would be raised, the horsepower required, the cost for this energy per year and per acre—are laid out in mind-numbing detail.[72] Schwarz had little to say about the people of the Okavango Delta, though he offered a list of "tribes" and their approximate population, and noted that they would need to be compensated for any lost land.

In a quest to remake the world, deserts have the advantage of being perceived as empty. As other scholars have noted, decisions about whose arid lands need to be redeemed or reclaimed and whose can be "waste-landed" both reflect and reinforce existing configurations of power that are frequently rooted in racial inequality.[73] In South Africa, the well-worn trope of empty land was repurposed for new geographical spaces and new times. For the most part, Schwarz and Karlson's proposals simply ignored local populations. When Schwarz stated that diverting the Kunene River would do no harm because the region was uninhabited, he rendered invisible the Himba pastoralists who relied on the river for seasonal food and grazing.[74] Karlson wrote that the population of the Bechuanaland Protectorate, except in the southeast, "is so scarce that there does not seem to be any practical difficulty for its colonization by the white man"—thus erasing the relatively dense population of the Okavango Delta.[75] Meanwhile, the question of who would help white settlers work their farms in the Kalahari was conveniently left unasked, continuing a pattern of ignoring the contradiction inherent in imagining a "White Man's Country" that ran on Black labor. The paternalism that offered Africans the benefits of white men's technology on white men's terms was could be paired with the conceptual annihilation of Africans when it was necessary to ensure logical coherence. In public debate, too, supporters and opponents alike focused on the land but ignored the people.

The fate of the Black majority was far less important to these engineering schemes than the well-rehearsed idea of an existential threat to whites as fear of a drying subcontinent combined with the fear of a fragile white population. A farmer in the northern Cape wrote that everyone who supported Schwarz's scheme was "helping his children in the fight they are going to have for existence."[76] Drought would ruin white men, who would then be driven from the land. Unless something like his scheme were built, Schwarz told a Queenstown audience, "they would have to clear out of South Africa."[77] Who was meant by "they" was understood.

Many Afrikaner nationalists opposed government-sponsored white immigration as a threat to their political power and a distraction from the problem of white poverty. But Schwarz's climate-engineering scheme did not force people to choose between aiding "poor whites" and increasing the white population. He promised to address both pillars of white insecurity, and this was a major reason why, in the early years of the debate, many National Party politicians supported him. A wetter climate and rivers flowing from the north across the Kalahari would open perhaps thirteen thousand square miles of land for settlement. There would be space for potentially millions of white families, and the resettlement of perhaps 75 percent of South Africa's poor whites.[78] By promising (racial and class-appropriate) abundance for all, environmental transformation

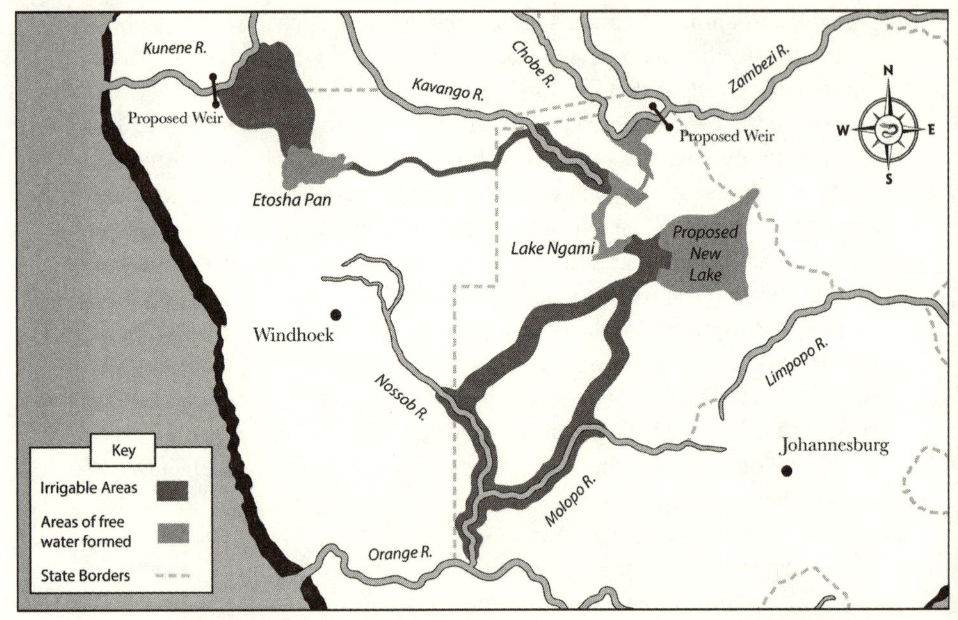

Map 7 Schwarz's scheme, with political borders shown. On Schwarz's original map, seen at the front of this book, the absence of political borders suggested the possibility of a unified "white man's land" that lay at the heart of his scheme. In reality, none of the newly created bodies of water would have been located within South Africa.

offered a way to avoid a future in which whites would be swept up in a "racial" conflict between British and Boer: a dystopian vision outlined by Karlson, who wrote that South Africa was facing "the black specter" and "the worse specter of the racialism among the whites themselves."[79] The rivers to the north would be "the salvation of South Africa."[80]

The northern rivers would facilitate the creation of new, territorially more expansive white man's land in southern Africa. Schwarz's map of the Kalahari Scheme included the lands, major rivers, and cities that lay south of the Kunene and Zambezi rivers, but it had no political borders. (map 7). Weidner imagined a "YOUNG SOUTH AFRICAN COMMONWEALTH" and dedicated his pamphlet to the leaders of South Africa, Southern Rhodesia, South West Africa, and the Bechuanaland Protectorate. (His dedication echoes that of the agronomist William MacDonald advocating the "conquest of the desert" a decade earlier: "To the million settlers of tomorrow on the dry and desert lands of South Africa: Welcome.") Karlson emphasized that the waters of the Okavango were located "300 miles from Pretoria" and promised "an entirely new state in the Union of South Africa." Because he correctly surmised that reservoirs would be necessary in Angola and Northern Rhodesia to make a major river diversion and irrigation scheme work, his imagined political union was even larger than those of Schwarz and Weidner.[81]

Hindsight makes South Africa's borders appear more inevitable than they appeared to those living in the early twentieth century. The 1909 Act of Union made provisions for the transfer of the three High Commission Territories—the protectorates of Basutoland, Bechuanaland, and Swaziland—to South African control. South Africa was awarded South West Africa as a mandated territory in 1919 and held it, in violation of international law, until 1990. And many white South Africans assumed that Southern Rhodesia's future lay with its fellow settler colony and were shocked when white Rhodesians voted not to join South Africa in 1922. While the historical literature has argued that the commitment to this expanded version of South Africa was largely confined to a few political elites, particularly Jan Smuts, in fact many white South Africans embraced the concept of territorial expansion, or at least assumed it would happen.[82] They did so not out of the self-confidence that historians have associated with politicians' plans to enlarge the South African state, but out of fear of what might become of them if they did not do so. The popularity of Schwarz's scheme reveals that in much of the popular imagination, territorial expansion and territorial partition could be two sides of the same coin: solutions to the problem of maintaining white dominance amid a disenfranchised majority. Expansion was understood as a sustainable path to macrosegregation or "total territorial partition."[83] It was an early iteration of partitionist sentiment that would later manifest itself in another form: the project of Grand Apartheid.

Blank Slates, White Fantasies

Schwarz had never seen the rivers he proposed to divert. Nor had he seen the Kalahari, the very space that he claimed was a threat to South Africa's future, and which he proposed to eradicate. As one of his biographers notes, Schwarz's Kalahari was "an imagined place, conceived from afar."[84] But Schwarz was not unique. Just as the lost lakes and vanished rivers had a vivid place in the white South African imagination, so too did the Kalahari. Legends about the Kalahari abounded: one of the most well-known was promoted by a Canadian working under the pseudonym of Guillermo Farini, who traveled the region in the 1880s and wrote a book claiming that he had discovered the ruins of a lost city there.[85] More prosaically, it was the birthplace of droughts and locusts and it was, many argued, encroaching on South Africa. The Kalahari was portrayed by the agronomist William MacDonald as "unsurveyed waterless desert."[86] Schwarz called the Kalahari "a wilderness of red, blown sand, covered with grass, thorny scrub, and in the north with forest."[87] The very title of his book, which referred to the "Thirstland," echoed a view of the Kalahari that dated back to Dutch colonists' first encounters with southern Africa's interior, in the nineteenth century.

And yet its seemingly magical transformation after the rains called into question whether the Kalahari was a desert at all. A German military officer wrote in 1906 of the "Kalahari 'desert' . . . a lush grassland, fertilized by rainfall and pearly dew," where "an industrious people could raise up a second Argentina."[88] Karlson called it a "so-called desert,"[89] and cited reports that it contained "80,000 square miles of the finest ranching country in the world."[90] Schwarz wrote, "The Kalahari after rain is the most beautiful country in the world."[91] In 1923, *United Empire* reported on a cattle trek through the Kalahari, "which was found to not be a desert at all, but 'a veritable grass garden.'"[92] The following year, the *Windhoek Advertiser* noted that an expedition seeking the breeding grounds for the locust swarms that plagued farmers had made "the somewhat surprising discovery that the Kalahari Desert is not the waste of sand that it is popularly supposed to be."[93]

The paradox of a place that could seem a desert one moment and a lush grassland the next explains why the Kalahari was designated as a "problem." Karlson titled his book *The Kalahari Problem*. Government reports referred to "the general problem of the Kalahari" and "the nature of the problem."[94] The Cape farmers asked the minister for agriculture to investigate "the Kalahari problems." The problem was both practical and existential. The presence of a purportedly unsettled and expanding "wasteland" on the doorstep of Africa's largest zone of white settlement was a threat to racialized rule based on white men's ownership of the land. These claims to power and property rested on making land "productive" in a way that Africans had supposedly failed to do. More immediately, the Kalahari was a problem because every suggestion for its development foundered on a basic lack of information. Was it a desert or wasn't it? Was it expanding or was it not? In 1923, *United Empire*'s editor expressed frustration over the conflicting information about the Kalahari: "Such discrepancies ought not to be possible today."[95] Schwarz cheerfully admitted this ignorance about the Kalahari in his book when he wrote that if his scheme were to be built, "it is impossible to state now where the water will flow."[96] And despite its importance as a labor reservoir for South West Africa's farms and mines, Ovamboland was little better known. It was placed under direct colonial control only in 1915. Kanthack's 1920 expedition on behalf of the Angola–South West Africa Boundary Commission was the first colonial survey of the Kunene River. And uncertainty remained about the origins of the floods that swept across the Ovambo plain during the rainy season, reaching Etosha Pan in years of very heavy rainfall—the inspiration for the western half of Schwarz's scheme.

Excitement over the Kalahari Scheme prompted a renewed search for information. In South West Africa, people focused on the western half of the scheme. In a series of articles for *Allgemeine Zeitung*, one man described the hydrology of the Ovambo plain and introduced readers to an entirely new river, which he called the Covelai (Cuvelai). These confirmed the long-forgotten findings of a French missionary named Duparquet, who had been stationed near the river's headwaters in southern Angola. Seasonal in all but its upper reaches, the Cuvelai in its lower course breaks into multiple channels which fill with water during the rainy season. The author provided readers with a name for these distinctive channels: *oshana* (plural *iishana*).[97] To the east, the government of the Bechuanaland Protectorate sought to find out more about Schwarz's "lost lakes." Ignoring the possibility that local Africans might contribute to this new body of knowledge, the governor general's office instead sought out the views of local white residents and former and current colonial administrators. This was a small pool of expertise indeed, since the entire Bechuanaland Protectorate operated with a skeleton British staff, and white settlement in the northwest was restricted to a handful of traders and missionaries. Their reports had remarkably different reactions to Schwarz's scheme. One local resident suggested that if a weir were built along the Chobe, "a large part of the country would be desolated." A former official in the Caprivi Strip, on the other hand, thought "it would confer a lasting benefit upon the country" but questioned whether it were possible. During the annual floods, "the Zambezi and Chobe are practically one sheet of water. . . . I consider it would be an impossibility to dam the Chobe without damming the Zambezi too."[98]

Most comprehensive was the report of Almar Stigand, the resident magistrate of the district known as Ngamiland. It was as damning as Kanthack's report on the viability of the Kunene half of the scheme. Schwarz had never visited the region, Stigand noted, and all his information came from a trader named Kays, who had exaggerated the area's potential in order to advance his own financial interests. Stigand insisted that Schwarz's information was based on "antiquated, useless and misleading maps together with inaccurate hearsay reports," and that building his scheme would require not a weir of mud and sticks but "huge engineering work." Hundreds of miles of channels would have to be straightened and broadened, and to what end? There was already a large body of surface water in the region, in the form of the Okavango Delta, which offered "little or no benefit to the local rainfall."[99]

Unlike Kanthack's Kunene report, Stigand also considered the impact of Schwarz's scheme on people. Schwarz's claim that his scheme would preserve African land rights was a dangerous delusion, the magistrate wrote. The delta's lagoons and backwaters provided the region's subsistence. The flora and fauna of the marshes—trees, water lilies, papyrus, water iguanas, and lechwe—were staples of the local diet that would vanish if the river were confined to a single channel aimed at delivering water to the "lost lakes." Local agriculture would also disappear, since it relied on seasonal inundations to fertilize and irrigate alluvial soils. By destroying this riparian ecosystem, the Kalahari Scheme "would inevitably reduce the country to starvation."[100]

Stigand is the only person in the written record to offer a defense of the inherent value of the Okavango Delta's wetland ecosystem and local people's use of it. While most observers saw a worthless and unhealthy swamp, Stigand insisted that the current environment was valuable and productive. The governor general's office stated that the resident magistrate's opinion carried particular weight. Yet the South African government—with the cooperation of the governments of the Bechuanaland Protectorate and Northern Rhodesia—ultimately chose to investigate the scheme anyway.

The Kalahari Scheme was sufficiently expansive to allow people to adapt it to their own utopian visions. If many imagined that it might enable the continued dominance of agriculture in a rapidly industrializing economy, others imagined different outcomes and customized the scheme accordingly. Richard Rothe, a store owner in Outjo, in South West Africa, was excited about the possibility of resource extraction and commerce: "The land between the Zambesi and the Chobe is a swampy one. . . . Only putting it underwater forever will make this part of Central Africa fit for the convenience of trade." He enumerated the region's supposed resources: "one of the biggest Copper Mines of the world" at Katima Mulilo; "big forests of Ebony Trees" that awaited cutting; "the great wealth of natural produce of the big Hinterland of Angola," including wild coffee and rubber.[101] A National Party politician from the Orange Free State stated confidently that Schwarz's scheme would avoid the problem of salinization that bedeviled other irrigation schemes because "the Etosha Pan has no salt."[102] The "thirstland" in need of redemption was a geographical abstraction that could be filled with virtually anything. The eastern Caprivi, which has no copper deposits, could be the site of one of the world's largest copper mines. The Etosha Pan, source of a regional salt trade for centuries and the largest salt pan in Africa, could be stripped of its salinity.

Schwarz had omitted the Zambezi River, southern Africa's largest, from his proposal in order to protect Victoria Falls.[103] His supporters had no such qualms. The mine engineer Blenkinsop misunderstood Schwarz's scheme as "the bringing of the waters of the Zambezi River into the Union" and supported it.[104] (He may have conflated Schwarz's and Karlson's schemes, since Karlson proposed constructing a diversion channel above Victoria Falls.) To Karlson, tapping the Zambezi was a necessary sacrifice. If his scheme were built, "in the winter the *falls will be dry*," he wrote. "South Africa cannot afford the luxury of keeping them."[105] A man living in the Karoo wrote to *Farmer's Weekly*, "Had I anything to do with the diverting of rivers, I would have Lake Ngami and all the depressions in the Kalahari flooded instead of allowing volumes of water to pour over the Victoria Falls as a theatrical display. Were this done, it must of necessity alter the climatic conditions of the country."[106] The German farmer Gessert agreed:

> Who would not cheerfully renounce a few hours of delight afforded by Nature in order to be rid of the tormenting idea that by hindering the scheme he has contributed to the misfortune of thousands of families, who under most unfavorable circumstances are gaining a scanty subsistence not worthy of a white man, and leading lives not much better than the life of a native?[107]

The possibility of diverting the Zambezi rather than its smaller tributary, the Chobe, also caught the attention of two industrialists. Kenneth Quinan was an American-born chemical engineer who had established a large munitions plant in the Cape. Charles Markham was the director of a British coal and iron company that had expanded into chemical manufacturing. Markham's information about southern Africa may have come from his daughter, Violet, who alluded to the debate about "drying up" in her book.[108] The two men dreamed of "a vast industrial project, involving the utilization of the waters of the Zambesi which now run to waste, for the generation of power, so essential to industry and also possibly for agriculture or grazing in the—at present—dry land of the Kalahari Desert." Markham did not "attach much importance to the filling up of lakes in the desert, except in so far as they may aid towards irrigation"; he was "an industrialist with an eye to the future." The "redemption of dry areas" was a secondary consideration.[109]

These were very different goals from those of the agrarian populists who were the core of Schwarz's support. But Markham and Quinan shared farmers' concern that Britain had failed to stop its territories' rivers from "running to waste." They also shared the same faith in monumental engineering projects to transform nature. Quinan thought the scheme should be investigated "with an eye to the ultimate though perhaps distant future of South Africa"—again using an expansive definition of the term—and to bear in mind "such engineering achievements as the Suez and Panama

Canals."[110] Using Quinan as an intermediary, in 1923 Markham anonymously donated one thousand pounds to Prime Minister Smuts for an investigation of the Kalahari Scheme.

Into the Kalahari

For years, South Africa's executive branch had resisted demands that it investigate Schwarz's scheme. But the Markham donation and the crescendo of popular calls for an investigation caused it to relent. Perhaps Markham's vision seemed more palatable than Schwarz's to the governing South Africa Party, or perhaps South Africa's leaders had concluded that the only way to put Schwarz's scheme to rest would be to investigate it. Regardless, by the time the Irrigation Department organized a "rapid reconnaissance" into the Kalahari, Smuts—the recipient of the donation—was out of power, and Hertzog's National Party had been elected on a promise to help average farmers.[111]

Scientists objected to the announcement in 1925 that the Kalahari Reconnaissance would investigate Schwarz's proposal. The meteorologist Sutton lamented, "The sad thing is that our Government is being persuaded into spending thousands of pounds over this scheme. A tenth of the money devoted to the needs of scientific meteorology would beget useful knowledge."[112]

The Kalahari Reconnaissance promised to establish whether the fears and fantasies of lost lakes and vanished rivers awaiting restoration had any scientific basis. But its mandate was far more limited, thanks to the interests of the donor. The South African government contributed three times as much to the cost of the expedition as Markham, but his preferences for engineering the Zambezi and building an industrial, not an agricultural, utopia dominated the agenda.[113] The expedition would thus learn little about the Kavango River and nothing about the Kunene River. (Nonetheless, administrators in South West Africa took a keen interest; in fact, Namibia's archives are the only place where much of the documentation around the expedition appears to survive.)

The expedition was led by the famed geologist Alex du Toit, whom we met in chapter 2 when he joined Schwarz at the Cape Geological Survey. During the year or so that they overlapped at the survey, du Toit and Schwarz traveled, surveyed, and wrote reports together. Now du Toit was tasked with assessing Schwarz's theories. The Department of Irrigation, cognizant of Schwarz's popularity, invited him to join the expedition. The staff included three engineers, a "meteorological assistant," a camp superintendent, a mechanic, and a surveyor who drowned in the Zambezi at the beginning of the expedition. No scientific journey into the African interior had ever functioned without African labor and knowledge, and this one was no exception.[114] But the presence and contributions of Africans were as invisible here, in a survey that sought to make their lands "knowable," as they were in debates over the causes of climate decline in South Africa itself. Only the report of the surveyor's death acknowledged their existence, noting that three unnamed African paddlers had also drowned and that another had managed to swim to shore—and blaming the incident on the African staff who had "lost their heads during a spell of inclement weather."[115]

African knowledge was equally invisible. In a series of press releases, the Department of Irrigation updated the public on the expedition's progress. These bulletins served to familiarize white South Africans with the region, adding to the body of popular knowledge about Ngami the names of lesser-known rivers and channels, such as the Savuti, Linyanti, and Thamalakane. But the public bulletins never indicated that the expedition drew on local knowledge. Instead, the goal was bring the region into a domain of universal knowledge. The director of irrigation announced that "what was most required was *measurement*."[116] A major goal of the expedition was to render this maze of land and water legible in Cartesian space: to fix the location of major waterways, the elevation of the terrain, and the direction of water flow.

But achieving such legibility proved daunting. Across the region that Schwarz proposed to transform, the land was so flat that its elevation changes were imperceptible to the naked eye, and difficult to measure even with the best instruments available. It would be decades before professional surveyors worked out the gradients. Water here behaved in ways that seemed astonishing. Some visitors to the Botletle River observed it flowing into Ngami while others observed it flowing out. F. H. Barber, recounting a visit to the Makgadikgadi Pan in 1875, recalled hunting on its dry surface one day and returning on another day to find it covered in a shallow sheet of water—the result, an "old Dutch hunter" told him, of the wind changing direction and pushing the water, since the country was "absolutely on a dead level."[117] That "dead level" also fascinated visitors who saw the Selinda Spillway, a channel linking the Kavango and Zambezi watersheds that could flow in either direction.[118]

In language reminiscent of colonial discourse about African peoples, Schwarz described Africa's rivers as lying outside the norms of the rest of the world. He called their courses "extraordinary" and described the Zambezi and Orange Rivers as "deceptive" in their apparent normalcy.[119] Of the Okavango Delta he wrote, "It is characteristic of the whole of this flat country, that the rivers seem not to flow in the ordinary sense of the word, but great masses of water infiltrate the country, filling all depressions, and then remaining stationary."[120] In "a normal country," he insisted, "evaporation balances rainfall."[121] These ideas were rooted in settler views of arid landscapes as abnormal, and Alex du Toit shared them. Two years after the expedition, du Toit referred to the hydrology of the northern Kalahari as "queer" and "decidedly not that of any normal river system."[122] Flatness denied visitors a "God's eye view" of the terrain—the ability to survey the entirety from an elevated vantage point.[123] Airplanes proved the closest thing to a divine perspective, and much of the investigation and mapping was conducted from the air. The use of airplanes fascinated the media and the public, and the department promoted it at every opportunity. But airplanes were as blind to minute differences in elevation as were the people on the ground.

The expedition also carried instruments to measure river flow, balloons to calculate wind currents, and a "wireless apparatus" that could calculate latitude and longitude. But modern technology was no match for the weather. Not for the first time, the mere mention of Schwarz's scheme seemed to open the heavens. According to one set of figures, 1925 was one of only three years between 1910 and 1938 that saw no districts declared drought-stricken in South Africa—and the others were 1918, the year in which Schwarz first presented his scheme, and 1921, six months after he published his book.[124]As MP de Jager noted in the April 1925 parliamentary debate, the latest campaign to get an investigation of Schwarz's scheme came at an unfortunate time because it had been a year of record rainfall across southern Africa. "At the moment, it almost seems as if we should drain South Africa," he told his colleagues.[125] The meteorologist Sutton, who accused Schwarz of launching a "crusade," observed sarcastically that "everybody [is] too busy drying their clothes to worry about drought."[126]

Rainfall was particularly high in the Okavango Delta: Ngamiland's annual total was 70 percent above average.[127] Arriving in the region in June, the expedition found itself in something like the landscape Schwarz dreamed of creating: a liquid world, where channels and pans that had not filled in decades were brimming with water. As late as September, near the end of the dry season, the land was still so flooded that they could not do any survey work. The Johannesburg *Star* reported that thanks to the heavy rains, "Ngami is promoted once again from a swamp to a lake!" For the first time in twenty-three years, it was possible to travel by boat to the Makgadikgadi Pan, one of the three "lost lakes" Schwarz sought to "redeem."[128] The resident magistrate wrote that for the first time, a white man had made the trip between Maun, in the Bechuanaland Protectorate, and Andara, in the Caprivi Strip of South West Africa, by boat—using channels that were recalled in oral migration traditions but which rarely see flowing water.[129]

Water was not the only obstacle the expedition faced in its quest to know southern Africa's interior land- and waterscape. The expedition group was limited in resources—time, money, and technology. Schwarz's plans covered an enormous area, most of it with no road infrastructure, and the group ultimately investigated only the part that lay within the Zambezi watershed, omitting the Kavango and Kunene Rivers—and the Cuvelai itself, whose existence is never acknowledged in the report nor in any other official document from this time. The aerial survey, conducted by the South African Air Force, was meant to address some of these limitations. It could view places where floods had excluded the ground expedition. It could travel farther and faster, and thus theoretically could take in the totality of Schwarz's vision. Alex du Toit, who wrote the expedition's report, considered the aerial reconnaissance key to making this region legible. It

> enabled information to be obtained of the nature of the country, of the evolution of its river systems, seasonal and progressive shifts of the courses of the channels, which could not have been gathered from the journeys alone. The flights also enabled a good idea to be acquired of the nature of the country which lay outside the narrow belts to which the party's field operations were naturally restricted.[130]

The air force surveyed not just the area where the ground expedition traveled, but also the Ovambo plain in South West Africa. The *Times* reported that the flight between the Kunene River and Etosha Pan "established the fact that the Kunene flows into the pan, so that it appears that millions of acres could be irrigated and brought under cultivation."[131] But the aerial survey established nothing of the sort. C. H. L. Hahn, the Ovamboland native commissioner who went along for the ride, claimed to observe "eastern streams originating from the Cuvelai, which is a tributary of the Kunene." He concluded, "The western water courses leading South from the Kunene . . . are filled with Kunene flood water."[132] But Hahn's descriptions of the journey reveal that he did not fly far enough west to see the Kunene. The supposed link between the Kunene and the water he saw remained grounded in the long-standing assumption that the Kunene fed the seasonal flood. The assumed superior authority of aerial surveys thus acted as a substitute for actually witnessing water leaving the Kunene and flowing into Ovamboland.

Du Toit and Schwarz remained on the ground. Their expedition diaries reveal similarities in their encounters with the region. Both men drew on the canon of Kalahari knowledge: Livingstone, Passarge, and the accounts of other travelers and local administrators. Both observed natural features in the landscape with the eye of a geologist, but did so within the framework of what Yusoff has called "White Geology": "a historical regime of material power" whose descriptive practices naturalize projects of extraction and dispossession.[133] Finally, both men relied on local knowledge to further this project of extraction—a fact revealed in their diaries but obscured in the press releases and the final report. "Native knowledge" filled in for places they could not access, and offered a time dimension to what was otherwise a snapshot of the region in a single season—an extraordinary season at that. Du Toit recorded that local people described the entire course of the Savuti River, which he could only follow for three miles.[134] He began to understand the riparian region as dynamic and the idea of distinct rivers as a fiction. Basins flowed into one another, and the connections changed from year to year. For example, informants told him that the "arms" of the river flowing south of Katima Mulilo had not flowed more than ten miles this year despite heavy floods, "which might have been due to channels being blocked, to others being opened and providing a freer connection, etc."[135] But indigenous knowledge could not resolve all uncertainties. Writing of the Savuti channel, du Toit reflected that "two Bushmen state that water came down at the upper drift, but others denied this."[136] Schwarz, too, recounted what local people told him. A local headman said that the Botletle flooded every year as it had in this year of high water—a fact that Schwarz ignored, since it did not fit with his theories about vanishing rivers and, doubtless, because it seemed impossible given the historic scale of the flood.[137]

In other respects, the two men's diaries are very different in the kind of information they collected, the observations they recorded, and how they interpreted what they saw. Du Toit experienced a sense of wonder at the distinctive environment, which he described as "absolutely amazing" country.[138] He took particular interest in the ephemeral rivers, whose channels had for years aroused controversy over which way they flowed and what they linked together. Did one river leading from "Lake Liambezi," itself an inaptly named floodplain in the Caprivi Strip, connect the Chobe to the Zambezi above the two rivers' confluence? Du Toit did not travel the channel's length, but concluded that such a connection "can hardly be doubted." The heavy floods of 1925 meant that du Toit saw the sheets of water that had prompted earlier visitors to wonder about connections between watersheds, and he stated that "no sharp boundaries can be drawn" between the two river systems. He argued that it would be difficult to map exactly how and where the Chobe "communicated" with the Zambezi, "because the water must be filtering through the swamp and sand into the open channels farther south. But during the flood anything could happen."[139]

This sense that the northern Kalahari's waterways did not follow hard and fast rules—that "anything could happen"—continued as du Toit explored the lands south of the Chobe, in Bechuanaland. Here two channels, the Selinda/Makgwena Spillway and the Savuti, had long captured the imagination of explorers, many of whom argued that they connected the Zambezi and Kavango Rivers and could flow in either direction, depending on the nature of the seasonal flood. Du Toit thought the Savuti was an abandoned riverbed, "a normal river channel that has dried up." Accepting Schwarz's general theory of recent hydrological changes at least in part, du Toit concluded, "It should be feasible to bring the Linyanti back toward the Savuti channel without difficulty."[140] Du Toit concluded his expedition at Lake Ngami, which had also been affected by the historic floods and had filled for the first time in many years.

Schwarz's Kalahari reconnaissance was markedly different in its route, its use of evidence, and its ultimate conclusions. He separated from the main party soon after it arrived in Northern Rhodesia and headed straight for the heart of the delta, where du Toit ended his journey. The heavy floods allowed Schwarz to travel via canoe to the very pans he proposed "restoring" as lakes—Ngami, Mababe, and Makgadikgadi. His diary is a jumble of place names, sketch maps, lists of numbers, geological observations, ethnography, and history. He identified evidence of desiccation despite the abundance of water around him. On 15 August he wrote, "Country desert-like. Many trees dead, flood has come too late for them." A few days later he described a "beautiful wide stretch with fine camelthorn trees. . . . One is looking at a country such as the Cape used to be 150 years ago or Kimberly 50." His diary refers to "drowned mealie-lands" and "country terribly dry" in quick succession, without explanation.[141]

Du Toit attached no apparent significance to his observations. He noticed old water channels everywhere, but offered no thoughts about what had become of the water that once flowed in them, or when it had disappeared. He saw the same sand ridges and dead trees that Schwarz saw, but rather than drawing conclusions that were different from Schwarz's, he drew no conclusions at all. By contrast, Schwarz's conclusions were sweeping, and reveal that his view of the region had been little changed by his direct encounter with it. His diary contains a draft of a letter to Frederic Creswell, head of the South African Labor Party, a supporter of white immigration, and minister of defense in the Pact government. Schwarz offered no introduction, indicating that the men had corresponded before. He wrote,

> What was painfully evident was that even in flood time—and this has been the biggest flood within the memory of any of the inhabitants—the rivers

> are miserable ghosts of their former selves. . . . Whatever the main expedition may report, the run of the flood waters has proved my contention. This great mass of water should have shocked South Africa for a generation, yet we have let ninety-nine hundredths run to waste to fill up the Atlantic and Indian oceans with the consequence that we have a long series of droughts in front of us.[142]

By playing to popular perceptions, Schwarz ignored evidence that most of this water did not, in fact, "run to waste" but was lost to seepage or evaporation. But he was right about the droughts. Schwarz would not be alive to see them, but they would reinvigorate public debate over whether his scheme could save South Africa's whites from extinction.

7 * Redemption Reimagined

The Kalahari expedition returned to South Africa in October 1925 and issued its report the following year. Alex du Toit rejected every aspect of the Schwarz scheme: its necessity, its feasibility, its cost-effectiveness, and its benefits. Rainfall in the Kalahari had not declined, and there was no "Greater Ngami" awaiting restoration. The cost would be orders of magnitude higher than Schwarz had stated, while any water that was impounded would be lost through evaporation, which would limit the potential for irrigation. And creating new bodies of water would have no effect on rainfall, since evaporation from the existing floodplains of the Chobe, Zambezi, and Kavango Rivers did not have any apparent effect on the region's climate.

Du Toit was not opposed to engineering southern Africa's waterscape to create a white man's land. He suggested that the Zambezi could be dammed and diverted at Katombora, a set of rapids below the Zambezi-Chobe confluence, which would allow parts of Markham's industrial and irrigation vision to be realized. But it would cost millions of pounds and it would not benefit South Africa.[1] Du Toit repeated the fables that explained Ngami's disappearance: the abandoned papyrus rafts and Moremi's interference with the lake's water supply.[2] He imagined a future in which "a portion of this waste space in southern Africa" could be "satisfactorily settled."[3] And he drew conclusions about the region's topography, agricultural potential, and water availability—conclusions about which he was surprisingly confident, given that they were based on a single visit during an unusual season. A decade later, another survey would cast doubt on many of his measurements.[4]

The victory of the National Party in 1924 left Britain wary of handing over the protectorates to South Africa. And so du Toit's alternative vision, even if it had been grounded in accurate information, was also irrelevant. When Hertzog, the National Party prime minister, publicly pressed for the incorporation of the Bechuanaland Protectorate, British officials argued that he was motivated by "the Dutchman's insatiable appetite for land"

and the need to address South Africa's poor white problem and inadequate African reserves. They appear to have interpreted the Kalahari Reconnaissance as a "peaceful penetration" designed to strengthen South African claims.[5] Meanwhile, the Bechuanaland Protectorate administration rejected even the most modest irrigation proposals, on the grounds that "the present native population probably does not stand in need of the benefits to be conferred."[6]

Some found du Toit's report persuasive. One man wrote to *Farmer's Weekly* that it proved the scheme was "dreams and dreams only." He continued, "I am sure most of us are really sorry for this, but as sensible people we must accept the indisputable fact that the 'bubble has been busted.'"[7] *Die Landbouweekblad*, which summarized the English-language report for its readers, had a similar reaction.[8] But public responses remained shaped by doubts about the objectivity of experts. In 1919, the *Star* had argued that any investigation of Schwarz's scheme required "engineers whose minds are open."[9] As the reconnaissance was preparing to begin its work, a National Party MP argued that Kanthack's 1920 survey of the Kunene as part of the boundary commission "was largely a shooting expedition, and . . . not really a survey of the merits of the scheme." His party colleague hoped that the Kalahari investigation "will be properly made and be unprejudiced. This is the great difficulty that Professor Schwarz has had in the past, namely, that the people who investigate the matter . . . had already their minds completely made up about the matter."[10]

Once the expedition began, the Irrigation Department issued regular bulletins designed to allay public skepticism that the department was conducting an impartial investigation. These press releases avoided discussing substantive disagreements between Schwarz and his scientist colleagues. One of them opined, "Professor Schwarz's enterprising and courageous journey in the face of the many difficulties and dangers associated with the Botletle River is deserving of the highest commendation." Minimizing the possibility that Schwarz might make serious scientific contributions to the expedition, the bulletin added, "He must have gathered much interesting information of a general nature. . . . The public may look forward to some interesting articles from Professor Schwarz."[11] Alex du Toit's final report, in 1926, had none of this diplomacy, however. It read like an exorcism of the frustration scientists had felt collectively during the years in which Schwarz's scheme had enchanted an adoring public. Du Toit referred to the "exasperating features of the Professor's writings," including "amazing errors of idea" and a "deplorable lack of evidence." Schwarz's calculations that rainfall would progressively increase once the lakes were formed "fairly take one's breath away," du Toit wrote, adding that the presumption of a closed system of atmospheric circulation was "outrageous."[12]

To Schwarz's supporters, and even to some who were undecided, such language proved that the investigation's real purpose had been to sideline Schwarz and find evidence for a predetermined conclusion. The choice of du Toit—Schwarz's former colleague and alleged professional rival—as leader indicated that the government had intended all along to discredit the scheme.[13] One man recalled that, prior to the expedition's departure, he had casually told a government engineer that while he had little knowledge of Schwarz's scheme, "the idea seemed possible and if so would be of immense value to our land. Talk about disturbing a hornet's nest! In two shakes of a duck's tail I was hanged drawn and quartered and, holding up both hands, I crawled out, breathless and astonished." This reaction proved, the man wrote, that the government had made up its mind before beginning the investigation.[14]

The white community's suspicion of experts, always an undercurrent, bubbled to the surface in these discussions. The most common criticism was that experts did not meet their own standards of objectivity. As one man stated, "The actual fact of the matter is that our 'experts' start with a prejudice. . . . The matter of the Kalahari scheme has been coloured by that prejudice, and the scheme accordingly abandoned."[15] L. A. Rose-Innes called the expedition a taxpayer-funded "holiday" for South Africa's scientific elite.[16] He mocked Captain Moubray, the man who had suggested that Schwarz's scheme was "dreams and dreams only," for "the distinguished appendages which he [had] affixed to his name" in his letter to the editor. (Moubray had noted his memberships in the Royal Geographical Society and the Institute for Mining and Metallurgy.) Moubray "professes to speak with the authority of the scientific expert," Rose-Innes wrote, calling him merely part of "a vast legion of what one might style 'the gratuitous destructive critics,' gentlemen who are for ever attempting to criticize and break down everything [but] have nothing to offer in place of what they propose to destroy."[17] This criticism was not entirely fair, since du Toit had suggested an alternative scheme. But the idea that experts were unconstructive, rejecting the knowledge and practices of "practical men" but offering nothing useful in their place, was one aspect of public hostility toward institutionalized expertise. Bureaucratic experts were also accused of being unaccountable for past conclusions that later proved to be false. One of the most vocal Schwarz supporters wrote, "Reports of Government experts and engineers in this country carry little weight with farmers; they have proved themselves wrong too often. Why should the Reconnaissance report be an exception?"[18]

Schwarz added fuel to this fire. In articles for South African and British publications and in talks and interviews around South Africa, he reported that the expedition had never gone near the Kalahari.[19] He hinted at conspiracy and coverup, claiming that the South African government had demanded that he turn over his field notes and photographs and, when he refused, had declined to reimburse him for his travel expenses—a charge the Irrigation Department denied.[20] And so the Kalahari Reconnaissance did not end debate on the Schwarz scheme. In the years after 1926, conversations about it ebbed and flowed with the cycle of wet and dry years. When farmers faced another drought in early 1927, one man suggested that even if Schwarz's scheme proved "a dismal failure," trying it "cannot possibly have any harmful effect on this country."[21] Schwarz's claims for climate modification amounted to a nonfalsifiable hypothesis, in that there was virtually no way to disprove them short of building his scheme. This uncertainty was, one man wrote, reason enough "to test Professor Schwarz's scheme thoroughly."[22]

Members of Parliament continued to feel pressure from their constituents. Conradie, a National Party representative for the arid Northern Cape and an early Schwarz supporter, admitted that du Toit's report had revealed "that the thing was not so easy as at first represented." His constituents, however, remained "very much intrigued with Professor Schwarz's scheme. They cannot estimate how far the Zambezi is from them; they do not appreciate that it is more than 1,000 miles away." What they did know was that water from the north recharged their groundwater supply, and that "in former centuries there was a tremendously large rainfall in the interior."[23] Conradie's reliable allies, the National Party member de Jager and the pro-British Byron, echoed his views. De Jager asked the prime minister to approve funds to survey the remainder of the Kalahari, concluding that it was their "holy duty" to do so.[24]

De Jager's justification for his request reveals that du Toit and other scientists had lost control of the narrative almost immediately. He told Parliament that the expedition had "absolutely proved what Professor Schwarz alleged."[25] In an article about the ongoing drought, the Dutch Reformed Church magazine *De Kerkbode* reported that "Professor Schwarz, who was appointed by the Government in 1925 to go through the Kalahari with an expedition to find out if there is a possibility to bring water into those arid regions, firmly believes that it is possible."[26] People were not only claiming that the expedition had proven the scheme's viability; they were casting Schwarz in the role of government expert! Increasingly, the press referred to him as "Dr. Schwarz"—an error Schwarz seems not to have attempted to correct.[27]

Schwarz engaged in his own rewriting of the past. He falsely claimed that he had first visited the Kalahari in 1895. He subtly shifted the focus of his scheme in response to the criticisms over his meteorological claims and the reduced likelihood that South Africa would annex the Bechuanaland Protectorate. Unlike his initial sketch of his plan, the map Schwarz provided to the *South African Mining and Engineering Journal* in 1927 had political borders on it.[28] He spoke less of increasing rainfall and more of bringing water into the Kalahari for irrigation. In July 1927, his populism turning petulant, he wrote, "I do not wish to push my scheme any farther if at every point I am met with misrepresentations."[29] But instead of withdrawing from the fray, Schwarz became more strident; at one point *Farmer's Weekly* removed one of his statements because it libeled du Toit.[30]

Schwarz also ventured into increasingly bizarre lines of speculation. His scrapbooks of newspaper clippings reveal a fascination with archeological and geological mysteries, including the controversy over a trove of artifacts found at Glozel, in France, and the origins of diamonds found along the Kei River. In a 1928 interview with *South Africa* magazine and in a lecture to the Royal Colonial Institute, he devoted far more time to his fascination with the Kalahari's people and their supposedly foreign origins than to his engineering scheme. In that year his new book *The Kalahari and Its Native Races* was published. It was a travelogue of his 1925 journey to the region, combined with a global history of "pygmies." In it, he described various Bushmen groups in the Kalahari, speculating that some had their origins in the Mediterranean while others were the result of intermixing with Malays or with Chinese sailors who long ago visited the coast of East Africa. He argued that they were descended from the same races that had given rise to fables of elves, gnomes, and fairies in Europe, as well as tales of little people in China, Mongolia, and Japan.[31]

While Schwarz turned his fertile imagination toward human origins, others continued to focus on his dream of a fertile Kalahari awaiting white settlement. A man writing to the *Star* under the name "Pioneer" recalled his experiences in the early 1890s. "I have traveled thousands of miles . . . but nowhere have I ever seen fatter cattle and sheep than I saw that winter on Khama's posts in the Kalahari." The following year, he had witnessed pans filled with water to the south of Ngami: "Birds were singing, thousands of frogs in every conceivable tone raised their voices, all nature rejoiced, and the so-called desert (what a misnomer!) was a paradise." Also hidden in the Kalahari, he speculated, were rich diamond deposits.[32] A man named Waldemar Teichmann urged the South African Agricultural Union to demand water and rail infrastructure in the Kalahari, "with a view to throwing that area open for land settlement." The resolution passed despite the protests of a representative from the Irrigation Department that boreholes were prohibitively expensive, given the depths of groundwater there.[33]

Farmers also continued to see the Kalahari as relevant to the fate of their homes. "Something is wrong," one Afrikaner farmer said. "The drought in the interior is as bad as it has ever been in the history of the country. . . . We may laugh at Professor Swarz's [*sic*] desert redemption scheme, but although I do not wholly agree with him his ideas contain the basis of further scientific investigation. It is the bounden duty of farmers to keep this matter before the Government." The farmer and his colleagues in the Cape had voted in favor of additional government investigations.[34]

The popularity of Schwarz's scheme outlived the man himself. In December 1928 he died of heart failure while on a geological expedition in Senegal. A colleague at Rhodes hinted that the fight for his scheme had taken a toll on his health.[35] The British geologist John W. Gregory, in an obituary for the journal *Nature*, noted that although many of Schwarz's ideas were speculative and the Kalahari Scheme had been rejected by his scientific peers, his "friendship and the stimulating originality of his views" would be missed.[36]

This chapter follows the fate of Schwarz's scheme after his death, through the Great Depression and into the late 1930s and early 1940s. Environmental historians have written about this period as a watershed in South Africa's pivot toward conservation, while agricultural historians have focused on increasing state intervention in the commercial farming economy. Major environmental and economic legislation was passed in both spheres. But the decade that began with drought and depression and ended with South Africa's entry into World War II was also one in which expert and popular narratives about aridity, climate, and the role of white farmers in the "redemption" of South Africa began to converge.

Drought and Depression

The year after Schwarz's death marked the beginning of a multiyear drought that coincided with the onset of the Great Depression. The arid and semiarid regions were hit particularly hard. In 1931, P. M. van der Westhuizen, who had written to the *South African Journal of Agriculture* in 1914 to explain how moisture on land drew down water from the sky, was threatened with losing his government lease on a two-thousand-acre farm because he could not pay his debts.[37] In the 1932–33 season, 10 to 15 percent of the livestock owned by whites died: an estimated 750,000 cattle and seven million sheep.[38] In 1925, when Schwarz had traversed the inundated lands of the northern Kalahari, South Africa's white farmers had harvested twenty-five million bags of maize. In 1933 they harvested eight million bags.[39] (As usual, the contributions of African farmers were ignored in such statistics.) News reports assumed apocalyptic tones. *Farm-*

er's Weekly described the "desperate plight" of farmers: pasturage and water supplies were exhausted across much of the country, and some of the most fertile districts had total harvest failures. The much-lauded Hartbeespoort Dam was nearly dry, and the poor whites who had been lured to settle on its irrigation scheme were "on the brink of starvation." Churches again held special services to pray for rain. Natal had been spared the worst, but a *Farmer's Weekly* correspondent reported that "even here in Natal many of us have been gripped by this demon as never before."[40] The MP for an arid Northern Cape district would remember that 90 percent of the farmers were "on the road" by 1933, seeking grazing for their stock or work for themselves.[41]

A man in the Transvaal argued that the drought was a far bigger problem than the economic depression. Indeed, it was "the greatest calamity (rinderpest, wars, etc., not excepted) that has faced South Africa since its occupation by White men."[42] In popular memory, the period from 1929 to 1933 was remembered as "the Great Drought," though in fact those years were not uniformly dry and it was not the worst drought in terms of the percentage of the country affected. Its impacts were compounded by the global depression and the collapse of agricultural prices.[43] In the words of one historian, "Starvation was rife. Many white farmers abandoned their lands to seek relief in neighboring small towns, leaving their Black labor tenants to survive as they might."[44] By the end of the drought in 1934, the head of the main agricultural college in the Orange Free State estimated that 90 percent of white farmers were insolvent.[45] The disparity in the performance of the agriculture, mining, and industrial sectors during those years deepened a sense of rural crisis. While agricultural prices collapsed, prices for gold remained very high, particularly once South Africa left the gold standard in 1932. In 1928, agriculture had contributed more to the economy than mining, and 50 percent more than industry. By 1933, industry was contributing more than agriculture, whose value had declined by more than 40 percent, and mining was generating almost twice as much income as farming.[46]

The pairing of economic depression and drought again seemed to portend the extinction of white civilization. The 1931 census showed that the loss of white population in the Cape's midlands and the Karoo had accelerated.[47] The genre of the Afrikaans *plaasroman*, or farm novel, which became popular at this time, described what such depopulation portended. In these novels, the continued existence of the white community depended on holding onto the land for the next generation, and losing one's land to drought resulted in a loss of identity.[48] Such fears were not limited to Afrikaners. A man farming north of Pretoria, on a farm called "Illawarra"—indicating that he had migrated from Australia, where a coastal region bears that name—argued, "The position of the country is fast becoming desperate, and if nothing is done to keep our moisture here, I am afraid the future will be black in more ways than one for South Africa."[49]

During the Great Drought, experts—and even scientists in unrelated fields—continued to insist on the possibility of a drought-proof farm. A Free State physician named J. D. Schonken, writing in the *South African Journal of Science*, argued in 1930 that farmers could "outwit all droughts" through good land use practices.[50] But as dams for irrigation ran dry, much of the focus began to shift toward pastoralism and extensive rather than intensive farming, and some experts began to cast doubt on the possibility of a densely populated white countryside. In 1934, Alex du Toit argued that very little of the land was suitable for cultivation and that pastoralism was the future.[51] An economist argued that South Africa suffered from a "chronic bucolic complex," and criticized purportedly wasteful government programs designed to subsidize uneconomic agricultural pursuits.[52] The agricultural economist Hubert Leppan also sought to dampen enthusiasm for South Africa's agricultural potential. "Patriotism cannot ignore the natural endowments of a country," he wrote.[53] The frequent comparisons between South Africa and California were misleading: the entire country had perhaps three million irrigable acres, while California alone irrigated ten million. "Closer settlement" was not the future, and prospective farmers needed to be comfortable with "a comparatively lonely existence" because farms had to be large to be profitable.[54]

Leppan accused politicians of overstating the possibilities of increasing the white agricultural population (ignoring the fact that some of his fellow government agronomists had done the same). "Back to the land" was a "hoary shibboleth" coined by those seeking popular support, he wrote. "The politician, wishing to preserve a mythical optimum urban-rural ratio of the population, pours vast sums of money into irrigation in trying to force the failures from the land back to the land."[55] But South Africa was not a land lying in wait for millions of new farmers. Despite all the money spent by the Land and Agricultural Bank over the preceding two decades, there had been no net gain in the rural white population.[56]

In 1933 the government began offering subsidies to farmers for anti-erosion works, such as dams and reservoirs for watering livestock. This focus on storing water had long united expert and popular conceptions of a fertile and productive land. A country filled with open water was, of course, a central aesthetic of Schwarz's scheme. But it was also one perpetuated in the pages of *Farmer's Weekly*, which had long featured photos of wetlands, farmers' dam reservoirs, and other water features that reinforced an ideal of agricultural fertility grounded in Northern European norms (figure 7.1). Water in the landscape, whether in the form of perennial

Figure 7.1 "Saving the land's life-blood from the sea": images of farm dams and irrigation on a white farm in the Bechuanaland Protectorate at the height of the Great Drought. The owner called himself "MoKalahari" in the pages of *Farmer's Weekly*, thereby claiming the authority of "a person of the Kalahari."

rivers or large reservoirs, was a precondition of "closer settlement" and a denser rural white population—what David McDermott Hughes has aptly named a "hydrology of hope."[57] Such photos were an especially poignant manifestation of hope when they appeared in 1933, at the height of the drought. But over the course of the 1930s, the focus of government subsidies shifted. In imagining a white agrarian future that was pastoral and extensive rather than crop-based and intensive, experts gradually ceased urging farmers to build dams and reservoirs, which had high rates of evaporation and siltation. Instead, they advised them to put boreholes on their farms and build up reserves of fodder. In 1936 the government began offering subsidies for silos to store maize for feeding livestock; it also began subsidizing boreholes.[58] In *Farmer's Weekly*, photographs of silos and haystacks began to replace photos of open water.

Such images of modern pastoral infrastructure lacked the emotional resonance of a shady, tree-lined pool, and many farmers maintained a preference for using government conservation funds to build dams.[59] When the rains failed again in 1935, *Farmer's Weekly*'s columnists railed against this preference and against farmers' supposed failure to store fodder. "Drought Gets Home Again despite Warnings: A Calamity That Is Not an Act of God but a Farming Disgrace," read the headline to one article that featured photos of starving cattle presided over by Black South Africans.[60] It was white farmers who were blamed, however. A farmer wrote on "the crime of farming without feed reserves," and accused fellow farmers of causing unnecessary animal suffering by failing to make plans for the inevitable droughts.[61] Another insisted that he had never had losses due to drought, and wrote of his system of feed storage, illustrated by photos of his silos and towering haystacks (figures 7.2a and 7.2b).[62]

Many farmers, meanwhile, interpreted the collapse of the agricultural economy during the Great Drought as further proof of the limitations of scientific farming prescriptions. Fred Nicholson, a Free State farmer, argued that "no amount of local conservation, tree planting, veld improvement, or other like schemes (desirable as they are in themselves) can possibly bring about an improvement in farming conditions." He blasted the editor of *Farmer's Weekly* for his "ridiculous suggestion of compulsory winter-feeding" and of committees to determine stock carrying capacity. Without rain the land's carrying capacity was nil, and it was impossible to grow fodder crops, much less store them.[63] A short story about the 1933 drought, published later in *Die Landbouweekblad*, expressed farmers' sense that experts did not understand their reality: "Some people speak so easily about drought! Of bore-holes, of maize, dry-feed and all such things but what does it help if it does not rain? As if humanity can survive without rain! . . . Drought! It is enough to drive the strongest man to insanity!" In the story, only the arrival of a rain shower saves the main character from dying by suicide.[64]

The rainmaker Charles Hall was among those who continued to push back against the idea that drought losses were entirely in the control of farmers. He criticized progressive farming prescriptions as unrealistic, accusing those who claimed to be relying on stored fodder reserves of buying grain to feed their stock. The *Farmer's Weekly* editor took him to task, citing examples of farmers who described their success at growing lucerne with almost no rain, as well as a survey in which farmers argued that "drought is our heritage" and that "crop production for sale is dangerous and uneconomic." The editor told readers that the path to prosperity lay in growing fodder for livestock. But this advice could offer new grounds for boundless optimism. One man in the poll suggested that "with the help of the scientist, the carrying capacity of the land should be almost unlimited."[65] An irrigation farmer in the Transvaal had a similar view: land was "the inborn heritage of the people" and was "inexhaustible if practically utilized, conserved, and reclaimed."[66] These claims demonstrated farmers' continued reworking of scientific orthodoxy to support a goal of endless expansion.

Figures 7.2a and 7.2b "The combination that spells safety: A full year's reserve feed supplies of legume hay and silage." Photos from *Farmer's Weekly*.

The vision of a countryside filled with white farmers retained its appeal, and the Great Drought brought new people into the pro-Schwarz camp. A man in Bloemfontein asked *Farmer's Weekly* to publish a summary of the scheme, noting that "there may be many who do not know of it," and the newspaper obliged.[67] During the drought's peak, in the first half of 1933, *Farmer's Weekly* published letters almost every week debating the merits of flooding the Kalahari. Supporters wrote from around the country, as well as from South West Africa and the Bechuanaland Protectorate. As one letter writer put it, "The sorrow and the suffering which one sees all around one today—such suffering and tragedy as I have never seen in all my thirty years here—is sufficient to make one eager to help in any way possible to bring about a change."[68] The Dominions Office also fielded queries. A man named Eric Nobbs, living in England, argued that the "yet untapped resources of the Kalahari should now be turned to account." In his opinion, the region was not suitable for the yeoman farming Schwarz had envisioned; instead, it should be opened to white ranchers—in fact reflecting what was becoming expert consensus about the best "productive" use of arid lands.[69] One can hear the note of exasperation when an employee of the Dominions Office, fielding another query about Schwarz's scheme, wrote to a colleague: "I do not know why so many people interest themselves in the Kalahari, or why so many people should want the waters of the Zambezi to be diverted in that direction."[70]

Some well-known progressive farmers became supporters of Kalahari redemption during the Great Drought. Perhaps this was because, as the head of the Orange Free State's main agricultural college noted with regret, progressive farmers had been affected the worst by the drought and depression. Taking the advice of extension agents, they had invested large amounts of capital on their farms and, where possible, they had focused on wool, the commodity whose price declined the most.[71] Temple Fyvie, a pith-helmeted *Farmer's Weekly* columnist who advocated for progressive farming practices, echoed many experts when he suggested that "we are suffering for the sins committed by our fathers and also in many cases for those committed by ourselves." But he also believed that "Professor Schwarz's scheme is capable of bringing incalculable benefits to the whole of South Africa."[72] In the eastern Karoo, long a center of Schwarz support, John Owen-Collett and his relative John G. Collett, established sheep farmers in the Cape Midlands and officers in local farm associations, became lifelong proponents of the Kalahari Scheme. The support of men like these continued a pattern among farmers of supporting both "expert" and popular environmental ideas. A decade later, such convergence would begin to occur among some experts as well.

The eastern Orange Free State, one of the country's most productive agricultural regions, became another center of Schwarz support. In a rare glimpse of how both information and enthusiasm moved along personal networks, the small farming town of Petrus Steyn became particularly crucial to sustaining Schwarz's vision. Afrikaner, German, and English farmers all threw their support behind Kalahari redemption. P. W. de Wet wrote, "I have always been an enthusiastic supporter of Professor Schwarz's scheme, and have lately spoken to a large number of persons about the scheme without coming across a single one against it."[73] Whether de Wet was the originator of Petrus Steyn's pro-Schwarz orientation or merely a representative of it, his claims were borne out in the pages of *Farmer's Weekly*. His letter was followed by others, including a rare letter from a woman, who urged her "fellow farmers' wives" to take an interest in a scheme that would allow "a return to the common decencies of life, children going to good schools."[74] A month later, her neighbor Adolph Kern wrote to add his support. Kern argued that most continents had naturally arid conditions on their western coasts, but that something was amiss in the Orange Free State, far from southern Africa's Atlantic coastal desert. "Why do the rivers, pans, vleis, and boreholes in the veld dry up? Why do trees and plantations of many years standing die from drought now, and why did they not in former years?"[75]

On 1 June 1933, Kern and Owen-Collett joined seventy other white men and one white woman—Schwarz's widow, Daisy—at a hotel in Bloemfontein to formally create the Kalahari Thirstland Redemption Society. Telegrams from supporters who could not attend were sent from around the country and South West Africa.[76] Newspaper accounts suggest that there were as many Afrikaans as English names among the attendees, who included farmers, scholars, and an editor from *Die Landbouweekblad*. The paleontologist Egbert van Hoepen, curator of the National Museum, and Thomas Dreyer, professor of zoology at Grey University College (the precursor of the University of the Free State) attended. So did W. F. Murray, superintendent of roads and public works in the Orange Free State, and delegates from the Bechuanaland Protectorate. Daisy Schwarz, who had taken up her husband's cause—even tangling with the prickly Weidner in the pages of *Farmer's Weekly*—was named an honorary president.

The group represented a range of views on the Schwarz scheme. Dreyer, for one, was "agnostic."[77] Van Hoepen accepted expert opinion that rainfall was not declining. Participants were aware of the criticisms of the scheme. But all supported an "impartial" government investigation, indicating a shared belief that the 1925 Kalahari Reconnaissance had been either biased or incomplete. They presented their own group of experts, including a Free State surveyor named F. le Roux who offered a ninety-minute presentation on his 1927 survey of the Kunene and Etosha region. Le Roux argued that a seventy-foot dam wall would be sufficient to move the water across the flat country. The German-born Kern told the gathering that in 1912, Germany had formed a secret commission to investigate the possibility of restoring the inland lakes to stop the process of drying up.[78] The conference members voted unanimously to advocate for a new investigation, but also agreed to cooperate with other organizations studying the impact of soil erosion and the conservation of "inland waters." They also announced a subscription drive to raise funds for a new investigation.

The possibility of simultaneously solving the problems of white poverty and small white numbers continued to generate enthusiasm for flooding and settling the Kalahari. But from 1933 on, the meaning of Schwarz's scheme changed in two ways. First, his claim that rainfall was declining subtly shifted to a claim that the distribution of rainfall had changed from the "gentle rains" of the past to downpours that now eroded soil and skewed annual averages to make rainfall appear better than it was. This reflected farmer sentiment, and had been embraced by Schwarz toward the end of his life. Indeed, some of Schwarz's most prominent opponents among the citizen scientists also began to embrace this argument in the wake of the Great Drought, though they continued to insist that human behavior was the cause. Weidner suggested that rainfall distribution had changed due to "the White man's unrestricted exploitation of this country." Echoing the beliefs of many of Schwarz's supporters, he argued that the soft "land" rains of yore had been replaced by cloudbursts.[79] Schonken, the Free State physician who wrote about South Africa's rain and droughts for fellow scientists and popular audiences, also suggested that rainfall distribution had changed. He blamed both rural and urban whites: "When the vegetation of a country is destroyed over large areas, the rainfall of that country becomes proportionally more and more erratic."[80]

In addition, support for climate change was no longer the sine qua non of identifying as a Schwarz supporter. E. G. Bryant, a longtime proponent who had met Schwarz, admitted that he had become skeptical of Schwarz's climate theories but argued that his other ideas still had merit.[81] As a government report would later note, by the 1930s Schwarz's supporters were advocating for something Schwarz himself had opposed: the diversion of the Zambezi River at Katombora, some sixty kilometers above Victoria Falls, for purposes of irrigation.[82] This scheme originated not with Schwarz but with his nemesis: it was the alternative scheme Alex du Toit had proposed in the report of the 1925 Kalahari Reconnaissance. The Kunene half of the scheme had subsequently fallen away, to be mentioned only by a few people based in South West Africa. And the focus on creating lakes for rainfall had grown decidedly more muted.

Members of the Kalahari Thirstland Redemption Society, later renamed the Schwarz Kalahari Society, were among those who conflated the Katombora scheme with that of Schwarz. John Owen-Collett, elected as the society's president, told the crowd assembled at Bloemfontein that anyone who doubted that Schwarz's scheme was feasible should read du Toit's report—the very report that had savaged the scheme. By 1933, the false claim that du Toit's report had proven Schwarz right had become accepted as truth. In Owen-Collett's rendering, the Kalahari Reconnaissance had concluded that damming the Zambezi would "control the water, give electricity to Johannesburg, supply the Falls with a regular amount of water, change the climate of that part, and so perhaps change our climate and make rain here more regular and precipitation much gentler and more frequent than is the case today."[83]

Du Toit's report had said very little of this. But in the 1930s, Schwarz's supporters came to equate his original dream with plans to dam the Zambezi River in part to generate power for southern Africa's white settlers. It was a dream that ultimately would result not in the restoration of the Kalahari's lakes but in the construction of the Kariba Dam—which created the world's largest man-made reservoir, a "great lake" Schwarz had never imagined.[84] The renewed focus on irrigation, which Schwarz had initially criticized, foreshadowed the expansion of popular demands and state ambitions for dam and irrigation projects within South Africa itself in the late 1940s and beyond. Schwarz supporters had shifted away from his particular engineering scheme, but continued to embrace something more fundamental in his imagined future: the dual promise of environmental security and white prosperity. For the next two decades, commitment to Kalahari redemption was linked less to the details of Schwarz's original vision than to the ideas that lay at the foundation of that vision: a refusal to accept expert pronouncements that white farmers were to blame for the impact of drought, and support for engineering the waterscape to create a white man's land of farmers.

Desert Dystopias and White Utopias

Once again, intensified discussion of watering the Kalahari coincided with abundant rainfall. In early 1934 the drought ended, and much of southern Africa experienced historic flooding. Along the Angola-South West Africa border, the Cuvelai and upper Kunene Rivers surged and Etosha swelled to huge proportions. In the Ovambo communities of the western Cuvelai plain, 1934 is known as the year of Shiwenge—the name of a rainmaker who had been called from Angola to break the stubborn drought. Along the Kavango River, people told of another famous rainmaker, Mbambangandu, who was allegedly forced by colonial authorities to travel to Windhoek to make rain. In retribution, he brought torrential rains down on the city until colonial officials paid him to stop.[85] Towns across South West Africa were flooded, and bridges, dams, and railways were washed away. In Upington, in the Northern Cape, farmers who had settled in the bed of the Molopo River over the previous several decades watched the river swell to ten feet and take their homes. Some lost their lives. The Molopo began flowing toward the Orange River—a sight not seen since 1894—and it filled dry depressions. "Millions of little fish are in the lakes," one farmer wrote. "What a marvelous transformation to have fish in the Kalahari!" The land had been remade: "The so-called desert turned into a tropical garden and lake country."[86] It was the world as Schwarz had imagined it.

As perennial rivers grew to gargantuan proportions, observers could see just how much water was being "wasted." A surveyor living along the Orange River calculated that the amount of water flowing past his town during a twenty-two-day period in January would have filled two lakes of fifty square miles each, to a depth of ninety feet.[87] Owen-Collett claimed that the Zambezi was discharging 180 million gallons of water every minute, and criticized the government's failure to investigate ways to use the water. "We little know what agricultural and horticultural riches lie hidden in the desert," he wrote, invoking Kalahari fantasies that were by now a century old. "It is hard for men . . . to sit still and see and feel the land of their adoption blowing and washing away, when half the Zambezi (at flood) might so change the climate and temperature of the Union and Rhodesia."[88]

The flooded rivers and greened Kalahari offered a vivid illustration of the conditions that Schwarz had insisted could be made permanent. But they also diminished the sense of urgency over South African desiccation, and reminded people that too much rain was also a bad thing. People drowned in rivers that had long been dry, and livestock died in flooded pastures. Discussion of the Schwarz scheme dropped out of the newspapers. The Kalahari Society's second annual meeting was poorly attended. Owen-Collett suggested that the end of the drought was to blame, and predicted that "as soon as climate conditions had changed for the worse, the clamor for an investigation would begin all over again."[89] In 1935 the Kalahari Society canceled its third meeting, and Dreyer resigned as secretary. The society had 2,500 members on its rolls and a number of supporters in Parliament, but little sense of momentum.

As drought gave way to floods, South Africa's political landscape was also transformed. In 1934, Hertzog and his political rival Smuts merged their parties into the United Party. The Fusion government marked a new era in state intervention in farming. It continued a quarter century's pattern of government support for white agriculture and the systematic exclusion of African farmers from economic opportunity. But the scale of state subsidies and support began to increase, particularly for "modern" and highly capitalized farmers. Agriculture's share of national income grew, fueled by subsidies that were paid for by the industrial and mining sector. Historians have noted the uniqueness of this pattern among modern economies, most of which became less agrarian over the course of the twentieth century.[90] In the late 1930s the government began to discard the back-to-the-land ideology of white uplift. The solution to white poverty was now seen to be in the cities, where industrial expansion offered new job opportunities. The economic boom produced by South Africa's entry into World War II essentially solved the poor white problem, though the National Party continued to invoke old statistics that indicated there were tens of thousands of poor whites.[91] World War II also caused the collapse of the Fusion government, as National Party members rejected the decision to align with the Allied cause.

Over the course of the 1930s and 1940s, there were increasing calls to do something about the "menace" that soil erosion posed to South Africa's social and ecological future. Historians have explored the rise of concern over soil erosion in South Africa in some detail, and have pointed out the influence of global debates and events such as the US Dust Bowl in the resulting conservationist legislation. In some places, fears of human-made soil erosion merged with continued global anxieties about desiccation. In West Africa, colonial officials continued to warn that the Sahara was encroaching on the Sahel, thereby threatening millions of colonial subjects with famine. In the 1930s, the colonial forester Edward Percy Stebbing introduced these ideas to the Anglophone world. Stebbing described a journey in which he found the ruins of villages "overwhelmed" by sand, and huge tracts of land where rainfall had declined to the point that crops could no longer be produced. Although he was vague about the mechanism by which human-created land degradation caused a decline in rainfall, he tacitly accepted the idea, still common in 1930s South Africa, that conditions on land affected what happened above. These ideas about Sahelian deforestation and human-influenced land degradation survived into the 1970s and 1980s, when catastrophic drought in the region was blamed not on climate variation but on local people's land use practices.[92]

The Sahara continued to be fertile ground for European imaginaries about expanding deserts and about the precarious future of the white race. In the late 1920s, the German architect Herman Sörgel proposed building a dam at Gibraltar to create huge quantities of hydropower while simultaneously lowering the water level of the Mediterranean. The newly exposed shorelines, combined with the diversion of water into the

Sahara's extinct lakes, would create a new geographical entity. Sörgel named this hypothetical land "Atlantropa"—a fusion of Africa and Europe that would create *Lebensraum* for an expanding white population. Sörgel was a techno-utopian motivated by the recent history of intra-European bloodshed. But Atlantropa was anchored in 1920s-era eugenicist debates over the "rising tide of color." In Sörgel's 1938 book, a world map shows numerous silhouetted Chinese caricatures dancing across "pan-Asia," and trails of money emanating from the "capital-rich" Americas. Europe, with its stagnant population, is represented by a single white silhouette—threatened by the "black continent" to its south, which contains several narrow black silhouettes. Africa was underpopulated, Sörgel wrote, but if its peoples ever learned to manage its tropical diseases, they would increase (in his calculations) three to five times faster than the white race.[93]

Like French colonial administrators of earlier decades, Sörgel believed that the Sahara had once been "the granary of Rome," and that the desert was expanding. And like Schwarz, he imagined his water diversion scheme not as a radical transformation but as a restoration of prior conditions. Recreating lakes in the desert would increase rainfall and open arid lands to white settlement.[94] Sörgel's proposal was grounded in the distinctive context of a Europe in purported decline as the rest of the world surged ahead. But his vision was surprisingly similar to that of Schwarz's supporters. His ultimate goal was "the reclamation of Africa" and "the domination of the black continent by Europe" via the settlement of whites on its ecologically transformed lands.[95] At some point, the architect stumbled across Schwarz's ideas and incorporated them into his own plans. By 1935, Sörgel was imagining the remaking of the entire African continent. With a Swiss engineer named Bruno Siegwart, he published an article describing plans to dam the Congo River and flood much of its basin, to enlarge Lake Chad, and to realize Schwarz's vision of a lake in the northern Kalahari.[96] Sörgel and Siegwart portrayed the flooding of the Congo basin—home to millions of Africans—as a net good for the white race in its struggle. Black populations had higher rates of reproduction, they argued, and unless kept in check, they would overpower—or "swamp," in the language of South Africans—white civilization.[97]

In Australia, too, plans to "drought-proof" the interior and change its climate reemerged in 1938 when John Bradfield, the engineer who had designed the Sydney Harbour Bridge, proposed diverting rivers inland. Supporters explicitly linked the Bradfield scheme to the White Australia policy. Ion Idriess, a journalist who popularized Bradfield's scheme, wrote, "We just hold this continent and we cannot . . . increase our population to a really safe extent, without water."[98] The rugby player Frederick Timbury added his support in a book called *The Battle for the Inland*. "Population makes for prosperity and safety," Timbury wrote. He argued that river diversion would allow the addition of one hundred million white people to Australia, consolidating the White Australia policy and safeguarding the country from foreign invasion.[99]

Atlantropa appears to have had little impact on the debate over Schwarz's scheme. Only Gessert, who remained engaged with developments in Germany, mentioned it.[100] That so few Schwarz supporters appear to have been aware of the Atlantropa and Bradfield schemes underscores the local roots of the Kalahari imaginary, despite its counterparts elsewhere. But the coincidence of these schemes also demonstrates the way that racial and environmental anxieties continued to be fused within European-descended societies in the interwar period—a pairing grounded in discourses about soil and climate degradation as well as eugenics.

The connection between Germany and South Africa came from another direction. In 1932, the German–South African prospector Hans Merensky invited Erich Obst, a German geographer, to investigate whether South Africa was drying up. Merensky's involvement offers a glimpse at the scope of support for Schwarz—support that endured despite the quieting of public debate during years of good rainfall. Like Charles Markham and Kenneth Quinan more than a decade earlier, Merensky never publicly wrote in support of the Kalahari Scheme. But, not content with expert pronouncements, he put money into testing its viability. His interests in geology and soil conservation seem to have facilitated, rather than conflicted with, his interest in progressive desiccation. And his pro-German sympathies probably accounted for his willingness to enlist Obst's expertise.

At the end of his six-month research trip during the peak of the Great Drought, Obst concluded that uplift in the earth's crust was causing South Africa's desiccation, and that it threatened the existence of whites. He also pronounced Schwarz's scheme technically viable.[101] Obst, like Sörgel, was heir to a lineage of German geographers who worried about desiccation and believed that believed that geographical expansion was necessary for Germany to flourish.[102] But while Atlantropa was a pacifist project meant to avoid future wars, Obst embraced the expansionist agenda of the National Socialists.

In 1935, Merensky funded Obst's return trip to South Africa. With fellow German geographer Kurt Kayser and several scientists from the University of Stellenbosch, Obst traveled to Mozambique, the Transvaal, Ovamboland, the Kunene River, the Kavango River, and Basutoland.[103] Given his procolonial and pro-Hitler sympathies, it is difficult to avoid the conclusion that his interest in southern Africa's climate and water supply was linked to his growing interest in reestablishing Germany's empire in Africa.[104] The question of progressive desiccation now took on geopolitical importance.

Obst was instrumental in blocking the publication of the work of Hermann Korn and Henno Martin, two young German geologists working in South West Africa who sought to disprove the theory of continental desiccation; he even tried to have them lured back to Germany to face arrest. The plan failed when the two men fled into the margins of the Namib Desert, where they lived in hiding for two and a half years. The anti-Nazi politics of Korn and Martin became entangled with the debate over climate change; one author argues that by creating an image of the subcontinent as a worthless, expanding desert, Obst was trying to increase Germany's chances of regaining South West Africa. At any rate, the stakes were high enough that the two geologists found themselves unable to return to Germany and their work banned even in South Africa, which succumbed to German pressure to censor them.[105]

Schwarz's supporters followed Obst's 1935 expedition with great interest. Obst told an audience in Cape Town that while he did not agree with all aspects of the Kalahari Scheme, he believed that much of what Schwarz had argued was correct and that "South Africa, up to the present, has not done full justice to this scientist."[106] Many local people were probably oblivious of how German support for "drying out" had become a marker of loyalty to Hitler, though perhaps not all were; a South African fascist group included in its platform the building of the Kalahari Scheme.[107] On the whole, however, people welcomed Obst's conclusions because they served local agendas. Minister for Agriculture Denys Reitz had to persuade the Cape Province Agricultural Association, which represented thousands of farmers, not to ask the government to fund Obst in drawing up a plan for the scheme. He asked "that the Government be trusted in the matter or irrigation and allowed to proceed along indicated policies."[108] Indeed, Obst's racial framing of the problem in subsequent publications within Nazi Germany would have been comfortable to most white South Africans. He argued that changes in rainfall due to geomorphological changes had particular salience for Africa, "the last colonial continent of the white race."[109] The damage could be reversed by "appropriate technical measures" designed to keep water on the continent: "Only then will Africa be able to provide food for the current population and for a much larger population in the future. Here lies one of the greatest tasks of the white race on the soil of the last colonial part of Europe!"[110] Obst's conclusions provided supporters of the Kalahari Scheme with an "expert" opinion to counter the continued opposition of South Africa's own experts.

Anxieties about white numbers continued to animate conversations around drought in the latter half of the 1930s. A farmer named E. Owen Wright wrote in *Farmer's Weekly*, "For better or for worse we have allowed a lot of Natives to become detribalized and to live among us, and it is now too late to put these Natives back in reserves and make South Africa a purely White man's country." He suggested that "the future of the White man in this country, particularly the farmer," required both stopping the movement of additional Africans into white areas and encouraging "increased emigration of Whites from the virile Northern nations."[111] The following month, an Afrikaner in Beaufort West, the largest town in the Karoo, argued for recruiting more whites from Holland. "To South Africa has been allotted the task of maintaining the White civilization in Africa," he wrote. As in the past, the United States was considered a model.

> The eminent position the United States holds among the nations today, is largely due to the tremendous influx of settlers during the early part of the 20th century. . . . Imagine a South Africa with, say, 20,000,000 Whites. We should then be able to live in complete safety.[112]

As in the past, the meaning of Black South Africans "living among" whites had less to do with a subordinate and supposedly transient labor force, and more to do with who qualified as a farmer, living on and cultivating land independently. It was autonomous Black "farmers," as opposed to "cultivators" or laborers, who were the real threat to the white man's land.

The revival of interest in the Kalahari Scheme came at a time when South African annexation of the neighboring British protectorates looked increasingly unlikely. This too, jeopardized white futures. Owen-Collett suggested that South Africa's lack of control over the protectorates was a problem. "Here we are sandwiched between Native areas over which we have no or very little say. Basutoland and the Bechuanaland Protectorate may be likened to an indolent old man who has a beautiful river of water passing his fertile lands, yet prefers to sit, sit, sit."[113] The man who imagined a South Africa populated by twenty million whites suggested that the safety this would bestow would come partly from the fact that "we should be able to dictate to the other African Governments what Native policy to adopt.[114]

In South Africa, support for some form of territorial expansion spanned the political spectrum. South Africa already had access to the labor and economic markets of the protectorates without the expense of controlling them, so what drove this support? Hertzog noted that farmers in particular pressured him over annexation, and he periodically focused on the Bechuanaland Protectorate as a priority—perhaps because of the interest the Schwarz scheme had aroused in the Kalahari's possible contribution to a white South Africa.[115]

But the British public was increasingly opposed to South Africa's campaign to incorporate the protectorates, in part because of draconian South

African racial policies. Legislation passed in 1936 eliminated all African voting rights (which had persisted in limited form in the Cape) and consolidated the 1913 Natives Land Act by adding land to the reserves so that they comprised 13 percent, rather than 7 percent, of the country. There were now tighter controls on land use within the reserves, including elimination of private African land ownership and, over the course of the 1940s, coercive soil conservation policies that sparked resistance. The segregationist state was further strengthened through the elimination of "black spots"—Black-owned land within areas designated for white occupation—and stronger enforcement of the provisions of the 1913 act, which prohibited independent African farming on white-owned land. The distinction between privately owned white farms and the "native" reserves was becoming starker.

These new restrictions on African economic and political rights unleashed new forms of African political protest, and helped spur the revitalization of the African National Congress. Within the protectorates the African elite, already opposed to incorporation by South Africa, increased the volume of their objections. In 1934 and 1935, colonial officials and Tshekedi Khama, the Bechuanaland Protectorate's most prominent chief, publicly debated the question of transfer in British newspapers.[116] Despite mounting opposition to incorporation, British colonial administrators in the Bechuanaland Protectorate allowed South Africa to again assess the viability of Schwarz's scheme, this time via an aerial survey of the Kalahari in 1935. Reitz—a Boer war hero and Smuts ally, and now minister of agriculture—reported that the survey showed that Schwarz's scheme would not increase rainfall.[117] In 1937 Reitz traveled to northern Botswana for three weeks, perhaps as part of the Ngamiland Waterways Surveys carried out by the South African engineer John L. S. Jeffares. That survey had been ordered by the British government in the Bechuanaland Protectorate, in part to establish the potential for development in the "swamps," including the restoration of Lake Ngami to nineteenth-century conditions. Jeffares noted the continued lack of information about the Okavango Delta and suggested that even the limited data on river flow, collected by Alex du Toit, was probably inaccurate. There was still no data at all on rates of seepage, absorption, transpiration, and evaporation, and this made it impossible to calculate the amount of water available for development. Jeffares's report noted that the primary mode of transport, the canoe, "confines the area over which observations may be made to very narrow limits."[118] The inability to gain a "God's eye view" in the flat landscape remained a problem.

Perhaps because Jeffares was hired by the protectorate's administration, which was committed to upholding African land rights, his report reads quite differently from most descriptions of the Delta by white South Africans. Jeffares accepted that rainfall had not declined, and he sought other explanations for changes in the delta's hydrology, including a decrease in large game populations and the practice of burning reeds. He speculated that even relatively minor engineering interventions, like the building of a dike to direct the flow of water, "might have very far reaching and unexpected effects."[119] Jeffares also recognized that local people had rational reasons for their patterns of resource use. They did not irrigate because the water supply was too unpredictable, and because interfering with the system of water dispersal could jeopardize vital grazing resources. He also argued that white naming practices for the region were fundamentally incorrect. "If all the districts or named areas were located and mapped [correctly], N'Gamiland would present a most civilized appearance."[120] Jeffares's insistence that the supposed lack of "civilization" was due to white misapprehension and not inherent conditions was radical in the context of a century of seeing the riparian region as a blank slate awaiting white transformation. This did not endear local people to him. "I had little cooperation from them on this survey," he wrote, "and did not have one item of information, good or bad, voluntarily given"—a rare and important piece of evidence that local people knew about and opposed plans for transforming their region.[121]

When Reitz returned from the Bechuanaland Protectorate, he again told the press that he had "no faith in Professor Schwarz's scheme for increasing the rainfall in the Union." He took pains to refer to Ngami as a "pan," noting that "it is not a lake at all; it is merely a vast, shallow depression." Similar attempts to refute the idea that Ngami was a great-lake-in-waiting had occurred sporadically over the years. In 1933, a man writing under the name John Kolobe Phoks told *Farmer's Weekly* readers, "Anyone with elementary knowledge of geology can say that the so-called Lake Ngami is not a lake but simply a pan or depression."[122] Such statements appear to have made little impact on white South African imaginations. And although Reitz did not imagine a future mega-lake, he did not give up all hope of reclaiming the area in the name of progress. He suggested that cutting channels through the "swamps" in order to make seasonal waterways into permanent ones could help develop "healthy cattle country."

It was not clear why South Africa's agriculture minister was involved in a survey meant to assess the potential for local development, and Reitz did not say for whom such a ranching industry should be developed.[123] But, in justifying their demand for incorporation, South African politicians persistently accused Britain of failing to develop these territories, implying that such neglect would not happen under their watch. This pressure found its target. A decade after Reitz's visit, the high commissioner for

South Africa wrote, "Most of Professor Schwarz's theories have been disproved, yet many South Africans still believe in them and there is in fact a Schwarz Society. . . . The Union Government may well ask us what we intend to do to bring into use the water which now runs to waste."[124] Schwarz cast a long shadow. Two decades after du Toit had criticized his scheme as unworkable, it continued to shape the geopolitics of the region and white South Africans' conceptions of how their place in southern Africa might be secured for the long term.

Establishing White Innocence

The idea of redeeming the Kalahari for white settlement remained popular, to the continued frustration of experts. They continued their campaign to discredit the Schwarz scheme and the climate ideas that underpinned it. In 1936 the agronomist W. R. Thompson published a book that sought to counter the "amazing amount of public interest" in Schwarz's scheme and to refute the belief that rainfall had decreased—a belief "handed down from one generation to another without any real proof." Thompson had no patience for Schwarz's "fantastic" scheme, which appealed to "laymen" but had no support in "responsible scientific circles." And he insisted that the evidence farmers produced for "drying up" was rooted in unreliable memories and "impressions."[125] In the book's foreword, the agricultural economist Leppan argued that Thompson's study was "a nail in the coffin of those who fatalistically aver a progressive decline in our rainfall and its inherent efficacy."[126] The two experts were remarkably naive in their optimism that this book would succeed where two decades of government reports and public pronouncements had failed.

Even as he sought to undermine public faith in Schwarz's ideas, however, Thompson inadvertently reinforced the popular view of abundant water just beyond the Thirstland—the very imaginary that made the Kalahari Scheme so appealing. Thompson's revised map reflected increased knowledge about the region's hydrology, but also gave equal prominence to intermittent and perennial rivers by representing both with thick, solid lines (figure 7.3). Beneath them, the map is blank. The viewer is left with the impression that northern Namibia and northern Botswana are densely laced with rivers that could redeem the waterless and empty space to the south.[127]

Schwarz's assertion that geology, not white settlers, were to blame for the apparent desiccation gripping the subcontinent had always been one of the most appealing aspects of his scheme. The irrigation director recognized this in a 1946 report, arguing that people supported the scheme out of wish fulfillment: hope for "an easier way out of the difficulty" of

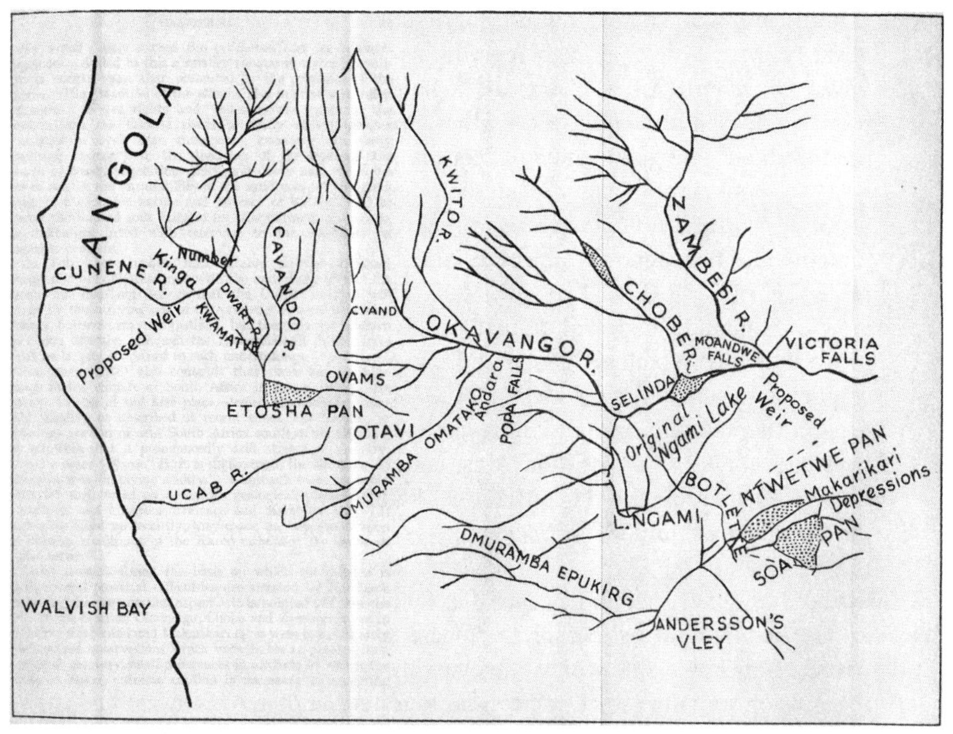

Figure 7.3 In this rendering of the area included in Schwarz's scheme, the semiarid lands of northern Botswana and Namibia appear to be laced with a dense network of rivers. Despite the author's claim to expertise, many of the names on this map were unknown locally, did not reflect phonemic convention, and seem to have been the result of typographical errors. Source: Thompson, *Moisture and Farming*.

droughts and agricultural precarity—easier than "expensive and tiresome programs" of soil conservation and agricultural improvement—as well as a desire to avoid "self-blame."[128] But the parameters of this debate began to shift after the Great Drought. Schwarz supporters spoke less of impersonal geological forces, and more about the precarity forced upon them by unpredictable rainfall and a lack of markets and transportation infrastructure. Government officials, in turn, more readily acknowledged the challenges of farming in a country with an unpredictable climate. They even began to admit that rainfall had declined, although they insisted that it was not a permanent trend. And Black South Africans began to emerge from their invisibility in these debates, now identified more frequently as a major cause of desiccation and land degradation.

In 1945, the *Times* of London suggested that Schwarz's scheme and soil conservation were mutually exclusive. "Respectable engineers and administrators" did not want to appear to support Schwarz's scheme, since it was "tainted with what many regarded as the quackery of rainmaking." It was white South African farmers who had kept the proposal alive. They were seeking "a miracle in the north to save them—an attitude which has infuriated the growing body of conservationists in the Union."[129] But by 1945, popular and expert ideas about South Africa's past climate and white agricultural future were beginning to converge. As we have seen, a number of prominent progressive farmers had become Schwarz supporters during the Great Drought. Many of them took a more nuanced view of climate change and argued for changes in rainfall distribution, rather than a wholesale reduction in the average. Few of these men saw a conflict between Schwarz's ideas about climate change and government experts' insistence that poor farming methods and land management practices contributed to desiccation. Most embraced both smaller conservation works on farms and the much larger Kalahari Scheme. Owen-Collett, the president of the Kalahari Society, was a case in point. He was also the president of the Midland Farmers and Wool Growers Association. He availed himself of government subsidies for anti-erosion works, and wrote articles on alternatives to contour trenching for controlling erosion. He wrote: "Whichever side we take, we are all for something. . . . Whether it is erosion stopping, stock control, contour netting first, or Schwarz and Government Reconnaissance Report suggestions first, we must all press for the lot."[130]

The support of men like these continued a pattern among farmers of supporting both "expert" and popular environmental ideas. But in the 1930s and 1940s, such convergence began to happen among some experts as well. The year after Owen-Collett became president of the Kalahari Thirstland Redemption Society, Alex du Toit became president of the South African Association for the Advancement of Science. In his presidential address, he admitted that fifty years' worth of records *did* show rainfall declines, "though whether of a progressive or a long-cyclic character remains unknown."[131] The following year, Theodor Schumann, South Africa's chief meteorologist, and the agronomy lecturer Thompson published a book that included a dozen line graphs plotting South Africa's rainfall over the previous half century. The graphs showed a general decrease since 1890, which the authors noted was "not in agreement with the findings of the Drought Investigation Commission." In their willingness to admit that rainfall had declined over the period farmers insisted it had, Schumann and Thompson were changing the conversation around climate and expertise. They argued that the reality of four decades of declining rainfall should be considered "when an explanation is sought in regard to drought losses, vegetal deterioration, and other matters dependent on moisture supply."[132] In other words, farmers' drought losses could not be blamed solely on farmer behavior.

In his own work, Thompson continued to challenge the orthodoxy created by the 1920 Drought Commission. The commission had confused cause and effect, he said. The loss of vegetation was not responsible for reductions in water supplies; rather, the loss of vegetation was a consequence of diminished rainfall. Farmers, therefore, could not be blamed for their hardships.[133] Thompson suggested that decreasing rainfall was the possible cause of white depopulation in the Karoo and Cape Midlands, and asserted that "practically the whole of South Africa" had "experienced a diminishing rainfall" since 1917.[134] But he argued that this decades-long decrease in rainfall was not permanent; rather, abundant rains in the early 1890s had given people unrealistic expectations of what was normal. Thompson did argue that farmers had contributed to their predicament: they had "gambled on occasional good years and ridiculed direct and indirect measures for efficient moisture preservation." This had set them up for devastating losses during extended dry periods.[135] Nonetheless, he was legitimating the arguments of those who criticized expert climate narratives and farming prescriptions, and was beginning to challenge the claims of an earlier generation of agricultural and climate experts.[136] Finally, Thompson acknowledged that without data on rainfall distribution as well as evaporation and transpiration, precipitation statistics had little meaning. It was impossible to counter the popular claim that rain now fell differently than it had in the past.[137]

This fusing of popular and expert ideas about climate and farming would accelerate in the 1940s as both Schwarz supporters and experts began to blend their ideas, albeit in distinctive ways. These were also the years in which state propaganda around soil erosion reached a fever pitch. It was described as "an enemy quite as dangerous as Hitler."[138] As the issue of soil erosion became increasingly embedded in the popular consciousness, the language around preserving the fertility of the soil increasingly sounded like the language Schwarz supporters had used when they talked about battling the encroaching Kalahari. Both carried existential implications. The first chairman of the National Veld Trust, a nonstate organization that worked closely with the government to educate the public about soil conservation, took to the radio to tell South Africans that they faced "the threefold challenge of the races, of population, and the soil." He invoked the kind of existential language long used by proponents of "drying out": "This South African civilization—let it be faced—is planted insecurely." If they did not meet the challenge of soil conservation, "then assuredly this southern civilization will pass, and, like [the precolonial state of] Zimbabwe, presently its ruins will trace a faint pattern on the wastes of Southern Africa."[139]

The Veld Trust spoke of *bestaansbeveiliging*—"securing existence." The popular Afrikaans press urged farmers to protect "the heritage of the nation" for the next generation. The government's Afrikaans-language agricultural publication, *Boerdery in Suid-Afrika*, argued that the soil was "the safeguard of our *volk*'s existence."[140] *Die Landbouweekblad* wrote, "Our descendants will never forgive us if they have to one day sit and perish in a barren desert, where once was fertile earth."[141] Readers picked up this theme as well, opining that "the survival of the nation" rested on the health of its soil. In 1943 a reader of *Die Landbouweekblad* wrote, "A badly fed white will inevitably be superseded and engulfed by the coloured."[142] In 1944, someone who saw a film about soil erosion called *South Africa in Danger* wrote a letter to *Die Landbouweekblad* arguing that those who abused their soil would be "wiped from the face of the earth." Leaders of the youth organization Landsdiens, founded in 1944, linked care for the soil to maintenance of "our European blood" and the continuance of civilization.[143]

Soil erosion was portrayed as a threat to the survival of the Afrikaner people themselves, and soil conservation as necessary to the future of the *volk*. But this was not an exclusively Afrikaner cause. Both English- and Afrikaans-speaking "progressive" farmers argued that healthy soil was a prerequisite to a healthy nation. Soil conservation, like Schwarz's scheme, became a unifying cause that connected the fears concerning white poverty and white population numbers. A British South African in the Transvaal argued that combating soil erosion required intensive effort: "It is my opinion that the country can never be properly controlled unless it is filled with the necessary man-power." This writer suggested that if ethnic politics made recruitment of settlers from Britain impossible, "then let us import 20 million Frenchmen."[144] Two unnamed livestock experts interviewed by *Farmer's Weekly* suggested that "degeneration" due to poor farming was affecting not just livestock but also South Africa's white children. "During the past quarter century we have seen humans, livestock, and farms in South Africa degenerating."[145] They concluded that unless South African farmers changed their methods to produce more nutritious food, degeneration "would convert the people of South Africa [again, implicitly white] into a race inferior both in physique and mentality to the present generation. The future of the nation lay in the hands of the farmer." They also insisted that "too much reliance was placed on the Native worker." Farmers should instead farm smaller parcels and "live in close relationship with the soil," like their counterparts in Britain, France, and the United States—strange advice coming from experts in pastoral production, but advice that reflected the spirit of the Kalahari Scheme and the world it would create.[146]

That spirit also was reflected in a growing techno-utopianism around water within the state bureaucracy. In 1943, *Time* magazine in the United States published an article proclaiming that Schumann, the chief meteorologist, was preparing "the first successful attempt in history to make rain artificially" by turning Table Mountain's cloud cover into thirty-one million gallons of water every day. *Time* was careful to note that Schumann had studied in the United States and Germany, including at Yale, thereby replacing him firmly on the side of respectable scientists. But his scheme sounded a lot like the plans of the schemers and dreamers who had written to the South African government in the decade after Union: "two parallel fences of wire netting," 150 feet high and 9,000 feet long, "with an electric potential of 50,000 to 100,000 volts between them."[147]

The modernizing aspects of Schwarz's vision—using technology and white ingenuity to radically change the land—were merging with the modernizing aspects of bureaucratic expertise. In this convergence, Schwarz could be lionized for his "vision" even by those who accepted that his scheme was not feasible. He was credited, rightly or wrongly, with being the first to imagine how taming rivers might save South Africa. Thus, the British high commissioner to South Africa wrote, in the year that Schumann sought to turn clouds into water, that while Schwarz's climate theories were incorrect, "if nothing is done I foresee nothing but further desiccation of the area and the gradual but steady advance of desert conditions." The biggest issue facing the southern African protectorates was therefore "the desirability or non-desirability of embarking on a large-scale undertaking . . . by harnessing for economic production and development the waters of the Okavango and Chobe Rivers," thereby putting "in the service of man" water that was "going to waste."[148]

The spirit of the Kalahari Scheme could also be seen in emerging notions of who bore responsibility for South Africa's environmental decline. In the early 1920s, Heinrich du Toit had accused white farmers of placing the country in peril. In the 1930s and 1940s, some experts were more sympathetic. Yes, the arrival of white settlers had resulted in environmental harm, but it was not their fault. South Africa's rainfall patterns predisposed the country to soil erosion.[149] Early farmers had no way of knowing this. Furthermore, the white men who did the noble work of opening up the land to settlement had been forced to engage in exploitative farming by circumstances over which they had no control. State economic policies had incentivized farmers to maximize production at the cost of soil fertility.[150] Combined with official recognition that rainfall had been unusually low for nearly forty years, such pronouncements partially absolved whites of their responsibility for rural precarity.

In 1944, the year after the Veld Trust was founded, an agriculture professor argued that "most dongas [erosion gullies] are dug by farm mortgages and poverty."[151] A Department of Agriculture employee named J. C. Fick wrote a book, published in both English and Afrikaans, arguing that "bad farming practices" were the cause of erosion. But he qualified his accusation: "*Nobody can be blamed for this.* It came to pass during the rapid development of a big and young country, with a large diversity of soil and veld, and precarious climatic conditions." Economic pressures had forced farmers to make difficult choices. In a passage that encapsulated decades of debate over the relationship between farmer choice, agricultural debt, progressive farming, and the absence of markets and infrastructure, Fick came down firmly on the side of the farmer. He wrote:

> People who are inclined to blame farmers for the burdens of debt on their farms should remember that most farms were undeveloped, and that the owners had to embark upon extensive improvement programs. Think of the thousands of miles of jackal-proof and other fences that had to be erected, the sheds, dams, silos, houses, etc. that had to be built. These were essential improvements which this and the former generation had to carry out, all without the assistance of steady rains or stable markets.[152]

In other words, the very forms of farming recommended by experts had created economic pressures that led to bad farming. Fick was essentially saying what farmers had said for a long time, his comments echoing the earlier words of the rainmaker and farmer Charles Hall, who had insisted—and been reprimanded by the *Farmer's Weekly* editor for insisting—that "the only farmers [*sic*] . . . able to run a farm as it should be run [is] the man that can afford the loss."[153] Fick's message was approved by the National Veld Trust, which distributed five hundred copies of his book.[154] In the same year, a committee on the reconstruction of agriculture offered a similar message. It argued that land settlement had not "proceeded along sound lines." Farms had been given out to those without the knowledge to use them wisely, and people had grown crops on land more suited to pastoralism. But the committee largely attributed this to farmers' inexperience.[155]

The establishment of white innocence was a precondition for a merging of popular and expert knowledge around key points that had once been pillars of the opposing identities of "experts" and "practical farmers." As scholars have noted, debates and action around soil conservation in South Africa were shaped by the context of a white supremacist society.[156] For all that "backveld" farmers and poor whites had long been spoken of in language reminiscent of that used for Africans, white and Black farmers received quite different treatment as the state implemented conservation programs. As Delius and Schirmer argue, the harmful farming practices of whites were attributed to their economic circumstances, which justified still more economic aid; similar practices by Black farmers were attributed to the inherent backwardness of Africans. While the Reconstruction of Agriculture report admitted that inexperienced white farmers had caused environmental damage, it said that far greater damage had been caused by Africans—due to insufficient land and "improvident farming practices" in the reserves, as well as the use of Black labor on white-owned farms, a practice that resulted in "extensive soil and veld abuse."[157] This was in notable contrast to the findings of the Drought Commission in 1920, which acknowledged African responsibility for environmental degradation but placed far more weight on the changes wrought by the arrival of white settlers. Despite decades of government proclamations about white responsibility for land degradation, in the 1940s it was African reserves, not white-owned farms, that saw coercive action around conservation.

It might seem obvious that Black South Africans would have been blamed for environmental degradation all along. Why resort to abstract geological forces when the subaltern offered a ready scapegoat? But narrative convenience encountered the problem of narrative coherence. One of the problems with blaming Black South Africans for the environmental stresses of the early twentieth century had always been the "empty land myth," which held that white settlers had encountered an unpeopled land rather than one filled with densely populated agricultural communities. In this rendering of history, popularized by George McCall Theal in the 1890s, Europeans and Bantu-speaking Africans had arrived in South Africa at the same time, with the Europeans coming from the south and the Africans coming from the north. As we have seen, whites also routinely denied that Africans had a long tradition of agriculture. They did not work the soil, but only "scratched" at it. White people mastered their environment, while Black people were subject to its whims. The notion of the simultaneous arrival of whites and Africans and the greater agency of whites over their environment not only legitimated white claims to South African land, but explained how accounts of men coming upon well-watered, fertile landscapes unsullied by human hands could make sense in the context of a country that, when these tales were recounted, had an African majority with a long tradition of agriculture. But it also implied that the loss of this paradise had coincided with the arrival of white settlers.

By the 1940s, the land could no longer be described as empty. In a speech on native policy, the high commissioner for South Africa, George Heaton Nicholls, offered a British audience a version of Theal's history, but with an ecological twist that carried it into the present. The "practically unpopulated" land of the seventeenth century was now populated by "Europeans and Natives in certain clearly-defined areas of their own choosing." Europeans with their livestock had occupied the semiarid regions, while Africans, "dependent on their crops," lived in the rainfall belts near the coast. Aridity was a white preoccupation because whites had naturally gravitated toward the drier lands.[158] The idea that Black South Africans lived on overcrowded rural lands not because whites had mandated it but because this reflected natural historical developments was another declaration of white innocence—one that would become more important in the 1960s as three million Black South Africans were forcibly relocated to homelands rife with poverty and malnutrition.

While whites had previously denied Africans the label of "farmer" on the grounds that they merely "scratched" at the soil, now Africans supplanted "backveld boers" as the primary villains in the saga of environmental degradation. Particularly in the wake of the Great Drought and Great Depression, when the notion of a drought-proof farm seemed risible, whites began pushing back on the idea that they were responsible for ecological devastation. A man from Southern Rhodesia wrote that he had "read most of the recent literature on erosion in Rhodesia and the Union, but, though looking closely for it has failed to find any mention of the Native squatter, or of the Native agriculturalist [again, never "farmer"!], whose unchecked destruction of all vegetation for the sake of three or four years' crops, and then passing on to new vegetal slaughter, is the outstanding cause of the erosion evil." The *Farmer's Weekly* editor insisted that there had been a great deal of discussion of "the destructiveness of Native farming methods," but in fact the correspondent was right. The tendency had long been to blame white farmers.[159] This began to change in the early 1940s. Farmers began to insist that "the biggest culprit" when it came to soil erosion was the African population. Africans were blamed for cutting down the trees that protected the country's water supply, and thus for the drying up of springs and streams. Readers described the Transkei as the most eroded part of South Africa, and openly suggested that, while the white population should not be compelled to adopt conservation methods, such compulsion should be applied to the African population. Even erosion on white-owned farms was not blamed on the actions of the farmers themselves, except inasmuch as they left their Black laborers in charge.[160]

Although white agricultural experts never bothered to collect any data, they confirmed white popular perceptions when they argued that the reserves were an ecological catastrophe in the making.[161] The populism of Agriculture Department officials and elected MPs who represented farming constituencies mitigated against any state intervention directed at white farmers, even as African farmers came under increasingly coercive regimes. As the director of the Division of Soil and Veld Conservation wrote later about the 1946 law, "The spirit of the Act is democratic, the intention from the outset being to encourage voluntary action by the farming community."[162] Redemption would be a voluntary effort.

In 1945, a man describing himself as "just a plain farmer" and signing with the name "Icarus" compared Schwarz to Galileo, "burned at the stake by the priests (the 'experts' of that epoch)." And he told readers, "It is said that the last words to his wife, as he lay on his deathbed, were that he knew his theory was right and practicable . . . and that she must carry on his gospel."[163] A popular mythology had emerged around the man who had died alone—not with his wife by his side—in West Africa. Schwarz's widow, Daisy, helped create this mythology. In the same year, she published an article in *The Outspan*, perhaps the most widely read magazine in South Africa, in which she purported to describe "the man behind the scheme." The headline, "Schwarz of the Kalahari," cast him as a sort of prophet of the desert. But the prophet had found his final rest in a working-class suburb of London (figure 7.4). Engraved on his striking headstone is the same title, while a sculpture of an open book displays two verses from the Book of Isaiah, the tale of another prophet: "I give waters in the wilderness and rivers in the desert," and "The parched ground shall become a pool, and the thirsty land springs of water." These verses, which predicted the redemption of Israel, placed Schwarz in a religious frame that he seems never to have embraced. But they also served to make him the equal of the men who worked for the Department of Irrigation, because the triumphal arch of the Hartbeespoort Dam also bears a quote from Isaiah: "I provided waters and rivers in the pathless desert so that I might give drink to my chosen people."

Schwarz had become a part of South Africans' ideas about their own redemption. The editor's introduction to the *Outspan* article states that while the Kalahari Scheme "may never come true . . . whatever development may take place in the future in this great section of the Continent will be due primarily to Professor Schwarz—a man who had the courage and daring to think and act big." Daisy herself could state, a quarter century after Schwarz had published his book, "Everyone in South Africa has heard of the Schwarz scheme for irrigating the Kalahari."[164] Even *Farmer's Weekly* joined in the mythmaking. Its editors had long been Schwarz skeptics. But when yet another expedition to investigate the Kalahari Scheme went north in 1945, the paper called it "one of the most romantic expeditions of modern times" and, in sensationalist prose, resurrected the Kalahari

Figure 7.4 "In Memory of Schwarz of the Kalahari," Willesden New Cemetery, London. Photo by Aubrey McKittrick-Troop.

fantasy: "For more than a quarter of a century the riddle of this desert and its supposed effects on weather conditions in southern Africa has been tantalizing Governments, Parliamentarians, scientists and farmers—in fact, all who have the progress of the sub-continent at heart." It concluded, "Whatever success may be attained in this connection will, in truth, be due to the pioneering work of the Prophet of the Kalahari."[165]

8 * Afterlives

In 1945, Andrew Meintjes Conroy, the minister of lands, led another expedition into the Kalahari. He was accompanied by nine hand-picked members of the South African Parliament—all from his own United Party—as well as two journalists and several employees of the Irrigation Department. The director of irrigation, Leonard Mackenzie, wrote that the expedition was a response to the "pressure of public opinion urging action in connection with the restoration of the Kalahari lakes." Repeated scientific attempts to refute Schwarz's ideas "appear merely to have developed a measure of obstinacy in the public mind," he observed.[1] One of the MPs on the trip told his colleagues in Parliament that he had received "hundreds of letters from farmers all over the country" asking for another investigation.[2]

Unlike the 1925 investigation, this group covered the entirety of the territory Schwarz's scheme encompassed: the Okavango Delta and the adjacent Chobe and Zambezi rivers; Etosha Pan and the Kunene River in northwestern South West Africa; the vast lands that stretched southward to the Orange River; and even the Molopo and Nossob, the dry tributaries of the Orange where South Africa, South West Africa, and the Bechuanaland Protectorate converged. This was an immense space, much of it controlled by South Africa. But the expedition generated no new information about the region and its rivers. One MP observed that the trip was "hurried," while the irrigation director reported that experts had actually refrained from commentary so that "impressions of the laymen accompanying the expedition might not be influenced."[3]

This was a new approach toward laying Schwarz's scheme to rest: taking experts out of the picture and letting the landscape doing the talking. The MPs saw with their own eyes what experts had told the public for years: there were already "vast water supplies, right inside a desert country," as one put it, and all that water did not produce rainfall in the desert. The travelers seemed awestruck by the absolute aridity of some of the places they saw, and sobered by the realization that most of this land could never support a dense population. They focused their hopes and dreams on the rivers themselves, much as Alex du Toit had done twenty years earlier. While they almost universally accepted that the Schwarz scheme would not change the climate, most supported extensive irrigation works in the Caprivi and Kalahari, in part to enable "a large white population" to settle there or, at minimum, to pave the way for European-led development of the area.[4] Contrary to the hopes of the MPs and their constituents, however, Mackenzie argued that that irrigation and ranching schemes were unlikely to be successful, because the riparian soils were surprisingly infertile and because there was less groundwater in the Kalahari than had once been supposed. "Those who fear the 'spread of the Kalahari' may rest assured that this is likely to follow only from man's occupation of the area," he wrote.[5]

Prior investigations had generated sarcastic commentary about taxpayer-funded safaris, but the 1945 expedition produced unprecedented vitriol and partisanship. The Fusion government that had merged the National and South Africa parties into the United Party in 1934 had collapsed over Afrikaner nationalist opposition to South Africa's entry into World War II. The rancor lingered after the war ended. The new, "reunited" National Party opposition in Parliament ridiculed the United Party expedition as a group of friends who "went on safari with evening dress and rifles to attack the Kalahari," and mockingly compared Conroy to David Livingstone. They repeated press reports of the group's provisions—which included sixty dozen eggs and five sides of bacon—and the "hardships" that included subzero temperatures that encased false teeth in ice and froze socks. National Party members—including former supporters of the Kalahari Scheme who had once accused experts of being biased against Schwarz—now accused the government of ignoring the findings of du Toit's 1925 investigation.[6] The attacks were sarcastic and intensely personal, as was Conroy's rebuttal, which included revealing the embarrassing university nickname of one of his harshest critics. One MP, who had switched from the United Party to the National Party a few years earlier, confessed his discomfiture that the public had witnessed the spectacle, which he called a "farce" that was unprecedented in his political career.[7] The mudslinging revealed that at the level of formal politics, opposition to the Schwarz Scheme had become an Afrikaner nationalist cause. At a popular level, however, this politicization was less evident. As the Conroy expedition departed, people wrote letters to both *Farmer's Weekly* and *Die Landbouweekblad*—now more willing to publish diverse reader opinions—in support of the scheme.[8]

Futures, Realized and Unrealized

In 1948, the year the National Party came to power on the platform of "apartheid," Wilfred Marais, a Cambridge-educated lawyer who had recently retired from farming on the highveld, published a pamphlet advocating for the Kalahari Scheme. Marais proclaimed white innocence, arguing that South African farmers were not responsible for the desiccation they saw around them—and that if they were, they would be "a nation not fit to own a good country." The South African farmer was "a true lover of the soil," and natural forces, not human action, were to blame for erosion and other environmental ills. Marais concluded that without addressing climate change, the newly elected government "will be wasting time with their elaborate preparations for Apartheid and Republicanism for a country destined for rapid and complete desiccation."[9]

Also in 1948, Daniel Kokot, an employee at the Department of Irrigation who had traveled with the 1945 Conroy expedition, published another book seeking to refute the claims of Schwarz and his supporters. *An Investigation into the Evidence Bearing on Recent Climatic Changes over Southern Africa* was the most extensive response yet to popular climate thinking. It ran to 160 pages; cited more than four hundred published and unpublished scientific studies, explorers' accounts, and government reports; and drew on research about southern Africa as well as the polar regions, Europe, North America, South America, Australia, Asia, East Africa, and the Sahara. Kokot argued that the features people took as evidence of desiccation—dry pans and riverbeds, signs of water erosion—were normal features of arid environments. "Much of the evidence, on which the belief that South Africa is drying up rests, falls away because the network of dry river beds needs no foreign climate to explain their existence."[10] While Kokot acknowledged that rainfall had been lower than average from 1918 to 1934, he wrote that after that point, "the rainfall increased in an almost spectacular manner," with nine of the next twelve years wetter than average.[11]

It is doubtful that Kokot changed many minds. The Schwarz Kalahari Society was reestablished in 1949, amid a drought that the *New York Times* called the worst in a century.[12] In the pages of *Farmer's Weekly* the following year, Kokot pointed out that twenty years after Schwarz's death, his most ardent supporters were fuzzy on what the Kalahari Scheme actually was. Men who claimed to support the scheme had focused much of their attention on diverting the Zambezi, an option Schwarz had refused to consider. And when challenged on the validity of Schwarz's climate theories, his supporters downplayed plans to change the climate. Marais, for example, had argued that Schwarz's opponents "lay too much stress on the question of additional rainfall."[13] But, as Kokot pointed out, increasing rainfall had been the primary purpose of the Kalahari Scheme. He expressed his frustration with the "offensive" assertions of Schwarz supporters that there had never been a "fair and unbiased" investigation of the scheme. These claims insulted Alex du Toit, "one of South Africa's most distinguished scientists," as well as the credentials of many others who had participated in the Kalahari expeditions. "It is open to speculation as to whether there is anyone in South Africa from whom a verdict against the Schwarz theories would be accepted," Kokot concluded.[14]

But almost at the very moment that Kokot insisted that support for Schwarz's scheme was impermeable to evidence, that popular support began to fade. The scheme continued to be discussed through the 1950s, but only sporadically. Men whose support remained unwavering were advancing in age. R. A. Hockly, an MP from the Eastern Cape and one of the original trustees of the National Parks in the 1920s, was the Schwarz Society's vice president when he died in 1945. Owen-Collett, Gessert, and Paulsmeier were in their eighties and each died in the 1950s. Other farmers who had been vocal in their early support were retiring from farming. Schwarz's opponents were also passing on. Alex du Toit died in 1948; Carl Weidner and Francis Kanthack in the early 1950s. A new generation took up the cause, including John G. Collett, a relative of Owen-Collett, as well as Marais. But it was a small group.

No more South African investigations were sent into the Kalahari. In 1969, John G. Collett sent a copy of a pamphlet he had published in 1949, as well as Marais's 1948 pamphlet, to the mining tycoon Harry Oppenheimer. Collett included a somewhat plaintive handwritten note: "Since 1949 I have distributed about two thousand of these pamphlets . . . to every prime minister and minister of agriculture and water affairs. None of them are interested."[15]

Schwarz's scheme, when it is mentioned at all, is now largely regarded as a historical oddity. Readers may be left unsatisfied, if also relieved, by its inglorious end. The torrents of debate, the flood of articles and reports, the spate of turbulent emotions among those for and against the scheme all dissipated, like the water of the Kavango River in the Kalahari's sands. But the imagined futures that animated debates over Schwarz's scheme did not disappear; their shape merely changed. The redemption of the Kalahari, and the ideas at its core, lived on in new forms to reflect new concerns over the maintenance of white supremacy.

Schwarz's scheme created the conditions for a populist embrace of white innocence, racial agrarianism, and salvation through hydrology, even among those who were not among the National Party faithful or apartheid ideologues. Since at least the early 1930s, allegiance to Schwarz had become less about the specifics of his plan than about support for four key beliefs that lay at the plan's foundation. First was a conviction that the country was "drying out," and that white actions were not the cause. Second was that white civilization was in a perilous state, and that farmers were the key to its survival. Third was that stopping water from "running to waste" was the solution to these threats because it kept farmers on the land. And last was that it was the responsibility of the state to manage the hydroscape, and thus to ensure the survival of white farmers. By the 1950s, however, these beliefs were hardly limited to Schwarz supporters. The decades of debate around the Kalahari Scheme had consolidated old truths and created new ones. Many rural whites shared these key beliefs, and they had filtered into the pronouncements of bureaucratic and political elites as well.

Let us begin with the question of white innocence in South Africa's "drying out." A 1955 schoolchildren's textbook called *Save the Soil*, written by two men trained in agriculture—one a high school teacher and one an agricultural inspector—shows how Schwarzian imaginaries had seeped into expert discourse. "South Africa has developed from its barbaric beginnings into one of the most civilized countries in the world," they wrote. But it was now on the same path as ancient civilizations that had been turned into deserts through human misdeeds. Using the same sources Schwarz had used thirty-five years earlier, they described a land that was once lush and well-watered—even in the time of students' grandparents. But the nature of its rainfall, coming in short, intense bursts, had made the country vulnerable to soil erosion once white settlement began. "Although the early travelers gave so many proofs of the well-wooded and well-watered nature of the country, South Africa is now, compared with some parts of the world, an arid land." While the world received forty inches of rainfall per year on average, South Africa received only seventeen.[16] These facts created the impression that rainfall was declining due to the destruction of forests, although the authors did not explicitly state this.

Unlike Schwarz supporters, the government agronomists suggested that white settlement was in some way responsible for South African aridity and its low rainfall figures. But they managed to assert the same white innocence for this state of affairs that Schwarz had offered his supporters. Like some of their colleagues in the 1940s, the authors of *Save the Soil* argued that while white farmers had overexploited their land, they had also borne the burden of placing modern improvements on farms, such as jackal-proof fencing and dams—the very infrastructure that marked scientific farming—"without the blessings of regular rains and established markets." They concluded, "To be charitable we ought, therefore, to forgive the sins of the past."[17]

Save the Soil is also striking for what it shows as opposed to what it says. It does not acknowledge that Black South Africans had been farming and raising livestock in the country for centuries before white settlement. But in its photographs, Black South Africans are no longer invisible. They are in native reserves, standing outside rondavels with their sheep on barren and eroded land, posing amid dongas and using cow dung as fuel rather than fertilizer, and cutting down mature trees. By contrast, on photographs of badly grazed white-owned land, the only people are "experts" and the captions use the passive tense, stating for example that soil "has been badly overgrazed."[18] In these photos, white men are not only innocent of malice toward the land; they are the key to its redemption. The point is not that all experts would have agreed with these two men. It was now possible for experts to embrace and even reproduce popular knowledge about rainfall, race, and responsibility while retaining the status of experts.

The second shared belief, in the precarity of "white civilization" and its links to a white countryside, is perhaps best demonstrated in a 1959 report on the racial demographics of the platteland (figure 8.1). The report lamented the white depopulation of the platteland as a threat to "white civilization." Farmers' individual incentives to employ cheap Black labor continued to conflict with what was supposedly the greater good: a territory inhabited and controlled by white men occupying and working the farms they owned. The commission stated that in the seven districts it had surveyed in the Orange Free State, only 75 percent of the farms were occupied by white owners. The remaining quarter were either unoccupied or managed by Black farmers and farm managers—a trend that had increased since 1945. In the near future, the report suggested, there would be one white person for every ten or eleven Black people on the platteland. This was "a serious danger for the future." White landownership brought social benefits such as feelings of equality and responsibility. "The possession of land brings with it spiritual stability and deep roots, an abiding love of one's own country and all the other conservative characteristics." This was especially true of Afrikaners, who were the majority of farmers.[19] Farmers were also the bulwark against the Black majority.

> In a country with a homogenous population the depopulation of rural areas would not carry dangers, but in this country with its heterogeneous elements the White platteland is largely the pivot of Western civilization. Should this pivot collapse, White civilization in the cities, too, would not be able to hold its own in the long run.

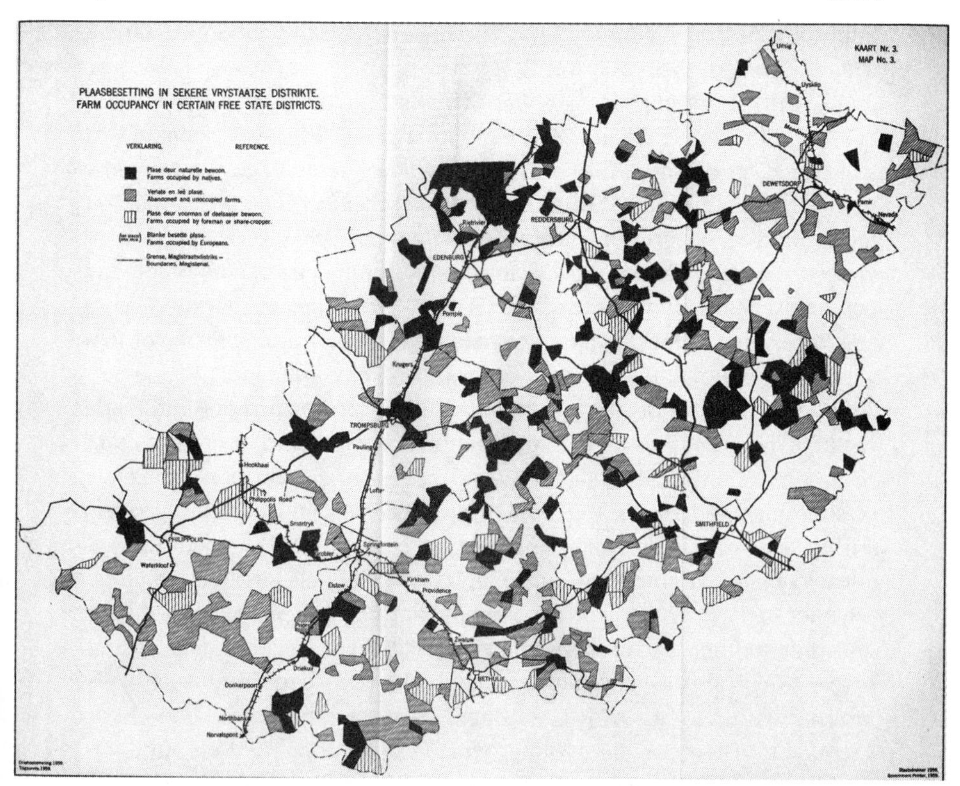

Figure 8.1 Map showing the prevalence of abandoned and Black-occupied farms (denoted by gray and black shading) in the purportedly "white" agricultural area of the Orange Free State. Source: *Report on the European Occupancy of the Rural Areas.*

This was not just a matter of cultural survival. If food production were in the hands of Black South Africans, "civilization would automatically have to capitulate, because those who produce the food of the nation, automatically exercise great influence on the course of events in any country."[20]

The report is revealing for what it does not say: that even if every farm were occupied by its white owner, the demographics of the platteland would change very little. As Foster notes, rural South Africa ultimately became a white aristocracy, not a society of yeoman farmers, because farmers were unwilling to do without Black workers. The country had become "a 'white man's land' in terms of ownership rather than labor," he writes.[21] There were also far fewer white South Africans, proportionately, on the land than there had been a half century earlier. In 1904, 47 percent of whites had lived in rural areas; by 1958, this had declined to 18 percent. Black South Africans had also urbanized, but 63 percent still lived in rural areas.[22] Yet, as others have observed, the platteland remained important to urban white people's imaginaries of a white South Africa, even as fewer of them had ties to it.

The third point of consensus was that water on the land was the key to keeping whites on the land. At the popular level, Schwarz's scheme was still sometimes discussed as a scheme to make rainfall. But it was now more commonly understood as an irrigation scheme, reflecting the shift in focus that had begun in the 1930s. The Kalahari Scheme promised more water—a vague promise that meant it had become just one among many possible schemes and was assessed as such. It also meant that most of its supporters and all of its opponents adopted the shared imaginary of irrigation when they imagined how to keep people on the land.

When members of the House of Assembly traded insults and snide remarks over Conroy's 1945 expedition into the Kalahari, it did not mark an absolute change in the temperature of partisan discourse. In that same session, members expressed cross-party consensus on engineering rivers within South Africa. A man who had gone on the 1945 expedition advocated for a national long-term plan "embracing all possible irrigation schemes in the country," and his National Party opponents expressed their gratitude to him. One of them opined that "before we can dream dreams of great Central African schemes" they had to take care of those already farming within South Africa. A United Party member later echoed this sentiment: "There are many things which can be done right on our doorstep to enable our people to make a living. Those things in the north are for future generations."[23] Meanwhile, another nationalist MP suggested that if more dams were constructed, "this country could carry at least double the present population." Such sentiments represented a political convergence on the terrain of South Africa itself: that focusing on engineering water within the contemporary borders of South Africa was the best path to white security, in terms of both rural white population and agricultural prosperity.

As South Africa's ambitions for domestic water engineering schemes grew, those who still believed that human interventions in the landscape could create rainfall came to support schemes closer to home. An Orange Free State farmer named Charles du Preez urged the state to take over the management of mountain catchments to protect their soil and trees. This would, he argued, "strengthen the fountains and increase the rainfall."[24] A man in Johannesburg accused irrigation engineers of being "the greatest enemies that we have in this country"—presumably because they tempered popular enthusiasm for large schemes—before advocating for the Schwarz scheme *and* for damming the Orange River to create an upstream lake, both with the goal of increasing rainfall. These arguments that covering the land with trees and water would increase humidity and thus rain demonstrate that such forms of popular knowledge continued to exist even as they were supplemented by other ideas about contouring, erosion control, and other "scientific" prescriptions.[25]

The near-universal enthusiasm for irrigation that emerged in the late 1940s was hardly an inevitable outcome. Schwarz's insistence that more rainfall, not irrigation, would guarantee security to all white men had fallen on fertile ground. Many members of the public had been highly critical of irrigation settlements' mixed track record at rehabilitating poor whites or offering farmers a secure livelihood, and those who were well-read repeated stories of irrigation failures elsewhere in the settler colonial world. Yet by the late 1940s and through the 1950s, members of Parliament pressured government ministers to build schemes in their districts, and agricultural journals were filled with letters and articles extolling the benefits of one irrigation scheme or another. The 1950s were a dam-building decade, with more than a dozen large projects constructed in the Transvaal and Orange Free State alone.[26] While many of these dams were multipurpose, providing water and power for cities and industry as well as farmers, it was the amount of acreage they could irrigate that was often the basis of their popular support.

Two things that accounted for the enthusiasm for dams and irrigation schemes were shortages and high consumer prices for food in South Africa. Globally, people talked about looming worldwide famine and the inability of farmers to continue feeding the world's growing population. Locally, the press, politicians, and the public spoke of the urgent need to increase food production. In this context, agriculture took on new meanings. It was not just the embodiment of a demographic and territorial claim to a white man's land and a more virtuous way of life. It was also key to the physical survival of both white and Black South Africans. Farmers had long called themselves the backbone of the country, but now their claim was newly persuasive.

The fourth point of common ground between Schwarz supporters and opponents was that engineering the white hydroscape and guaranteeing white rural prosperity were responsibilities of the state. Whereas in the past, farmers had advocated for small dams on every farm, now the kind of infrastructure envisioned to stop water from running to waste was beyond the capacity of individuals or private groups. Over the course of the 1950s and 1960s, furthermore, government subsidies for agriculture increased even as farmers continued to complain that the state policies favored cities over the rural areas. White farmers benefited from state largesse whether they practiced conservation-driven farming or not—one of many facts demonstrating that political, not environmental, concerns drove government policies around land rehabilitation.[27] That state support, combined with mechanization and the increased use of fertilizer, caused rapid increases in white agricultural productivity, especially for crop production. White farm incomes also rose, nearly tripling between 1936 and 1956. By 1959, the country imported almost no food, and people were eating more diverse diets.[28] Large concrete dams dotted the countryside, and many more were planned. Photos of these dams resemble scaled-up versions of the photos white landowners had taken of their reservoirs and farm ponds half a century earlier: a new "hydrology of hope."

The hydro-utopianism inherent in Schwarz's scheme continued to be evident in the schemes readers still proposed. In 1954, a man suggested using atomic power to pump water from the Zambezi at Katombora into the Kalahari. Three years later, another man told readers of the *Rand Daily Mail* that a "syndicate of international financiers" was backing a plan to build a canal in the northern Kalahari that would connect the Zambezi and Kunene Rivers, thereby linking the Atlantic and Indian Oceans and opening up the Kalahari for ranching. In 1959, an engineer said he had come up with a scheme to divert water from Port Elizabeth to the Sundays River via a seven-mile tunnel, and that the Department of Water Affairs was interested in seeing his plans.[29]

Other features at the core of Schwarz's scheme would resurface in South Africa long after general discussion of watering the Kalahari had faded away. But they took longer to manifest themselves. In the 1960s, a newly emboldened South African state pursued key aspects of Schwarz's vision in new forms. They turned a land with a Black majority into a white-majority country, embarked on a massive scheme to unite distinct river systems, and looked beyond their borders for water and security.

Watering White Supremacy

In the year that Marais and Kokot published their treatises for and against the Schwarz scheme, white South Africans went to the polls and narrowly elected the National Party to power on the platform of apartheid. The NP won just 40 percent of the popular vote, and prevailed only because the rural constituencies that were the center of its support were given more weight in the electoral system. The election of 1948 has often been seen as a watershed in South African history, matched in importance only by that of 1994, when South Africans of all races voted for the first time. Historians have suggested, however, that it was less clear at the time how it would change the country. The National Party had been in power before. And before the election, apartheid was more of a political slogan than a coherent set of policies. Its meaning was discussed mostly among Afrikaner intellectuals who sought to give substance to a word that merely means "apartness" or "separateness," in a society already organized along the lines of *segregasie*, or segregation. Those theorists argued that separate development would be more just than segregation, because it would benefit both groups rather than subordinating the needs of one group to another. Within the context of what was already a white supremacist society, they wrestled with the competing demands of the survival of the Afrikaner *volk* and what they deemed to be the fair treatment of Black South Africans. Most suggested that whites would need to make a significant sacrifice for apartheid to work, whether this meant giving up land, Black labor, or both.[30]

But if in its early stages the construction of apartheid was an elite project, it did not remain that way. The term was given meaning and substance through the interactions of the political elite and white voters who embraced a "crude vernacular racism."[31] The notion of absolute territorial separation—partition—and white "sacrifice" was considered seriously only by apartheid's intellectuals. As Daniel François Malan, the prime minister, wrote to an American minister in 1954, "Theoretically the object of the policy of Apartheid could be fully achieved by dividing the country into two states, with all the Whites in one, all the Blacks in the other."[32] But it was clear that outside of the world of academic theorists, no white South African seriously contemplated doing without Black labor or ceding large quantities of fertile land to Black South Africans. In keeping with Neil Roos's call to think about the distinctive lives and worldviews of ordinary whites, it is important to recognize how little popular support the idea of radical partition had.[33] While the agricultural press occasionally profiled white farmers who managed without Black labor, there was little evidence that most readers cared to emulate them. Beyond easy access to inexpensive Black labor, white farmers wanted Black people as a group—personal relations aside—to simply not matter. As one such farmer wrote in response to a 1947 National Party questionnaire, "I have neither the time nor inclination to bother with kaffirs, coolies, and hotnots (Coloured people)."[34] A historian who analyzed responses to the questionnaire noted that farmers and townspeople were far more concerned with the separation of the races and social control than ministers and academics were.[35] The white man's country would be not be consolidated through radical partition, but through a combination of legal fictions, coercive laws, and state largesse toward whites.

Hermann Giliomee argues that apartheid was a "survival plan," and this idea resonates with the language of extinction and survival that runs like a thread through the early twentieth century. But the question is—as Saul Dubow notes—what "survival" meant in the context of the 1940s.[36] By this time, white poverty had almost disappeared. Afrikaans language and culture had been secured through a half century of nationalist activism. What then was the threat? Dubow suggests that white survival came to be equated with continued white supremacy in the face of growing African nationalism, global decolonization, and communism, particularly in the 1950s and beyond. Initially, however, most whites did not seem to sense that a Black revolution was imminent. What rural whites did continue to talk about was their future as farmers in an arid land, and the responsibility of the state to secure that future. The state would respond, though not by building the Schwarz scheme.

The radical partition that Schwarz's scheme had proposed—where Black residents "need not come in contact with the white settlements at all"—was no longer even a hypothetical option by the 1950s. The vision of apartheid's theorists gave way to *baaskap*: the overt domination of the Black majority by the white minority. Partition was achieved not through the demographic expansion of the white minority into new lands, but through legal fictions, spatial engineering, and brute force. As Dubow notes, perhaps 3.5 million people—more than 10 percent of the population—were forcibly relocated into ethnically defined "homelands" between 1960 and 1982, a level of population redistribution unprecedented in world history outside wartime. The goal was to create ethnic minorities out of a racial majority, "to deprive blacks of their South African citizenship and to transform whites from a political minority into a sovereign majority."[37] The economy continued to rely on Black labor, but Black workers were now erased as members of the civic community, defined as temporary residents in a white man's land.

The state's all-out assault on Black South Africans met with a dramatic increase in Black resistance over the course of the 1950s. In 1960, police in Sharpeville fired on a crowd of unarmed people protesting South Africa's pass laws, killing sixty-nine and shooting many of them in the back. As protests spread around the country, the government arrested thousands of people and eventually banned the African National Congress and the Pan Africanist Congress. Its response catalyzed the decision of the formerly nonviolent movements to begin armed resistance. It also drew international condemnation in the form of a United Nations resolution that passed thanks to the votes of many former colonies that had become newly independent, neutralizing the "no" votes of South Africa's sister settler colonies and the European colonial powers.

In 1961, South Africa withdrew from the British Commonwealth and declared itself a republic. Two years later, Theodor Schumann wrote a book called *The Abdication of the White Man*. He praised the spread of the white race to the world's temperate zones as "one of the greatest events in the cultural history of man." Part history primer and part anticommunist screed, the book argued that most of the world's white populations had abdicated their natural role as the world's civilizers, thanks to the psychological insecurities inflicted by two world wars. Only South Africa remained confident in its destiny.

Schumann was an heir to some of the men introduced at the beginning this book: people like Charles Pearson and Isaiah Bowman, who looked into the future and saw a world in which white people were at the mercy of the "rising tide of color." But he was neither politician nor scholar. When we met him in chapter 7, he was the chief meteorologist of South Africa, defending farmers' perceptions of rainfall against the earlier pronouncements of his fellow experts. He was also a techno-optimist in the manner of Schwarz himself, as evidenced by his scheme, profiled in *Time* magazine, for harvesting water from Table Mountain. Now, in 1963, Schumann was writing as a purveyor of transnational white supremacy. He lashed out at a changing world that was condemning South Africa—a situation made possible by the fact that European colonial powers had "abdicated" their responsibility by allowing their African colonies to become independent under Black majority rule.

Schumann's intransigence in the face of international opprobrium was not unique. In the wake of the Sharpeville Massacre and the economic panic brought on by international condemnation, the National Party government abruptly announced that it would build a massive scheme on the Orange River, South Africa's largest river. Prior to the 1960s, the Orange River had been utilized on a small scale by a few irrigation settlements, such as Kakamas, and by mavericks such as Carl Weidner. Its potential for irrigation was limited by the mountainous terrain around it. In 1928 the irrigation director had suggested that Orange River water could be piped through tunnels to the Eastern Cape to help struggling farmers along the Sundays River, who had failed in their attempts to acquire artificial rain. In subsequent decades, farmers along both the Sundays and Orange Rivers continued to clamor for dams and irrigation works, arguing that the rivers could be the "salvation" of South Africa.[38] Throughout the 1950s, the National Party government rebuffed the farmers' requests, on the grounds that the project would be expensive and technically difficult. The government's about-face in 1960 revealed that white supremacy, not economic logic, was driving the project. Schwarz supporters had insisted that cost should be no object in considering the social benefits of his scheme; now, the state took the same attitude toward the Orange River Project, which would increase South Africa's irrigated land by 40 percent at a time when overproduction was already depressing crop prices. White farmers growing staples such as maize on government irrigation schemes already had 80 percent of their costs subsidized by the state.[39] The early twentieth-century experts who had argued that white farming needed to proceed on economic grounds had lost the debate.

In truly Schwarzian language, the government argued that the Orange River Development Project would "transform the desert into a paradise."[40] Its prime agricultural beneficiaries would be Cape farmers who were usually numbered among the National Party's opponents. Drawing on long-standing fears of the depopulation of the interior, supporters of the project argued that the Cape province "was traditionally the white man's country, and the north-west Cape his hinterland. Inhabited by hardened, proud, freedom-loving farmers, and comparatively free of Africans, it held great possibilities of development and could not afford to lose its white population."[41] The prime minister, playing to these same fears, promised that the project would increase the white rural population. People living within the Orange River's watershed were also offered benefits: flood control, a large reservoir and hydroelectric scheme, and water for towns and cities. But concern over white rural depopulation meant that most of the water harnessed by the Orange River Project would be devoted to agriculture, though water devoted to farming yielded a fraction of the economic return generated when the same amount of water was directed toward mining and industry.[42]

In 1963, experts attending the annual meeting of the South African Association for the Advancement of Science raised concerns about the project. They had relatively few technical objections—not least because the project had been designed so quickly that its flaws would not become apparent until construction was underway. Rather, most of the speakers were concerned about plans to spend so much money to expand an already unprofitable, highly subsidized sector and to add a few thousand rural whites to the population when the future of the white population and the economy clearly lay in the cities. The state addressed the technical concerns and ignored the rest.[43]

The project cost far more than its original estimates, but it was immensely popular. It was also so ambitious that, as one study notes, "it exceeded even the dreams of its most fervent proponents."[44] The dam was completed in 1972 (figure 8.2). When a fifty-two-mile tunnel linking the Orange and Fish Rivers was finally opened in 1975, it was the largest of its kind in the world. In its grandiosity and its symbolism, the project was a realization of the future imaginaries that animated support for Schwarz's scheme.

A newly emboldened South Africa also turned its attention back to the rivers that the Schwarz Kalahari Thirstland Redemption Society had dreamed of transforming. In the late 1960s and early 1970s, it would participate in the engineering of the Kunene and the Zambezi rivers—not

Figure 8.2 The Gariep Dam, formerly the Hendrik Verwoerd Dam, was the centerpiece of the Orange River Development Project. Conceived in the 1950s and 1960s and completed in 1972, the dam's primary purpose was irrigation, although it also provided water and electricity for industry and urban areas. Photo by Aliwal2012. CC license.

for climate control, but for hydropower and the security of white supremacy in a rapidly changing world. The Katombora diversion scheme on the Zambezi became impossible when Southern Rhodesia completed the Kariba Dam in 1959. But South Africa attached itself to another Zambezi dam scheme. In 1969 it agreed to help Portugal fight an African liberation movement in Mozambique in exchange for most of the electrical power generated by the proposed Cahora Bassa dam, downstream from Kariba. The dam and its lake, designed partly to slow the liberation movement, were "a security project masked as a development initiative," in the words of the historian Allen Isaacman.[45]

At the same time, South Africa tightened its claim over South West Africa, in violation of United Nations demands for the colony's immediate independence. In response, an armed liberation movement formed in the northern part of the colony in 1966. The Cuvelai plain, where Schwarz had photographed the remains of famine victims and imagined a land of water and fertility, became a war zone. The Kunene River, which he had planned to divert to Etosha in order to improve the rainfall, was put to other purposes. Portugal granted South Africa construction rights along the river, and surveyors again came into the region to assess the Kunene's potential to save white society. In the 1970s, South Africans constructed a series of dams along the Kunene to regulate its flow, generate hydropower, and pump water into a network of canals.[46] Electrical transmission lines carried most of the power to the white-dominated lands to the south. Ovamboland, although much closer to the river, had very little electricity infrastructure into the early 1990s.

The first phase of water diversion from the Kunene brought much-needed water to the Cuvelai plain, though the open, cement-lined canals never came close to meeting demand. Later phases of the project, which were never built, involved running water through a pipeline to white towns and industries. As an analysis of the scheme noted at the time, even the infrastructure for Ovamboland was meant to benefit the white minority by stabilizing the supply of cheap migrant labor. The Portuguese prime minister complained that international opponents of the scheme were criticizing South Africa and Portugal for bringing development to "the nearly empty lands of the African continent."[47] It was water that would make it possible to populate and develop those purportedly empty spaces. Water, both within and beyond South Africa, remained crucial to imaginaries of a white man's land in Africa, and plans for its use remained tied to white racial anxieties and aspirations.

EPILOGUE

This book has argued that the multiple ways in which whites imagined their futures mattered to history, even when those futures did not come to pass. Paying attention to the environmental fears of whites in early twentieth-century South Africa and exploring how those fears were entangled with well documented racial fears does not invalidate other, older explanations for the entrenchment of apartheid. Increasingly radical moves toward consolidating white supremacy had many causes, including tensions around urbanization and capitalism and, especially, the growing African resistance to segregation. But in the quest to secure white dominance, the variety of social imaginaries—including territorial expansion, racial partition, and turning deserts into rain-rich lands—helps to explain the path that dominance took as it consolidated. Apartheid was itself once an imaginary, albeit one that eventually had brutally real consequences. The idea that a society could be perfectly engineered along racial lines is no less ludicrous than the idea that an arid land could be engineered into the home of millions of yeoman farmers. Neither future came to pass, but both nurtured intellectual currents that shaped the conditions of possibility in the 1950s and beyond.

Over the course of this book, readers may have wondered what the "real" story of rivers and climate is in southern Africa. And they may have observed that a story of disdain for expert consensus on climate, of enthusiasm for techno-utopian schemes, and of an embrace of populist white nationalism has some apparent parallels a century later. This book concludes by briefly addressing both those points.

Schwarz's vision of what these river systems looked like in the past was fundamentally correct. The topography of the southern third of the African continent, with its high interior plateau and coastal ranges that drop sharply toward the ocean, lends itself to the kind of river capture that Schwarz described. Once upon a time, the Kunene River did flow into Etosha Pan, which was a much larger but shallow lake. However, Schwarz's chronology was off by at least two million years.[1] Much more recently, the upper Zambezi River flowed southward into an enormous "paleo-lake" that encompassed Lake Ngami, the Makgadikgadi Pan, and other low-lying areas. At some point the lake overtopped its banks, pouring into the Zambezi River at Katombora—the place Schwarz supporters had identified for diverting the river into the Kalahari. The overtopping caused a catastrophic flood, and the paleo-lake emptied. This process was much more recent than the diversion of the Kunene; it happened perhaps 150,000 years ago. Humans would have witnessed these dramatic changes, and perhaps would even have been caught in the flood.[2]

As for the climate, analyses of daily rainfall records demonstrate that at least since 1921, parts of South Africa have experienced more episodes of very intense rainfall, which are consistent with the claims of Schwarz's supporters. While there is currently no definitive evidence that rainfall was progressively declining in South Africa in the first half of the twentieth century, some studies indicate that a progressive decline in rainfall began in the western half of the country by the latter part of the century. Scientists assume that this is not part of a cyclical variation, but indeed a harbinger of the future. Southern Africa is considered a climate change hotspot, and models suggest that precipitation will decline further as a result of human activity.[3] The assertions of some South Africans in the early twentieth century, even if they are not borne out by the scientific data we have now, may well prove to be correct for the current century. But the reliance on models—many of which do not fare well when tested against conditions in the recent past—underscores that even today, experts struggle to quantify how much any given natural disaster is the result of global climate change and how much it is simply part of the natural climatic variability one might expect. South Africans faced much the same dilemma a hundred years ago, but without data-packed climate models.

In addition, recent work on range ecology in southern Africa suggests that another claim white farmers made, about the relationship between rain and the health of local ecologies, also has some merit. While many experts blamed farmers for land degradation, the farmers insisted that it was rain, not good management, that was lacking, and they cited environmental changes even in places beyond the zone of agriculture. In southern Africa, scholars working in communal and white commercial farming areas, as well as in the Namib Desert, have argued that in fact, precipitation is the major determinant of vegetation cover in these arid environments, in many cases dwarfing the impact of humans.[4]

Fears that South Africa is "drying out" have taken on a new meanings and a contemporary urgency in recent years. In early 2018, the world waited to see whether Cape Town would become the world's first major city to run out of water. "Day Zero" never came, though the city's water supply remains vulnerable to drought. Meanwhile, Schwarz's home city of Grahamstown, renamed Makhanda in 2018, struggles with persistent, citywide water outages made worse by the rolling blackouts that have recently put South Africa in the world news again. These are problems driven partly by climate, but also by the legacies of apartheid-era infrastructural inequalities and by poor planning and state corruption in the decades since apartheid ended.

When we look at global debates over climate change today, we see that the positions of experts and their skeptics are reversed from South Africa's debate a century ago. Now it is scientists who assert that there is consensus on the reality of climate change, and lay people who challenge them. This is, once again, not a debate between those who identify as proscience and those who identify as antiscience. As a recent book argues, many US climate-change skeptics are aggrieved not because they dismiss the validity of science but because they feel scientists deny them full—and equal—"participation in public debate." Like Schwarz and his supporters, they perceive themselves as "a group of misunderstood truth seekers who are undervalued and persecuted by society at large."[5]

Tellingly, however, the role of white innocence follows the same lines that it did in Schwarz's day. Experts insist that people are responsible: that rich countries—mainly Europe and its former settler colonies, especially the United States—set contemporary climate change in motion. Skeptics argue either that there is no progressive change or, more recently, that any change is due to "natural," nonhuman causes—or, still more recently, that China and India (notably, "nonwhite" countries) are to blame. The role of techno-optimism in these divisions is also similar. In Schwarz's day, lay people embraced techno-optimist schemes to "fix" South Africa's climate and arid environment, insisting that those schemes could solve a range of problems faced by white society. Today, geoengineering is embraced by "visioneer" billionaires and right-wing activists who promote climate denialism, at least in the sense of human-caused climate change. In his 2014 study of contemporary climate-engineering initiatives, Clive Hamilton argues that a commitment to denying the human role in climate change and an embrace of engineering-based solutions is not merely about money; it is also about different understandings of the relationship between humans and the nonhuman world. "For some, instead of global warming's being proof of human failure, engineering the climate would represent the triumph of human ingenuity."[6] Similarly, Schwarz's supporters argued that "drying out" was not the result of human failure, and that his scheme would demonstrate that white men controlled their own destiny. As in an earlier era, virtually all of the modern visioneers are white men, suggesting that race and masculinity remain deeply entangled in this way of seeing the world.

The idea of moral hazard also features in these parallel stories of climate change debates. In 1948, the irrigation engineer Daniel Kokot stated openly what his fellow experts had only hinted at: "So long as the farmers of South Africa believe that the drought problem can be solved by the easy method of carrying out the Schwarz scheme, they will not give due attention to the real causes of desiccation, and to those remedial measures which may be carried out on practically every farm throughout the land."[7] Similarly, most climate scientists have opposed geoengineering research on the grounds that it would reduce incentives to reduce emissions of greenhouses gases, a concern validated by the fact that fossil fuel companies have occasionally suggested that geoengineering could substitute for reducing CO_2 emissions.[8]

But today, some scientists are beginning to embrace geoengineering in a way that was never embraced by Schwarz's scientific peers. In 2021, the National Academies of Science, Engineering, and Medicine (NASEM) recommended that the US government create a solar geoengineering research program. The University of Oxford, the University of Chicago, and Harvard University, among other elite institutions, already have centers dedicated to researching geoengineering technologies. Proposed techniques would use technologies similar to those that white men in South Africa proposed, but on a grander scale: launching particles into the atmosphere, using mirrors to direct sunlight, tinkering with the oceans, and covering the land with some sort of sheeting. Meanwhile, some international NGOs have suggested that these new technologies could be tested in Africa, on the argument that they could protect the poorest and most vulnerable countries from the impacts of climate change.[9]

Critics of these initiatives have noted that, as in South Africa a century ago, there is little thought to who will bear the environmental risks of geoengineering. The NASEM report "acknowledges that some types of uncertainties can only be reduced through 'actual deployment'"—a position strikingly similar to that of Schwarz supporters who suggested that it couldn't hurt to try his scheme, rendering the risks to Black residents of the affected regions invisible or irrelevant.[10] In the words of one African climate scholar, modern-day geoengineering schemes echo "some of the worst aspects of colonialism," making Africa and other parts of the Global South "a test case for an unproven technology."[11]

Finally, the early twenty-first century has seen a resurgence of white nationalism, driven in part by fears of demographic change that are remarkably similar to those expressed by many white South Africans. From conspiracy theories about "white genocide" in South Africa to a "Great

Replacement Theory" that circulates globally, populist leaders and grassroots communities in the twenty-first century have embraced and elevated white racial anxieties, with deadly consequences around the world.

What are we to make of the fact that the politics of racial anxiety, scientific knowledge, and climate change seem so intertwined both in early-twentieth-century South Africa and in the world's European and settler colonial societies a century later? I have argued at the beginning of this book that both race and the environment were seen as existential issues for whites in South Africa, who perceived human and nonhuman threats to their survival in similar terms and timescales. Both of these threats were grounded in perceptions of scarcity: of not enough white people, not enough rain, not enough fertile land. Scientists, who denied that such scarcity existed, were also a threat—to people's deeply held understanding of their experiences in the world. Hence the charge that they were "theoretical men" completely out of touch with "practical men." It was only when experts came to validate those experiences that popular and expert knowledge met on common ground.

This is, of course, not an argument for catering to racist and environmental worldviews that jeopardize the well-being of the planet and the people on it. But two features of the debate over the Schwarz scheme complicate this story and perhaps offer it as a cautionary tale. The first is that scientists, too, were engaging in an identity politics of knowledge, insisting on the validity of their claims well beyond what the evidence would support. The second is that populist and scientific knowledge were not entirely distinct. They shared a number of fundamental assumptions about the world: that whites should naturally dominate, that Indigenous knowledge was inferior to white knowledge, and that whites had come to an empty land at the same time as the Black majority had.

Scholars, journalists, and much of the general public struggle to explain people's loyalty to messages peddled by "merchants of doubt" in the face of facts that are easily verified. This book does not offer all the answers. But it suggests that we should pay attention to the imaginaries that underlie people's political commitments, and explore connections between seemingly unrelated beliefs to see what forms of popular knowledge and visions of the future reveal themselves.

ACKNOWLEDGMENTS

The longer a project takes, the longer the list of people to thank becomes. This book took far too long to finish, and I have incurred a proportionately long list of debts. Its long germination also means that I have surely overlooked some of those who helped me along the way, and I apologize to anyone not named in these pages.

The seeds of this book were planted during a long-ago stint at the Woodrow Wilson International Center for Scholars, where I had the luxury of time to delve into the histories of countries and topics that were new to me. An overlapping National Endowment for the Humanities fellowship gave me the opportunity to begin research in Botswana, where I found the first traces of Schwarz, and in Namibia, where he popped up again. Subsequent grants from Georgetown University funded return travel as well as multiple visits to archives in South Africa, the United Kingdom, Germany, and the United States.

The raising of three children and a side adventure in farming meant slow progress on any academic project that required sustained concentration over long periods. But they offered other lessons and compensations. Aubrey, Jack, and Tobias saved me from any workaholic inclinations I may have had. An Appalachian valley turned out to be an inspiring place to teach myself the field of environmental history. The process of coaxing life from the earth and waiting for rain—not to mention the many things that could and did go wrong on our farm—gave me a measure of empathy that was otherwise often difficult to muster for these particular historical subjects. And my rural community, where my surname was enough to mark me as an outsider, offered me a real-world education on how distrust of elites and ideas about race and climate were entangled.

As I traced Schwarz and his acolytes through the historical record, I monopolized far more than my share of the time of Georgetown research librarians Maura Seale and Francesca Kang and their interlibrary loan colleagues. I'm grateful to them for all their help, and hope the interlibrary loan budget is recovering now that this book is done. I also benefited from the time and knowledge of staff at multiple archives and libraries, who were remarkably patient with a researcher who always seemed to show up with limited time and an iPhone in hand: the National Archives of Namibia, the National Archives of South Africa, the Western Cape and Provincial Records Service, the Grootfontein College of Agriculture, the Albany Museum, and the National Archives of the United Kingdom, as well as the National Library of Namibia, the National Library of South Africa in Cape Town and in Pretoria, Stellenbosch University, the University of the Free State, Goethe University Frankfurt, the New York Public Library, and the US National Agricultural Library. Frikkie Mouton gave me a tour of Ferdinand Gessert's farm; and the Catholic sisters at Pella housed me, fed me dates, and arranged to take me to the ruins of Carl Weidner's Orange River outpost. The Owen-Collett family kindly hosted me in Cape Town, and regaled me with tales of the family farm and Charles Hall's rainmaking adventures. Khadija Khan digitized the Alex du Toit papers for me while she studied abroad in Cape Town. Kasonde Mukonde took time from his PhD work at Wits to survey William Scully's papers and, while on his own research trip, searched for traces of Schwarz in the Rhodes University archives. Bernie Moore, Rob Gordon, and Julia Tischler were generous with their archival sources. Wesel Visser obtained copies of government reports from South Africa, and Dag Henrichsen's vast archival knowledge has pointed me in the right direction many times. Robert Raderschatt provided additional information on Hermann Korn and Henno Martin, and Paul Shaw kindly shared a copy of the unpublished Jeffares report.

It took me a while to figure out the historical significance of Schwarz's scheme, and I had a lot of help along the way. The work and comments of colleagues at the South African Empire Conference in Basel in 2013 and at the Global Deserts conference at the University of Arizona in 2015 shaped the direction of this project, as did feedback from colleagues at the Northeastern Workshop on Southern Africa. Michael Schnegg and Julia Pauli gave me an opportunity to present the almost-finished product at the Ethnology Institute's colloquium at Hamburg University. I've also benefited from the ideas and feedback of Kate de Luna, Nancy Jacobs, Thom McClendon, Bernie Moore, Kathy Olesko, Neil Roos, Sandra Swart, and Julia Tischler. Anonymous reviewers of related work published in the *Journal of Southern African Studies*, *Environmental History*, and *History of Meteorology* also helped me develop some of my ideas.

The first half of this book was written during a year at the Rachel Carson Center in Munich, where I was surrounded by an impressive range of scholars working in the environmental humanities. I'm grateful to Christof Mauch for providing me with the time and space to focus on writing, to Arielle

Helmick for helping with the details of moving a human and canine family overseas, and to my officemates Ruth Morgan, Anna Pilz, Julia Tischler, and Monica Vasile for their good company. Anna organized our morning writing group, Julia offered perceptive thoughts on the book's main arguments, and Ruth answered questions about Australian parallels. Many fellows participated in a seminar where I presented an early version of chapters 2 and 3, and I'm particularly grateful to Astrid Bracke, Floor Haalboom, and Rory Hill for their detailed comments.

The second half of this book was written in circumstances that were diametrically opposite to the above, during the general upheaval and isolation of the COVID-19 pandemic. The Zoom writing group organized by Michelle Wang connected me to academic life beyond my home office, and got me to resume writing in between homeschooling stints. I'm grateful to Georgetown University for a sabbatical that let me finish the manuscript and for the subsequent research fellowship that gave me time to revise it, and to Carole Sargent for helping me navigate the publication process. At the University of Chicago Press, Karen Merikangas Darling, the series editors Adrian Johns and Joanna Radin, and two anonymous readers saw value in the book and, when I was lost in the weeds, sometimes comprehended its larger significance better than I did. I'm grateful for the time they devoted to making it better.

The collegiality of Georgetown's History Department is a tremendous asset to a scholar moving well beyond her comfort zone. Feedback in a department seminar helped me shape this project in its early stages. My faculty and graduate student colleagues have talked to me about how their fields approach topics that paralleled mine, pointed me to archives, and answered obscure queries about the parts of the world they research. John McNeill read an early version of the project's prospectus, and his comments made it better. Kathy Olesko found biographical information on every German scientist I couldn't trace. Chandra Manning talked to me about race, citizenship, and yeoman farming in the US South. Dagomar Degroot told me about the link between Mars and desiccation debates. Katie Benton-Cohen's friendship and her no-nonsense advice to just finish the damn thing already have been invaluable. Kate de Luna is an endless font of creativity and wisdom about all things historical—not to mention parenting, cooking, gardening, and teaching. Carpooling and consuming wine (not at the same time) have never been more productive. Matt Johnson went looking for traces of Fournier in Brazilian archives, and Dale Menezes helped with Portuguese and Afrikaans translations. Matthew La Lime, Tracy Mensah, and Trishula Patel kept me thinking about race, migration, land, and belonging. When I was wrestling with the epilogue, members of the Georgetown Environmental History Workshop gave generously of their time and intellects to offer me feedback even as they were racing to finish their own semesters' work. Thank you to Bryna Cameron-Steinke, Francisco Centola, Silvia Danielak, Yuan Gao, John McNeill, Jackson Perry, Andrew Ross, and Rachel Singer.

It has sometimes been hard writing a book about distrust of scientific expertise, the consolidation of white supremacy, and fears of a climate apocalypse while the world around me is riven by culture wars over scientific expertise, hosts a newly emboldened white supremacy, and displays ever-clearer evidence that climate change is reshaping our planet. The community of friends and family around me is the sanctuary amid this turbulence. Chris, Lera, Clare, Erica, Katie, Andrea, Meredith, Jill, Pam, Pilar, and all the good people in each of their orbits have kept me grounded, laughing, and sane (more or less). Leta Brown passed away as this book was entering production, and the world feels emptier without her friendship and her Tennessee drawl. Thanks to each of them—and to my family, who enrich my life every single day. Aubrey, Jack, and Tobias grew up alongside this project. On the research trip where I discovered the Schwarz scheme, Aubrey lost her first tooth and befriended baby goats while Jack ecstatically drove toy tractors in the world's largest sea of sand. Now they are young adults finding their own paths in the world. Tobias was not yet born when I started this book; now he has his own informed opinions about its dominant themes. The weekend after I completed it, we rode waterslides together at the Kalahari Sands waterpark, a reminder that the Kalahari continues to be a potent—and infinitely flexible!—imaginary today.

No aspect of my very fortunate existence would be possible without Don, my partner for half my life. He almost certainly had no idea what he was getting into when he married an academic. He has single-parented when I traveled for work, followed me to Munich for a year, tolerated my grim musings about the environmental cost of the avocado that just went bad, and used his talents to enrich our shared life in thousands of ways large and small. Don has endured Schwarz's presence in our lives for a very long time. I hope this book justifies the intrusion.

NOTES

A note on sources: The Albany Museum, part of Rhodes University in Grahamstown, South Africa, contains a small collection of Schwarz's papers that was crucial to the writing of this book. The collection, however, reflects Schwarz's preference for big ideas over tedious detail. His field notebooks have no particular organization. They are filled with doodles, personal observations, mysterious columns of numbers, transcribed quotations from German and English publications, and even the partial draft of a novel. Schwarz carefully tracked media coverage of his scheme, and his two large scrapbooks of newspaper clippings are invaluable since they represent more than two dozen publications. But they are an organizational nightmare. He pasted articles on the pages in no particular order, often without noting their publication dates or even the titles of the newspapers. When this information is missing and I have been unable to trace it, I have noted in the endnote that the article comes from Schwarz's personal papers.

Introduction

1. "Remedy for Droughts: Our Lost Lakes: How They Can Be Restored," *Star*, 30 January 1918.
2. "Remedy for Droughts," *Star*, 30 January 1918.
3. Schwarz, "Desiccation of Africa."
4. The term is from McCray, *Visioneers*. While the United States led the world at this time in water engineering and "reclamation" schemes, there were no major proposals to change the climate.
5. Engerman, "Introduction"; Gordin et al., *Utopia/Dystopia*, 10–11.
6. Lake and Reynolds, *Drawing the Global Colour Line*.
7. Stoler, *Along the Archival Grain*, 21.
8. Coen, *Climate in Motion*, 8–9. Among the excellent studies of popular environmental and racial knowledge in white settler colonies, most deal with the pre-twentieth-century world, perhaps because of this assumption of encroaching scientific universalism. See, for example, Valencius, *Health of the Country*.
9. Geary et al., *Global White Nationalism*.
10. This often appears in the context of recovering Indigenous agency and arguing for the influence of Indigenous knowledge on colonial knowledge. But see also Weisiger, *Dreaming of Sheep*, for an excellent account of parallel Indigenous and expert narratives of the same dryland environment.

11. Conway and Oreskes, *Merchants of Doubt*; Doel, "Comments," 21. Work from South Africa has a similar elitist orientation.

12. Coen, *Climate in Motion*; Beattie and Morgan, "Engineering Edens"; Davis, *Arid Lands*; Beattie et al., *Eco-Cultural Networks*; Tilley, *Africa as Living Laboratory*; D'Souza, "Water in British India"; Beinart, *Rise of Conservation*; Phillips, "Lessons from the Dust Bowl"; Dunlap, *Nature and the English Diaspora*; Grove, *Green Imperialism*.

13. Beattie et al., *Eco-Cultural Networks*; Veracini, "Imagined Geographies"; Ballantyne, *Orientalism and Race*; Zimmerman, *Alabama in Africa*.

14. Indeed, they are part of the story of the Kalahari Thirstland Redemption Scheme. Some of Schwarz's most prominent critics had themselves been educated in the United States or had worked on irrigation projects in India. And the scheme itself, though much criticized by these global scientists, had its roots in imperial enthusiasm for moving water across the landscape and between river systems.

15. Lester, "Imperial Circuits," 135.

16. Du Bois, "Souls of White Folk," 924.

17. Lake and Reynolds, *Drawing the Global Color Line*; Stovall, *White Freedom*; Lake, *Progressive New World*; Powell, *Vanishing America*; Tischler, "Cultivating Race." Unlike most of this work, Tischler does not focus exclusively on Anglos, but includes Afrikaners and Africans in her transnational networks.

18. More recently, there has been a shift toward exploring the rural aspects of this global color line. Tischler, in "Cultivating Race," argues that the progressive project of rural agrarian reform also constructed transnational racial identities.

19. Powell, *Vanishing America*, 11.

20. Van Sittert, in "Supernatural State," argues for recognizing "the irrationality embedded in European epistemologies," including scientific ones (931).

21. Swart, "'Bushveld Magic.'"

22. McKittrick, "Race and Rainmaking."

23. Crosby, *Ecological Imperialism*. For a sampling of the literature that shows how European perceptions and environmental conditions shaped the contours of settler colonialism, see Zilberstein, *Temperate Empire*; Davis, *Resurrecting the Granary of Rome*; Dunlap, *Nature and the English Diaspora*.

24. Belich, *Replenishing the Earth*.

25. This is not a one-time event—in the form of genocidal violence, for example—but an ongoing process (in Patrick Wolfe's much-cited formulation). The very success of these projects of elimination obscures, for many people, their settler colonial origins. Not all US historians see themselves as scholars of settler colonialism; in countries like the United States, Canada, and Australia, a subset of scholars has had to insist on making visible the exterminationist origins of these societies.

26. Veracini, *Settler Colonialism*, especially 35–49.

27. Veracini, "'Settler Colonialism': Career of a Concept," 315.

28. Bowman, "Pioneer Fringe," 50–51.

29. Lave et al., "Arid Lands," 35.

30. Veracini, "Imagined Geographies," 183.

31. Belich, *Replenishing the Earth*, 339, quoting Wrobel, *Promised Lands*, 28.

32. Reisner, *Cadillac Desert*, 35–39; Kutzleb, "Rain Follows the Plow."

33. Dunlap, *Nature and the English Diaspora*, 175–78.

34. Worster, *Rivers of Empire*, 150.

35. Quotation in Lake, "White Man's Country," 351.

36. Schwarz, *White Man's World*, 11.

37. Wooding, "Populate, Parch, and Panic," 60.

38. Stoler, *Along the Archival Grain*, 22.

39. For example, Fleming, *Historical Perspectives on Climate*; Coen, *Climate in Motion*; Fleming, *Inventing Atmospheric Science*; Edwards, *Vast Machine*; Weart, *Discovery of Global Warming*; Stehr, ed., *Eduard Brückner*.

40. Fleming, *Historical Perspectives*; Fleming, *Fixing the Sky*; Zilberstein, *Temperate Empire*; Valencius, *Health of the Country*; Douglas, "'For the Sake of a Little Grass'"; Anderson, *Predicting the Weather*. Exceptions that do span the 1900 dividing line are Beattie, "Science, Religion, and Drought"; and Legg, "Debating the Climatological Role."

41. Fleming, "Civilization, Climate, and Ozone," 229–30.

42. In 1946 the director of irrigation observed that attempts to refute the idea that South Africa was progressively drying out "appear merely to have developed a measure of obstinacy in the public mind." MacKenzie, *Report on the Kalahari Expedition 1945*, 1.

43. Edwards, *Vast Machine*, xiv.

44. South Africa, *Report from the Select Committee on Droughts, Rainfall, and Soil Erosion*, 41–42.

45. Sayre, in *Politics of Scale*, notes how ecological variability "confounded the needs and expectations of the state and capital alike," and limited the predictive power of science (24–27). For attempts to establish rainfall cycles, see O'Gorman, "'Soothsaying' or 'Science'?"; also Stehr, ed., *Eduard Brückner*, ch. 8; Hutchins, *Cycles of Drought*.

46. Coen, *Climate in Motion*, 7.

47. For parallel processes outside of Western European colonization, see Maya Peterson, *Pipe Dreams: Water and Empire in Central Asia's Aral Sea Basin* (New York: Oxford University Press, 2019).

48. Bowman, "Pioneer Fringe," 62–63.

49. McDermott Hughes, *Whiteness in Zimbabwe*, xii; Foster, *Washed with Sun*, 227.

50. Koselleck, *Futures Past*.

51. Jasanoff, "Future Imperfect," 4.

52. Lehmann, "Infinite Power"; Goswami, "Imaginary Futures"; McCray, *Visioneers*, 22–23.

53. Connelly et al., "Forecasts, Future Scenarios," 1434.

54. Lehmann, "Infinite Power," 99.

55. Schwarz, *Kalahari*.

56. The word would have been familiar to virtually all whites in the form of the "Dorsland trekkers"—several groups of Afrikaners who left the Transvaal in the 1870s, crossing the "Thirstland" of modern-day Botswana and Namibia and enduring heavy casualties before ultimately settling in southern Angola. But the term appeared in written records by 1840. In an agreement delimiting the bounds of the state led by the Boer leader Andres Potgieter that year, the reference to lands stretching northwest to the Dorsland—Thirstland—does not define the term, thus indicating that it was in common usage at that point. See *South African Archival Records*, 351.

57. Mauldin, *Unredeemed Land*.

58. Schwarz, *The Kalahari*, 152. Bryce, in *Impressions of South Africa*, states that even "the most competent specialists" thought the gold seams would continue to produce for fifty years, or "possibly even for seventy or eighty . . ." (301–2).

59. Krikler, "Re-Thinking Race and Class," 144. See also Koorts, "'Black Peril.'"

60. Quoted in Krikler, "Re-Thinking Race and Class," 147.

61. Roos, "South African History," 127.

62. Hyslop, "Imperial Working Class"; Krikler, *White Rising*; Morrell, *From Boys to Gentlemen*.

63. Beningfield, *Frightened Land*, 76–77. For other work that deals with rural whites and the creation of the racial state, see Higginson, *Collective Violence*; and van der Watt, "'It Is Drought.'"

64. Dubow, *Racial Segregation*, 31; Koorts, "'Black Peril.'"

65. Coetzee, *White Writing*, 5; Beningfield, *Frightened Land*, 90–98.

66. Hoffman et al., "Land Degradation."

67. Mukerji, *Impossible Engineering*, 37. Weiner, in *Little Corner of Freedom*, 416, notes that the Siberian river diversion scheme had a similarly abstract quality: as long as it was in the "realm of fantasy," it wasn't criticized.

68. Hyam, *Failure*, 104.

69. The League of Nations mandates were established after World War I as the basis for administering the former colonies of Germany and the Ottoman Empire. All mandates were supposed to be administered as a "sacred trust of civilization." Class C territories were regarded as the least developed, and were to be administered as integral parts of the governing territory.

70. Schwarz, *Kalahari*, 157.

Chapter 1

1. Barnard, *Encountering Adamastor*, 48.

2. T. G. Bonney, obituary; Prof. J. W. Judd, *Nature*, 9 March 1916, pp. 37–38.

3. Darwin, *On the Structure*, 270.

4. Dalrymple, *Age of the Earth*, 38.

5. Dalrymple, *Age of the Earth*, 13.

6. Information from the 1896 *Register of the Associate and Old Students of the Royal College of Chemistry, the Royal School of Mines, and the Royal College of Science*, courtesy of Bryony Hooper, assistant archivist and records manager, Imperial College London.

7. Beinart, *Twentieth-Century South Africa*, 28.

8. Bryce, *Impressions*, 304–5.

9. 1896 Selborne Memorandum, quoted in Ross et al., *Cambridge History Vol. 2*, 94.

10. Letter confirming membership, 30 October 1895, Schwarz papers, Albany Museum. The South African Association for the Advancement of Science, where Schwarz would later introduce the Kalahari Scheme to his peers, was created in 1903. Barnhard, *Encountering Adamastor*, 20. Dubow, *Commonwealth of Knowledge*, details the creation of South Africa's "scientific societies."

11. Cape of Good Hope, *First Annual Report*, 2.

12. Price, *Catalogue of Publications*; Corstorphine biography, http://www.s2a3.org.za/bio/Biograph_final.php?serial=584, accessed 18 October 2023. The two men had very different dispositions. Rogers was careful and cautious; Schwarz impulsive and

almost compulsively curious. http://www.s2a3.org.za/bio/Biograph_final.php?serial=2385, accessed 18 October 2023.

13. Belich, *Replenishing the Earth*, ch. 3.

14. Hall, *Drought and Irrigation*, 3–4; Cunniff, "The Great Drought," 78–94.

15. Davis, *Arid Lands*, 79.

16. Drylands can, of course, have groundwater that bubbles up in springs. These might feed lakes or small rivers, but they also have high rates of evaporation loss.

17. Quoted in Davis, *Arid Lands*, 132–33.

18. Sweeney, "Wither the Fruited Plain.'"

19. Widney, "Colorado Desert," 44. Widney is best known as the first president of the University of Southern California.

20. Linton, *What is Water?* 123–24.

21. Gregory, *Dead Heart*, 12; Limerick, *Desert Passages*, 29.

22. Quoted in Towner, "Lake Eyre," 68.

23. Rennell, *Geographical System*, 667. Nineteenth-century views of a previously wet Sahara are explored in Heffernan, "Bringing the Desert to Bloom."

24. Blanford, "On the Nature," 498. Blanford was a naturalist and geologist, and brother to the head of the Indian Meteorology Department.

25. Henry Hubert, quoted in Martel, *Nouveau Traité*, 735.

26. Adhikari, *Anatomy*; Penn, *Forgotten Frontier*.

27. For the mapping of these rivers, see McKittrick, "Making Rain."

28. Brown, *Hydrology*, 212.

29. Resident commissioner Ovamboland to secretary for the protectorate, 27 February 1918, NAN, SWAA vol. 598, folder A66/1.

30. Quoted in Grove, "Scottish Missionaries," 166.

31. Bernhard Schwarz, quoted in Kienetz, "Nineteenth-Century South West Africa," 582.

32. Scully, *Between Sun and Sand*, 5. Scully was only posted to the region for three years in a long career, but his experience made a deep impression on him.

33. Schwarz, diary entry (journey to South West Africa), 30 November 1917, Schwarz papers, Albany Museum.

34. Livingstone, *Missionary Travels*, 75.

35. Gordon-Cumming, *Five Years*, 234.

36. Livingstone, *Missionary Travels*, 75–76.

37. VanderPost, "Early Maps," 203.

38. Potten, "Aspects," 64–65.

39. Because they form on dry land, inland deltas are more correctly termed alluvial fans.

40. Schulz and Hammar, *New Africa*, 251.

41. Von François, "Bericht," 212.

42. Hauptmann Friedrich von Erckert, "In der Kalahari," *Deutsche-Südwestafrikanischen Zeitung*, 5 December 1906, p. 5.

43. Quoted in Endfield and Nash, "Drought, Desiccation," 39–40.

44. Brown, *Hydrology*, 27.

45. William Charles Scully, "The Desiccation of South Africa Part 3," *Cape Times*, 14 September 1922. A very similar point is made in an unattributed article in the *Cape Argus*, 10 August 1922; Brown, *Hydrology*, 94–95.

46. Livingstone, *Missionary Travels*, 76, 79–80.

47. Schinz, *Deutsch-Südwest-Afrika*, 435.

48. Passarge, *Kalahari*, 103. Passarge gave scientific respectability to the narrative of regional desiccation, and introduced it to an international audience. See Barnard, *Encountering Adamastor*.

49. Typescript of March 1918 Senate debate, pp. 7, 20. Schwarz papers, Albany Museum.

50. Schwarz, *Kalahari*, 98; for details on these travel accounts, see McKittrick, "Making Rain."

51. South Africa, *Debates of the House of Assembly*, 3 (1925): 1975.

52. The Orange River is today more commonly known as the Gariep—the Afrikaans version of the Khoekhoegowab name, !Garib.

53. C. E. Stewart, letter to the editor, *Northern News*, 14 October 1922. See also MacDonald, *Conquest of the Desert*, 14; and letter from W W. M. Percy Fraser, *Farmer's Weekly*, 3 September 1919, p. 63.

54. South Africa, *Debates of the House of Assembly* 3 (1925): 1982.

55. F. H. Barber, letter to the editor, *Agricultural Journal of the Cape of Good Hope* 36 (1910): 167.

56. Schwarz, "Kalahari Project," 305.

57. MacDonald, *Conquest of the Desert*, 64.

58. William Jacobs, letter to the editor, *Farmer's Weekly*, 10 October 1923, p. 457.

59. Schwarz, "Kalahari Scheme," 212.

60. Schwarz, "Kalahari Project," 305.

61. W. G. Collins, letter to the editor, *Cape Times*, 11 September 1920.

62. Oswell, *William Cotton Oswell*, 203.

63. Schulz and Hammar, *New Africa*, 20.

64. Blanford, "On the Nature," 500.

65. Tate, "Post-Miocene Climate," 53. Tate repeated many of the same ideas a decade later in his inaugural address as president for the Australasian Association for the Advancement of Science. See Tate, "Inaugural Address."

66. Passarge, *Kalahari*, ch. 34.

67. Widney, "Colorado Desert."

68. Araripe, "Historia da Provincia do Ceará," 9–10; Cunniff, "Great Drought," 60–61.

69. Davis, "Restoring Roman Nature," 70–74; Davis, *Resurrecting the Granary of Rome*.

70. Mackenzie, *Flooding of the Sahara*, 217.

71. Martel, *Nouveau Traité*, 735.

72. Quoted in Endfield and Nash, "Drought," 39–40.

73. Chapman, *Travels Vol. 2*, 61–62, 64.

74. Reid, "Journeys," 582.

75. Livingstone, *Missionary Travels*, 76; also 16, 62, 125. The accounts of past travelers are detailed in Brown, *Hydrology*; and Schwarz relied heavily on many of the same accounts.

76. Anderson, *Twenty-Five Years*, 97. Anderson was in Griqualand West, near the Kimberley diamond fields.

77. Dalrymple, *Age of the Earth*, 26.

78. This trajectory was embraced by a number of scientists. See, for example,

Arrhenius, "The Fate of the Planets"; Martel, *Nouveau Traité*, 722; Lowell, *Mars*; Davis, "Coming Desert."

79. Dalrymple, *Age of the Earth*, 26.

80. Martel, *Nouveau Traité*, 717; "Must Humanity Perish," 305–7; Jackson, *Causes of Earthquakes*, 258.

81. Kropotkin, "Desiccation of Eur-Asia, 725, 733–34; responses from Holdich, 735; Seely, 738; Conway, 737; Mill, 740; Blanford 736.

82. Gregory, "Reported Progressive Desiccation," 341, 343.

83. Quoted in Kokot, *Investigation*, 86.

84. Kokot, *Investigation*, 85–87; Stigand, "Ngamiland," 405–8.

85. Stigand, "Ngamiland," 408.

86. Hodson, *Trekking the Great Thirst*, 153.

87. Livingstone, *Missionary Travels*, 566–67; Wilson, "On the Progressing Desiccation," 108.

88. Kokot, *Investigation*, 87.

89. Brown, *Hydrology*, 22, 27–28.

90. McKittrick, "Making Rain."

91. Kanthack, "Notes," 322. In fact, the rivers are quite different from each other.

92. Nitsche, "Aus dem Deutsch-Südwestafrikanischen Schutzgebiete," 217.

93. "Among the Mirages of Makarikari," *Star,* 12 June 1926.

94. Schwarz, *Kalahari*, 133; Schwarz, "Kalahari Project," 306.

95. Wilson, "Water Supply," 119.

96. Wilson, "Water Supply," 117. Wilson admitted that "even colonists of European descent" burned pastures in the Cape to encourage the growth of new grass. But he held Africans primarily responsible.

97. Wilson, "Water Supply," 120–21.

98. Grove, *Green Imperialism*.

99. Grove, "Scottish Missionaries," 163–87.

100. Brown, *Hydrology*, 158–91. Brown does not appear to commit to a timeline for African deforestation practices or their arrival in the region.

101. Sim, "Modification," 323.

102. Barber, letter to the editor, 167–68. Passarge and Stigand both told versions of this story. It remained the favored theory of some scientists and historians into the 1980s. See Shaw, "Desiccation of Lake Ngami."

103. Weidner, *Fallacy*, 11–13. Stories of chiefs causing Ngami's desiccation first appeared in published sources in 1925, almost thirty years after the stories of local people with their papyrus rafts.

104. McKittrick, "Capricious Tyrants"; Hamilton, *Terrific Majesty*; McKittrick, "Making Rain."

105. Endfield and Nash, "Drought," 40–41.

106. Chapman, *Travels Vol. 2*, 62.

107. McKittrick, *To Dwell Secure*, 28; Endfield and Nash, "Drought," 35–39; Moffat, *Missionary Labors*, 217–18; Carnegie, *Among the Matabele*, 38–42.

108. F. Ellis, letter to the editor, *Farmer's Weekly,* 29 March 1933, p. 87.

109. "The Problem and Danger of Exploiting the Kalahari," *Farmer's Weekly,* 8 December 1937, p. 906.

110. Jeffares, *Report.*

111. There were a limited number of Afrikaans-language sources that published readers' letters on such a wide range of subjects. But Afrikaans speakers were also sources for such stories—their role as originators often, but not always, disguised. We must not assume that the disproportionate number of English-language sources on Ngami's disappearance reflects anything other than who had access to publication outlets.

112. "Thirstland Redemption: A Reply to Dr. Sutton," *Diamond Fields Advertiser*, 25 April 1925.

113. South Africa, *Debates of the House of Assembly*, 3 (1925): 1975. De Jager's source of information appeared to be the "old timers"—white men who had lived in the region in the 1890s.

114. Cape of Good Hope, *First Annual Report*, 21.

115. The article, in the *Cape Times*, 26 July 1895, is referenced in the inventory for the Schwarz papers, Albany Museum.

116. Cape of Good Hope, *First Annual Report*, 29.

117. Schwarz, *Kalahari*, 150.

118. Cape of Good Hope, *First Annual Report*, 21.

119. Roe, "Notes," 26. The author was probably the well-known Graaff-Reinet photographer, but also could have been his son, who assisted him.

120. Roe, "Notes," 26.

121. Cape of Good Hope, *First Annual Report*, 23.

Chapter 2

1. Schwarz to Judd, 24 October 1899, Imperial College Archives, KGA/JUDD/3, folder 13.

2. Chetty, *Africa Forms the Key*, 19–22.

3. http://www.s2a3.org.za/bio/Biograph_final.php?serial=813; accessed 18 October 2023.

4. Chetty, *Africa Forms the Key*.

5. Barnhard, *Encountering Adamastor*, 19–20; Dubow, *A Commonwealth of Knowledge*.

6. Chetty, *Africa Forms the Key*.

7. Barnhard, *Encountering Adamastor*, 22.

8. Barnhard, *Encountering Adamastor*, 51.

9. Barnhard, *Encountering Adamastor*, 51; Wright, "Grahamstown of Today."

10. Cannon, *General and Physiological Features*, 19, 30.

11. Schwarz, "In the Footsteps of an Alien Race," *Cape Times*, 6 February 1926; Schwarz, "Did the Chinese Build Zimbabwe?" *Star*, 20 September 1924; Green, *Thunder on the Blaauwberg*, 163–64; Schwarz, "Former Land Connection"; Schwarz, "Lost Land." The idea of Lemuria later spread into occult circles; meanwhile Fournier, who proposed engineering the climate of the *sertão*, himself later became deeply involved with occult movements.

12. Barnard, *Encountering Adamastor*, 61; the house is referenced in the Schwarz papers at the Albany Museum.

13. Schwarz, "Three Paleozoic Ice-Ages."

14. Schwarz, "Only Moss on Mars," *Star*, 14 February 1925; "Slavery in Portuguese West Africa," *South African Review* 67, no. 1801 (26 November 1926), 5; "No Slavery in Angola," *Star*, 25 November 1926.

15. Schwarz, “Rivers of Cape Colony,” 272.

16. Geologists today accept the idea of interior rivers that were captured by coastal streams, but reject the idea that this process poses any imminent threat to the Orange River, since they date these changes to between 180 million and 300 million years ago. Catuneanu et al., “Karoo Basins.”

17. Schwarz, “Rivers of Cape Colony,” 272.

18. Schwarz, “Rivers of Cape Colony,” 278.

19. Gilmartin, “Imperial Rivers,” 86.

20. H. J. Lewis, “Populating the Vast Inland,” *The Mail* (Adelaide, Australia), 10 October 1925.

21. Beattie and Morgan, “Engineering Edens,” 39.

22. Beattie and Morgan, “Engineering Edens”; Gilmartin, “Imperial Rivers”; Beinart et al., “Experts and Expertise.”

23. Peter MacDonald, letter to the editor, *Farmer’s Weekly*, 10 September 1913, p. 50.

24. South Africa, *Report of the Select Committee on Drought Distress Relief*, 40.

25. R. M. Bowker, letter to the editor, *Farmer’s Weekly*, 26 February 1913, p. 2191.

26. Edward Shingles, letter to the editor, *Farmer’s Weekly*, 17 September 1919, p. 338.

27. Schwarz, “Uitenhage Springs,” 9–10, Cape Town Archives Repository, 3/UIT vol. 80, ref 198.

28. “Professor Schwarz’s report on Boring for Water: Sandfontein Farm,” 1914, p. 15, Cape Town Archives Repository, 3/UIT vol. 52, ref. 194.

29. Grove, *Green Imperialism*.

30. Davis, *Arid Lands*, 79.

31. Cunniff, “Great Drought,” 60–61.

32. See, for example, Hough, *Report upon Forestry*.

33. Davis, *Arid Lands*, 137.

34. Marsh, *Man and Nature*, 128–29.

35. Marsh, *Man and Nature*, 132.

36. Davis, “Restoring Roman Nature,” 70–74; Davis, *Resurrecting the Granary of Rome*.

37. Fairhead and Leach, *Misreading the African Landscape*, 239–41; Moloney, *Sketch*, 224–27, 240–42; Bovill, “Encroachment,” parts 1 and 2; Stebbing, “Encroaching Sahara.”

38. Marsh, *Man and Nature*, 196. He remained doubtful in his later edition, titled *The Earth as Modified by Human Action*.

39. Davis, *Arid Lands*, 79.

40. Davis, *Arid Lands*, 118; also Grove, *Green Imperialism*. Davis and Grove underplay this fact by focusing on a few voices and geographic locations, and by limiting their lens to “powerful people,” rather than those with whom those people were conducting a debate.

41. Blanford, “On the Nature and Probable Origin,” 500.

42. See Kropotkin, “Desiccation,” 725; Martel, *Nouveau Traité*.

43. Roudaire, *Rapport*, 346.

44. Blanford, “On the Nature and Probable Origin,” 500.

45. Holdich et al., “Desiccation,” 737–38.

46. South Africa, *Report from the Select Committee on Droughts, Rainfall, and Soil Erosion*, 49.

47. Grove, *Green Imperialism*. In the nineteenth century, a diverse range of scientists sought to demonstrate the impact of forests on rainfall. Woeikov, in “Der Einfluss

der Wälder," clearly influenced the German farmer Ferdinand Gessert when he argued that it is the capacity of forests to stop the wind, rather than their effect on moisture or temperature, that explains their effect on rainfall.

48. Humboldt, *Ansichten der Natur*, 1st ed., 18.

49. Humboldt, *Ansichten der Natur*, 3rd ed., 17.

50. Widney, "Colorado Desert," 47.

51. Tate, "Post-Miocene Climate," 55.

52. Passarge, Kalahari, 664–68.

53. Widney, "Colorado Desert," 47. Widney thought the Colorado River was shifting back toward its former basin.

54. Laflin, Salton Sea, ch. 1.

55. Rolle, *John Charles Frémont*, 251. President Díaz's response is not recorded. Widney's proposed scheme was eventually carried out, though by accident rather than design. The Salton Basin partially refilled in 1905–7, when the Colorado River overwhelmed diversion canals meant to bring irrigation water into the basin. A 1915 study by a team of engineers concluded that the creation of the sea had not changed the area's rainfall. Cory et al., *Imperial Valley*, 1416.

56. Marsh, *Man and Nature*, 525–26. Marsh goes on to speculate that Mediterranean plants and animals would fill in "the new home which human art had prepared for them"—but also that a massive increase in hydrostatic pressure due to the new basin could "produce geological convulsions the intensity of which cannot even be conjectured." (526)

57. Roudaire, *Rapport*, 346. Roudaire was building on earlier French ideas about recreating the sea, though without any anticipated climate benefits. See Heffernan, "Bringing the Desert to Bloom," 97; Mackenzie, *Flooding of the Sahara*," 222–23.

58. Marsh, *Earth as Modified by Human Action*, 593.

59. See Heffernan, "Bringing the Desert to Bloom," 98; Lehmann, *Desert Edens*.

60. Roudaire, *Rapport*, 347; Lehmann, *Desert Edens*.

61. Charles Stewart, letter to the editor, *Brisbane Courier*, 19 July 1892.

62. Cunniff, "Great Drought," especially 106–7.

63. Powell, "Trees on Arid Lands," 170–71.

64. Rafferty, *Oceans and Oceanography*, 242; http://www.eeo.ed.ac.uk/public/JohnMurray.html, accessed 16 December 2016.

65. Murray, "On the Total Annual Rainfall," 65.

66. As just one example, Murray used figures for Asian rivers, such as the Indus and Ganges, that were around half and one-fifth of their actual flows.

67. Paul Edwards writes that Brückner was an "'issue entrepreneur,' promoting political action on the basis of scientific evidence . . . a remarkable precursor to modern climate politics." Edwards, *Vast Machine*, 67.

68. Brückner, "Über die Herkunft." Brückner was part of a larger conversation that sought to calculate the earth's water balance. See Lvovitch, "World Water Balance."

69. Brückner, "Über die Herkunft," 96.

70. Passarge, *Kalahari*, 664.

71. Walther, *Gesetz der Wüstenbildung*, 45.

72. For a more comprehensive look at the impact of Murray and Brückner's ideas, see McKittrick, "Theories of 'Reprecipitation.'" Sections of this chapter are adapted from that article, first published in the open-access journal *History of Meteorology*.

73. Fournier, *Problema*, 18.

74. Fournier, *Problema*, 27–28.

75. Fournier, *Problema*, 14–15, 54.

76. Schwarz, *Causal Geology*, 42.

77. Schwarz, *Kalahari*, iii.

78. Passarge, *Kalahari*, 666.

79. Von Weber, "Ferdinand Gessert," 37; South African Association for the Advancement of Science biographical database at s2a3.org.za, accessed 9 July 2016.

80. Steinbach, "Carved Out of Nature." Wallace, in *History of Namibia*, 194, discusses the creation of "Germanness" in the context of Namibia.

81. One of the judges was Alexander Supan, editor of *Petermanns Geographische Mitteilungen*, the oldest German geographical journal. Supan largely accepted Brückner's ideas about land-based rains, and had himself written about reprecipitation.

82. Kienetz, "Nineteenth-Century South West Africa," 585, and ch. 12.

83. Kienetz, "Nineteenth-Century South West Africa," 592–602.

84. Kienetz, "Nineteenth-Century South West Africa," 600.

85. Gessert, "Reise längs der Flussthälter des südwestlichen Gross-Namalandes," 190.

86. Kienetz, "Nineteenth-Century South West Africa," 887–88.

87. "Aus dem Schutzgebiet," *Deutsch-Südwestafrikanische Zeitung* 6, no. 26 (29 July 1904): 2.

88. Von Weber, "Ferdinand Gessert, Farmer in Sandverhaar"; author interview with Frikkie Mouton, current owner of Sandverhaar, June 2017.

89. Gessert, "Reise längs der Flussthäler," 190.

90. Gessert, "Seewind," 297.

91. Coetzee, *White Writing*, 44.

92. Gessert, "Seewind," 297.

93. It is unclear who first suggested diverting the Kunene River into north-central Namibia, but it is referenced in Brincker, "Bemerkungen," 80.

94. Gessert, "Seewind."

95. Gessert, "Klimatische Folgen"; Gessert, "Mutmasslichen klimatischen Folgen."

96. He also echoed Woeikov and Supan, without attribution, when he wrote that reprecipitation accounted for the heavy rainfall in the Amazon River basin.

97. Gessert, "Über Rentabilität."

98. McKittrick, "Making Rain."

99. Von Weber, "Ferdinand Gessert."

100. Gessert, "Zur Aufforstungsfrage"; Gessert,"Über Rentabilität."

101. Schmidt, ed., *Albert Schmidt*, 118–19.

102. Bridgman, *Revolt*, 104; Wallace, *History of Namibia*, 161–62.

103. Gessert, "Zur Aufforstungsfrage," 135.

104. Wallace, *History of Namibia*, 162–65; Bridgman, *Revolt*, 112.

105. Gessert, "Über Rentabilität."

106. Gessert, "Auf der Flucht," 81.

107. Gessert, "Über Rentabilität," 341.

108. H. Paulsmeier, letter to the editor, *South African Agricultural Journal* 7 (January–July 1914): 737–39.

Chapter 3

1. "Die Bewässerung der Kalahari," *Der Farmwirtschaftliche Ratgeber* (supplement of *Allgemeine Zeitung*), 12 August 1920, p. 1.

2. South Africa, *Report from the Select Committee on Droughts, Rainfall, and Soil Erosion*, iii.

3. Some material for this chapter was first published in McKittrick, "Talking about the Weather," in *Environmental History* (copyright holder Oxford University Press).

4. Li, "Beyond the 'State,'" 386.

5. "Notes on the Northern Cape," *Farmer's Weekly*, 29 January 1913, p. 1807.

6. *Farmer's Weekly*, 20 December 1911, p. 585.

7. "Notes on the Northern Cape," *Farmer's Weekly*, 29 January 1913, p. 1807.

8. *Farmer's Weekly*, 10 January 1912, p. 731.

9. *Farmer's Weekly*, 10 January 1912, p. 730.

10. A. Crezesky to governor general, 20 January 1912, NAR, GG vol. 105, folder 3/823.

11. "Rain" (editorial), *Farmer's Weekly*, 14 February 1912, p. 973.

12. E. S. Morgan to administrator, Free State, 12 September 1912, NAR, PM vol. 1/1/247, folder 113/18.

13. Spes Bona, "Notes from the Border," *Farmer's Weekly*, 5 February 1913, p. 1884.

14. "The Weather," *Farmer's Weekly*, 17 December 1913, p. 1207.

15. "The Drought," *Farmer's Weekly*, 21 January 1914, p. 1823.

16. "The Weather," *Farmer's Weekly*, 24 February 1915, p. 1699.

17. "The Weather Prospects," *Farmer's Weekly*, 31 January 1915, p. 237; "The Maize Crop and the Need for Statistics," *Farmer's Weekly*, 31 January 1915, p. 262.

18. *Farmer's Weekly*, 16 February 1916, p. 2280.

19. Cole, *South Africa*, 58.

20. Levinkind, "Droughts," 84.

21. Levinkind "Droughts," 85.

22. Keegan, *Rural Transformations*, 166.

23. This was standard practice around the British Empire. See O'Brien, "Imported Understandings."

24. Kienetz, "Nineteenth-Century South West Africa as a German Settlement Colony," 427, 782.

25. South Africa, *Final Report of the Drought Investigation Commission*, 4.

26. "Jonas Klip," letter to the editor, *Farmer's Weekly*, 10 February 1915, p.1560.

27. "Griqua," letter to the editor, *Farmer's Weekly*, 11 March 1914, pp. 82–83.

28. A. Kern, letter to the editor, *Farmer's Weekly*, 29 March 1933, p. 85.

29. *Farmer's Weekly*, 6 January 1915, 1226–27.

30. "Cnoc-Na-Ra," letter to the editor, *Farmer's Weekly*, 1 March 1916, p. 2460.

31. W. Vogel Hughes, letter to the editor, *Farmer's Weekly*, 4 February 1914, p. 2058.

32. E. J. Hughes, letter to the editor, *Farmer's Weekly*, 24 May 1933, p. 538.

33. "Jonas Klip," letter to the editor, *Farmer's Weekly*, 10 February 1915, p. 1560.

34. South Africa, *Final Report of the Drought Investigation Commission*, 180.

35. George van Zyl, letter to the editor, *Farmer's Weekly*, 29 March 1933, p. 87.

36. E. G. Bryant, letter to the editor, *Farmer's Weekly*, 22 March 1933, p. 29. See also R. E. Edelmann and G. W. van Zyl. letters to the editor, *Farmer's Weekly*, 23 November 1927, n.p.

37. "In Search of Veld," *Farmer's Weekly*, 1 March 1916, p. 2492.

38. "A Desert in the Making," *Farmer's Weekly*, 9 July 1913, p. 1744.

39. O. Webber, letter to the editor, *Farmer's Weekly*, 5 March 1933, p. 143.

40. Plummer, *Rainfall and Farming*, 8.

41. The idea of drought as a sign of divine displeasure probably represented a significant section of Afrikaner opinion, but one poorly represented in *Farmer's Weekly*. But see M. C. Pretorious, letter to the editor, *Farmer's Weekly*, 29 March 1933, p. 87.

42. Van Sittert, "Nation-Building Knowledge," 96.

43. "A Desert in the Making," *Farmer's Weekly*, 9 July 1913, p. 1744.

44. D. Hodgson, letter to the editor, *Farmer's Weekly*, 24 May 1933, p. 537.

45. Resident commissioner Ovamboland to secretary for the protectorate, NAN, SWAA vol. 598, folder A66/1.

46. South Africa, *Report from the Select Committee on Droughts, Rainfall, and Soil Erosion*, 40.

47. Schwarz, *Kalahari*, 5.

48. F. Ellis, letter to the editor, *Farmer's Weekly*, 29 March 1933, p. 87.

49. South Africa, *Report from the Select Committee on Droughts, Rainfall, and Soil Erosion*, 40.

50. W. B. Phillips, reprinted speech, *Farmer's Weekly*, 3 September 1913, p. 2642. A sluit is a dry gully formed or deepened by water erosion. While experts would attribute such changes to poor land management, Phillips suggested that they were all due to declining rainfall.

51. South Africa, *Final Report of the Drought Investigation Commission*, 5.

52. "Die Reenval van Suid Afrika," *Die Landbouweekblad*, 24 March 1920, p. 1453.

53. "Is Suidafrika aan Opdroë?" *Die Landbouweekblad*, 21 April 1920, p. 1651.

54. "Jonas Klip," letter to the editor, *Farmer's Weekly*, 10 February 1915, p. 1560.

55. Lehmann, in "Average Rainfall," also notes the importance of sensory experience in colonial depictions of climate.

56. South Africa, *Debates of the House of Assembly* 3 (1925): 1976–77.

57. South Africa, *Report from the Select Committee on Droughts, Rainfall, and Soil Erosion*, 43.

58. South Africa, *Report from the Select Committee on Droughts, Rainfall, and Soil Erosion*, 44–45.

59. "Karroo," letter to the editor, *Farmer's Weekly*, 23 November 1927, p. 999.

60. E. G. Bryant, letter to the editor, *Farmer's Weekly*, 22 March 1933, p. 29.

61. South Africa, *Final Report of the Drought Investigation Commission*, 179.

62. Indigenous concepts also anthropomorphize the west wind, albeit in distinct ways. See Schnegg, "Life of Winds."

63. Gessert, "Seewind."

64. South Africa, *Final Report of the Drought Investigation Commission*, 180; Philip Townshend, letter to the editor, *Farmer's Weekly*, 24 December 1919, p. 2238; "Karroo," letter to the editor, *Farmer's Weekly*, 23 November 1927, p. 999.

65. Philip Townshend, letter to the editor, *Farmer's Weekly*, 24 December 1919, p. 2238.

66. South Africa, *Final Report of the Drought Investigation Commission*, 182.

67. Anderson, "Looking at the Sky."

68. South Africa, *Report from the Select Committee on Droughts, Rainfall, and Soil Erosion*, 40. Nicholson, "ITCZ," notes the history of this concept and its limitations, even as it has dominated understandings of African rainfall for a century.

69. Schwarz, *Kalahari*, 61.

70. Schwarz, "Kalahari and Its Possibilities," 10.

71. South Africa, *Report from the Select Committee on Droughts, Rainfall, and Soil Erosion*, 44–45.

72. H. Langley, letter to the editor, *South African Agricultural Journal* 7:735.

73. P. M. van der Westhuizen, letter to the editor, *South African Agricultural Journal* 7:897.

74. Sim, "Modification of South African Rainfall," 322.

75. R. B. Chase, letter to the editor, *South African Agricultural Journal* 6 (August–December 1913): 693–94.

76. Jas. P. van Zijl, letter to the editor, *South African Agricultural Journal* 6:1028.

77. W. Akkersdyk, letter to the editor, *South African Agricultural Journal* 7:736.

78. O. T. Farthing, letter to the editor, *South African Agricultural Journal* 7:261.

79. W. W. Steers, letter to the editor, *South African Agricultural Journal* 7:736–37.

80. D. Pretorius, letter to the editor, *South African Agricultural Journal* 6:1029.

81. South Africa, *Report from the Select Committee on Droughts, Rainfall, and Soil Erosion*, 8, 14.

82. Tischler, "Cultivating Race."

83. "Burning the Veld," *Farmer's Weekly*, 12 March 1913, p. 2410.

84. W. B. Phillips, reprinted speech, *Farmer's Weekly*, 3 September 1913, p. 2642. For similar views in Australia, see Beattie, "Science," 140.

85. "Burning the Veld," *Farmer's Weekly*, 12 March 1913, p. 2410.

86. South Africa, *Report from the Select Committee on Droughts, Rainfall, and Soil Erosion*, 44, 21.

87. South Africa, *Report from the Select Committee on Droughts, Rainfall, and Soil Erosion*, 39–47.

88. South Africa, *Report from the Select Committee on Droughts, Rainfall, and Soil Erosion*, 41–42.

89. The controversial Yale geographer Ellsworth Huntington, who argued that climate determined the nature of civilizations, was among those who had moved away from his earlier belief in progressive planetary desiccation by 1915. See Gregory, "Professor Huntington." See also Ward, *Climate*, 338–45.

90. South Africa, *Report from the Select Committee on Droughts, Rainfall, and Soil Erosion*, iii–v.

91. Beinart, *Rise of Conservation*, 250; South Africa, *Final Report of the Drought Investigation Commission*, 3.

92. G. W. van Zijl, "Colesberg en Reen," *Die Landbouweekblad*, 4 February 1920, p. 1177.

93. "Ladybrand Distrik," *Die Landbouweekblad*, 4 February 1920, p. 1175.

94. "Boerderij te Wepener," *Die Landbouweekblad*, 4 February 1920, p. 1177.

95. "Stof of Reen?" *Die Landbouweekblad*, 11 February 1920, p. 1195.

96. See Beinart, *Rise of Conservation*, 250–257. Tischler, "Cultivating Race," elaborates on the trend toward the bureaucratization of agricultural improvement over the course of the twentieth century.

97. Beinart, *Rise of Conservation*, ch. 7, deals extensively with du Toit.

98. Du Toit, letter to the editor, *Farmer's Weekly*, 6 December 1920, n.p.

99. South Africa, *Final Report of the Drought Investigation Commission*, 24, 27.

100. Stoler, *Along the Archival Grain*, 29.

101. Ashforth, *Politics of Official Discourse*, 254.

102. South Africa, *Final Report of the Drought Investigation Commission*, 3.

103. South Africa, *Final Report of the Drought Investigation Commission*, 5.

104. South Africa, *Final Report of the Drought Investigation Commission*, 64.

105. South Africa, *Report from the Select Committee on Droughts, Rainfall, and Soil Erosion*, v–vi.

106. South Africa, *Final Report of the Drought Investigation Commission*, 63.

107. South Africa, *Report from the Select Committee on Droughts, Rainfall, and Soil Erosion*, 42. Marsh, *Man and Nature*, 547, writes: "The possibility that the distribution and action of electricity may be considerably modified by long lines of iron railways and telegraph wires . . . in fact rests on much the same foundation as the belief in the utility of lightning rods, but such influence is too obscure and too small to have been yet detected."

108. W. H. Scholtz, letter to the editor, *Farmer's Weekly*, 2 March 1921, p. 3256.

109. South Africa, *Debates of the House of Assembly* 3 (1925): 1984.

Chapter 4

1. Hancock and van der Poel, *Selections from the Smuts Papers*, vol. 1, 484–85.

2. Gessert, "Über Rentabilität," 341.

3. F. E. Geldenhuys, "Toespraak gehou aan die Modderrivier naby Bloemfontein," in *Die Landbouweekblad*, 20 December 1922, p. 769, quoted in van der Watt, "'It Is Drought,'" 165–66.

4. Beinart, *Rise of Conservation*, 257; "The Agricultural Department," undated *Cape Times* article, Schwarz papers, Albany Museum.

5. Krikler, "Re-Thinking Race," 144.

6. Jeeves and Crush, *White Farms*; Ross et al., *Cambridge History of South Africa Vol. 2*, 211–53.

7. Swart, "Five-Shilling Rebellion."

8. Mamdani, *Citizen and Subject*. As some historians have pointed out, however, the other half of Mamdani's binary—Africans who lived as "subjects" under chiefly authority and communal land tenure regimes—was a reality that only emerged over the course of the early twentieth century, and likely reflected African preferences as well as colonial templates. See e.g. Beinart and Delius, "Historical Context."

9. Beinart and Delius, "Historical Context," 671.

10. Quoted in Tischler, "Cultivating Race." For English-speakers and agricultural modernization, see Keegan, *Rural Transformations*.

11. Darwin, *Empire Project*, 254.

12. Lewis, *Economics*, 22–23.

13. Wellington, *South West*, 409; Brownell, *Collapse of Rhodesia*, 3.

14. Etherington, *Great Treks*, 1.

15. On the genocide of South African hunter-gatherers, see Adhikari, *Anatomy*; Prada Samper, "Forgotten Killing Fields"; Penn, "British and the 'Bushmen.'"

16. President's annual address, in *The Transactions of the South African Philosophical Society* vol. 1, part 2 (Cape Town: J. C. Juta 1879), xxii.

17. Browne, *South Africa*, 107. Notably, German colonial soldiers were attempting to do precisely this in South West Africa at the time the person wrote.

18. Bell, "John Stuart Mill."

19. Lake and Reynolds, *Drawing the Global Colour Line*, 80.

20. South Africa, *South African Native Affairs Commission*, vol. 3, 173.

21. Browne, *South Africa*, 106.

22. Macdonald, *Trade, Politics*, 41.

23. Macdonald, *Trade, Politics*, 41; Wybergh, "Imperial Organization," 807–8. For connections between climate, race, and citizenship, see Ikuko Asaka, *Tropical Freedom;* Lake and Reynolds, *Drawing the Global Color Line.*

24. Giliomee, *Afrikaners*, 297. Africans had some voting rights in the Cape Colony and retained them until 1936.

25. Keegan, "Gender, Degeneration," 461.

26. https://www.sahistory.org.za/archive/white-mans-task-jan-smuts-22-may-1917. Accessed 10 October 2023.

27. Krikler, "Re-Thinking Race and Class," 144. The fear of "being outnumbered and overwhelmed by the black majority" was, in the words of an anti-apartheid activist, "a phobia as old as the colonization of South Africa." Gray, "Race Ratios," 140.

28. Both quoted in Giliomee, *Afrikaners*, 288.

29. Koorts, "'Black Peril,'" 563.

30. Creswell, "Transvaal Labour Problem," 456. Creswell was a mining engineer who resigned in 1903 to campaign for Chinese exclusion; he was elected to Parliament in 1910.

31. South Africa, *South African Native Affairs Commission*, vol. 4, 129.

32. "Pride of Race," *Farmer's Weekly*, 29 November 1911, p. 436.

33. L. Reyersbach, quoted in Bozzoli, *Political Nature*, 53.

34. Krikler, "Re-Thinking Race and Class," 146.

35. "Erosion and its Consequences," *Farmer's Weekly*, 20 December 1911, p. 587.

36. Quoted in Keegan, "Gender, Degeneration," 462.

37. W. A. M., "The Case for the Settler," *Farmer's Weekly*, 9 September 1925, p. 2791.

38. Herschensohn, *A Great South Africa*, 16.

39. South Africa, *Report from the Select Committee on Closer Land Settlement*, viii.

40. Karlson, *Kalahari Problem*, 19.

41. Hertzog, *Segregation Problem*, 12, 15.

42. Keegan, *Rural Transformations*, 183.

43. Spies et al., *Hertzogtoesprake*, vol. 2, 249.

44. "Sour Dough," letter to the editor, *Farmer's Weekly*, 21 February 1912, p. 1035.

45. Recounted in Keegan, *Rural Transformations*, 181.

46. W. F. van Coller, letter to the editor, *Farmer's Weekly*, 27 March 1912, p. 156; "Free State Farming," *Farmer's Weekly*, 27 December 1911, p. 643; Giliomee, *Afrikaners*, 307; Denoon, *Settler Capitalism*, 118; Van Onselen, *Seed.*

47. Keegan, *Rural Transformations.*

48. Van Onselen, *Seed Is Mine*, 116–17.

49. Keegan, *Rural Transformations* 182, quoting *The Friend* from 1912.

50. Spies et al., *Hertzogtoesprake*, vol. 1, 134–35.

51. Beinart and Delius, "Historical Context," 668, 674; Swart, "Five-Shillling Rebellion," 95.

52. Koorts, "'Black Peril,'" 564.

53. Markham, *South African Scene*, 341.

54. Karlson, *The Kalahari Problem*, 19.

55. D. F. Malan, *Die Groot vlug: 'N nabetragting van die Armblanke-Kongress, 1923, en van die offisiële sensusopgawe* (Cape Town: National Press, 1923), 7. At this point, in contrast to his later position, he argued that the threat affected both whites and Coloureds.

56. Ross et al., *Cambridge History of South Africa Vol. 2*, 260.
57. "Land for Our Sons," *Farmer's Weekly*, 27 October 1926, p. 653.
58. South Africa, *Final Report of the Drought Investigation Commission*, 179.
59. Stals, 12 August 1924, quoted in Seekings, "'Not a Single White Person,'" 382.
60. South Africa, *Debates of the House of Assembly* 3 (1925): 1983.
61. Money and van Zyl-Hermann, *Rethinking White*, 7.
62. Eric MacDonald, letter to the editor, *Farmer's Weekly*, 17 July 1912, p. 1256.
63. MacMillan, *South African Agrarian Problem*, 46.
64. Van Duin, "Artisans and Trade Unions," 104.
65. Reginald Luck, letter to the editor, *Farmer's Weekly*, 4 September 1912, p. 1725.
66. Giliomee, *Afrikaners*, 334.
67. Krikler, "Re-Thinking Race and Class," 147.
68. Krikler, *White Rising*, 128–29.
69. Giliomee, *Afrikaners*, 336.
70. South Africa, *South African Native Affairs Commission*, vol. 2, 707.
71. South Africa, *South African Native Affairs Commission*, vol. 3, 323.
72. Markham, *South African Scene*, 341.
73. Giliomee, *The Afrikaners*, 323.
74. Adam and Giliomee, *Ethnic Power Mobilized*, 150; Seekings, "Not a Single White Person," 379.
75. Giliomee, *Afrikaners*, 339.
76. Headlam, ed., *Milner Papers,* vol. 2, 459.
77. South Africa, *Report of the Select Committee on Drought Distress Relief*, 99.
78. Hancock and van der Poel, *Selections from the Smuts Papers*, vol. 2, 328, 331.
79. Quoted in Giliomee, *The Afrikaners*, 328.
80. H. F. Wirsing to Earl of Athlone, 5 November 1924, NAR, GG vol. 941, folder 17/997.
81. Fred T. Milsom to Earl of Athlone, 5 November 1924, NAR, GG vol. 941, folder 17/1000.
82. Fred T. Milsom to Earl of Athlone, 5 November 1924, NAR, GG vol. 941, folder 17/1000.
83. The focus on the governor general may reflect the nature of the archive, or may indicate that English-speaking farmers hoped to get a more sympathetic hearing from the governor general than from the Afrikaner prime minister or from a Lands Department widely deemed unsympathetic to struggling farmers.
84. A. G. Godwin to governor general, 9 October 1926, NAR, GG vol. 941, folder 17/1048.
85. Mrs. A. E. Roberts to Prince of Wales, July 1925, NAR, GG vol. 941, folder 17/1015.
86. F. G. Staples-Cooke to governor general, 23 October 1925, NAR, GG vol. 941, folder 17/1026.
87. A. M. Staples-Cooke to governor general, 26 October 1925, NAR, GG vol. 941, folder 17/1026.
88. C. H. Rotherforth to governor general, 16 August 1925, NAR, GG vol. 941, folder 17/1019.
89. Department of Lands to governor general, 31 October 1925, NAR, GG vol. 941, folder 17/1026.

90. Minute dated 26 November 1924, NAR, GG vol. 941, folder 17/1000.

91. Prime minister to governor general, 29 October 1925, NAR, GG vol. 941, folder 17/1019.

92. C. H. Rotherforth to governor general, 26 October 1925, NAR, GG vol. 941, folder 17/1019.

93. Giliomee, *Afrikaners*, 344; Clynick, *Reformers*.

94. South Africa, *Final Report of the Drought Investigation Commission*, 16; Davie, *Poverty Knowledge*, 82–91. By climatic conditions, du Toit meant the tropical and malarial conditions of northeastern South Africa, an area often deemed unsuited to white men.

95. Schwarz, *Kalahari*, 160–62.

96. South Africa, *Report of the Select Committee on European Employment*, 100.

97. Temple Nourse, letter to the editor, *Farmer's Weekly*, 20 October 1920, p. 1022.

98. South Africa, *Report of the Select Committee on Drought Distress Relief*, 30, 37, 89, 102.

99. J. X. Merriman to Jan Smuts, 20 December 1915, in Hancock and van der Poel, *Selections from the Smuts Papers*, vol. 3, 329.

100. South Africa, *Report of the Select Committee on Drought Distress Relief*, 61.

101. *South Africa, Report of the Select Committee on Drought Distress Relief*, 72.

102. H. J. T., letter to the editor, *Farmer's Weekly*, 10 October 1923, p. 451.

103. H. J. T., letter to the editor, *Farmer's Weekly*, 10 October 1923, p. 451.

104. "In Search of Veld," *Farmer's Weekly*, 1 March 1916, p. 2492.

105. H. J. T., letter to the editor, *Farmer's Weekly*, 10 October 1923, p. 451.

106. South Africa, *Debates of the House of Assembly* 2 (1917): 240.

107. South Africa, *Debates of the House of Assembly* 7 (1922): 179.

108. "Unity," letter to the editor, *Farmer's Weekly*, 27 March 1912, p. 156.

109. Fred Strauss, letter to the editor, *Farmer's Weekly*, 20 December 1911, p. 568.

110. Jas. B. Theron, letter to the editor, *Farmer's Weekly*, p. 1725.

111. South Africa, *Report of the Select Committee on European Employment*, 62; Davie, *Poverty Knowledge*, 89.

112. Lang, *White, Poor, and Angry*, 147–48.

113. South Africa, *Report of the Select Committee on European Employment*, v–vi.

114. South Africa, *Report of the Select Committee on Drought Distress Relief*, 37.

115. Davie, *Poverty Knowledge*, 82.

116. Quoted in Clynick, "Reformers," 113–14.

117. Quoted in Giliomee, *Afrikaners*, 297.

118. Money and van Zyl-Hermann, *Rethinking White Societies*.

119. South Africa, *Report of the Select Committee on Drought Distress Relief*, 97.

120. South Africa, *Report of the Select Committee on European Employment*, 95.

121. W. Temple Nourse, letter to the editor, *Farmer's Weekly*, 20 October 1920, p. 1022.

122. Fedorowich, "'Foredoomed to Failure,'" 196.

123. "Die brennende Frage," *Allgemeine Zeitung*, 31 July 1920, p. 1; Kennedy, *Islands of White*.

124. Tischler, "Cultivating Race"; Coetzee, *White Writing*, 75–81.

125. "Hou Vas aan die Plaas!" *Die Landbouweekblad*, 11 January 1922, p. 926.

126. Editorial, *Farmer's Weekly*, 13 March 1912, p. 26.

127. Peter MacDonald, letter to the editor, *Farmer's Weekly*, 17 December 1919, p. 2102.

128. Claude Lowe, letter to the editor, *Farmer's Weekly*, 5 April 1933, p. 143.

129. South Africa, *Debates of the House of Assembly* (1925): 1983.

130. W. G. Collins, letter to the editor, *Cape Times*, 11 September 1920.

131. Coetzee, *White Writing*, 75–81.

132. Beningfield, *Frightened Land*, 76–77.

133. Byron, *Back to the Land*, 1.

134. Wolmarans, letter to the editor, *Farmer's Weekly*, 20 April 1921, p. 661.

135. Byron, *Back to the Land*, 1.

136. Fedorowich, "'Foredoomed to Failure'"; Byron, *Back to the Land*; MacDonald, *Settler*.

137. Wallace, *History of Namibia*, 219.

138. Schwarz, "Menace to Ovamboland," 253.

139. Letter reprinted from *East London Despatch* regarding farming in Thaba 'Nchu, *Farmer's Weekly*, 27 December 1911, p. 643.

140. "Unity," letter to the editor, *Farmer's Weekly*, 27 March 1912, p. 156.

141. "Unity," letter to the editor, *Farmer's Weekly*, 27 March 1912, p. 156.

142. Quoted in Seekings, "'Not a Single White Person,'" 382. This was a circular sent out by Hertzog to government departments. Note that the Economic and Wage Commission report he quotes says that a civilized standard of living is that of being able to employ a native servant.

143. Temple Nourse, letter to the editor, *Farmer's Weekly*, 20 October 1920, p. 1022.

144. G. L. Coventry, letter to the editor, *Farmer's Weekly*, 24 April 1912, p. 420.

145. R.W. Berney, letter to the editor, *Farmer's Weekly*, 24 April 1912, p. 420.

146. Zollman, "Negoatiated Partition"; Posel, "Meaning of Apartheid."

147. Giliomee, *Afrikaners*, 299; Dubow, *Racial Segregation*, 6. Dubow, among others, notes the vagueness of the concept of segregation into the 1920s, arguing that there was really no segregationist ideology prior to World War I and that many of the segregationist laws were not rigidly enforced (4).

148. Dubow *Racial Segregation*, 43.

149. Posel, "Meaning of Apartheid."

150. South Africa, *South African Native Affairs Commission*, vol. 3, 173.

151. R.W. Berney, letter to the editor, *Farmer's Weekly*, 24 April 1912, p. 420.

152. Editorial, *Farmer's Weekly*, 13 March 1912, p. 26.

153. White women could not vote until 1930.

154. Tischler, "Cultivating Race," ch. 8. Tischler notes that the United States in the 1920s and 1930s also prioritized "re-agrarianization" despite its economic contradictions.

155. Van der Watt, "'It Is Drought,'" 32, citing letters from *Die Landbouweekblad* in 1922.

156. South Africa, *Report of the Select Committee on Closer Land Settlement*, 135.

157. South Africa, *Report of the Select Committee on Closer Land Settlement*, 23.

158. "Sour Dough," letter to the editor, *Farmer's Weekly*, 21 February 1912, p. 1035. On claims that, like animals, Africans merely "disturbed" the soil, see van der Watt, "'It Is Drought,'" 33, citing letter from *Die Landbouweekblad* in 1944.

159. Scully to Nora Scully, 5 January 1902 and undated, Wits University Historical Papers Research Archive, papers of William Charles Scully, Fonds A1312.

160. Hershensohn, *A Great South Africa*, 40.

161. South Africa, *Report of the Select Committee on Closer Land Settlement*, 14, 155.

162. South Africa, *Report of the Select Committee on Closer Land Settlement*, 12.

163. "Flooding of the Kalahari," *Queenstown Daily Representative and Free Press*, 19 December 1924.

164. Evans, *Cultures of Violence*, 19.

165. Coetzee, *White Writing*, 5. See also Beningfield, *Frightened Land*, 90–98.

166. Zollman, "Negotiated Partition," 407.

167. E. O'Connor, letter to the editor, *Farmer's Weekly*, 2 December 1925, p. 1273.

168. South Africa, *Report of the Select Committee on Drought Distress Relief*, 65, 67.

Chapter 5

1. Beattie and Moran, "Engineered Edens," 44.

2. Grove, "African Environment," 193.

3. Gilmartin, "Imperial Rivers," 92; Beinart and Hughes, *Environment and Empire*, 132–34.

4. Weidner, "Lost Lakes of South Africa," 28.

5. Union of South Africa, *Final Report of the Drought Investigation Commission*, 45.

6. Kalb, *Environing Empire;* Lau and Stern, *Namibian Water Resources*; van Vuuren, *In the Footsteps of Giants*, 30–31.

7. "Overstocking," *Farmer's Weekly*, 3 September 1913, p. 2642.

8. South Africa, *Final Report of the Drought Investigation Commission*, 15.

9. W. W., letter to the editor, *Farmer's Weekly*, 25 February 1920, p. 3487.

10. C. W. de Kiewet, quoted in Jeeves and Crush, *White Farms*, 9.

11. Ross et al., *Cambridge History of South Africa Vol. 2*, 250; Jeeves and Crush, *White Farms*, 5.

12. Jeeves and Crush, *White Farms*, 1–2.

13. "The Merino Sheep," *Farmer's Weekly*, 5 June 1912, p. 840.

14. "The Merino Sheep," *Farmer's Weekly*, 5 June 1912, p. 840.

15. "Dry-Farming and Mr. Campbell," *Farmer's Weekly*, 23 February 1916, p. 23.

16. "A Progressive Free Stater," *Farmer's Weekly*, 5 June 1912, p. 845.

17. Krikler, *Revolution*, 77.

18. On du Toit's career and influence on agriculture see Beinart, *Rise of Conservation*.

19. Phillips, in "Lessons," details American influences on South African agriculture and soil conservation.

20. Department of Agriculture, *Report for the Period 31st May 1910 to 31st December 1911*.

21. Phillips, "Lessons."

22. Van der Watt, "'It Is Drought,'" 32.

23. "Farmer George," letter to the editor, *Farmer's Weekly*, 27 July 1921, p. 2041.

24. South Africa, *Report from the Select Committee on Droughts, Rainfall, and Soil Erosion*, 12.

25. South Africa. *Report of the Select Committee on Drought Distress Relief*, 62.

26. *Agricultural Journal of the Cape of Good Hope* 34, no. 6 (June 1909): 647.

27. Van der Watt, "'It Is Drought,'" 38.

28. Quoted in Krikler, *Revolution*, 77.

29. "Backward Farmers," *Farmer's Weekly*, 20 March 1912, p. 118.

30. "Notes from the Northern Cape," *Farmer's Weekly*, 10 April 1912, p. 308. White farmers, like Black farmers, resisted dipping in part because it sometimes proved fatal to livestock. While Black farmers had no recourse in such cases, newspa-

pers reported lawsuits over white farmers' cattle being poisoned through mandatory dipping.

31. "Ignorant Farmer" and "A Seeker of Water," letters to the editor, *Farmer's Weekly*, 22 May 1912, p. 686.

32. Article reprinted in *Farmer's Weekly*, 10 January 1912, p. 730.

33. Ross et al., *Cambridge History Vol. 2*, 217.

34. Beinart et al., "Experts and Expertise."

35. W. Thornhill, letter to the editor, *Farmer's Weekly*, 24 April 1912, p. 430.

36. For an early example of this, see "The Road Question," *Farmer's Weekly*, 10 March 1915, p. 29.

37. Farmer George, letter to the editor, *Farmer's Weekly*, 27 July 1921, p. 2041.

38. For a description of these, see Keegan, *Rural Transformations*, 113–18.

39. See, for example, *FW*, "What Is a Farmer?" *Farmer's Weekly*, 20 January 1915, p. 1382.

40. W. H. F. Hughes, letter to the editor, *Farmer's Weekly*, 18 February 1914, p. 2310.

41. Keegan, *Rural Transformations*, 118.

42. South Africa, *Report from the Select Committee on Droughts, Rainfall, and Soil Erosion*, 10.

43. "Notes on the Economic Position in South West Africa," December 1921, p. 10, NAR, PM vol. 1/2/69, folder 21/15.

44. "In Search of Veld," *Farmer's Weekly*, 1 March 1916, p. 2492.

45. "Drought and Distress: Losses in the Cape Midlands," *Farmer's Weekly*, 28 Jan. 1920, p. 2971.

46. *Farmer's Weekly*, 22 October 1919, p. 1043.

47. *Farmer's Weekly*, 15 October 1919, p. 927.

48. See van der Watt, "'It Is Drought,'" p. 160.

49. MacDonald, *Conquest of the Desert*, 117. The expert consensus was that very little of the country was suitable for irrigation.

50. Thomas Parker, "Drought-Proof Farm," *Farmer's Weekly*, 23 April 1913, p. 573.

51. South Africa, *Final Report of the Drought Investigation Commission*, 56.

52. McDonald, letter to the editor, *Farmer's Weekly*, 17 July 1912, p. 1256.

53. Edward Shingles, letter to the editor, *Farmer's Weekly*, 17 September 1919, p. 338.

54. South Africa, *Final Report of the Drought Investigation Commission*, 82.

55. South Africa, *Final Report of the Drought Investigation Commission*, 15.

56. South Africa, *Report of the Select Committee on Drought Distress Relief*, 41.

57. Quoted in van der Watt, "'It Is Drought,'" 31.

58. Kanthack, "Irrigation Development," 646–47.

59. James Field, letter to the editor, *Farmer's Weekly*, 20 September 1911, p. 50.

60. South Africa, *Report of the Select Committee on Drought Distress Relief*, 89–90.

61. "The Weather," *Farmer's Weekly*, 17 December 1913, p. 1207.

62. South Africa, *Report of the Select Committee on Drought Distress Relief*, xvii.

63. See, for example, "A Lesson for South Africa," *Farmer's Weekly*, 7 January 1920, p. 2319.

64. *Farmer's Weekly*, 27 August 1913, p. 2498.

65. Du Toit, *Dry Farming*, 19.

66. Eric McDonald, letter to the editor, *Farmer's Weekly*, 17 July 1912, p. 1256.

67. MacDonald, *Conquest of the Desert,* 113, 116.

68. "A Rainless Wheat," *Farmer's Weekly*, 9 July 1913, p. 1744.

69. "The Weather," *Farmer's Weekly*, 17 December 1913, p. 1207; "Notes from the Border," *Farmer's Weekly*, 5 February 1913, p. 1884; also J. G. James, "Agricolorum," and "Northern," letters to the editor, *Farmer's Weekly*, 22 January 1913, pp. 1686, 1687, 1688.

70. Beinart, *Rise of Conservation*, 247.

71. Krikler, *Revolution*, 79.

72. Worster, *Rivers of Empire*, 150.

73. Quoted in Thompson, *Moisture and Farming*, 50.

74. E. R. Bradfield, letter to the editor, *Farmer's Weekly*, 21 January 1914, p. 1824.

75. MacDonald, *Conquest of the Desert,* 5.

76. For example, Senate, *Report from the Select Committee on Droughts*, vi.

77. Kanthack, "Irrigation Development," 650. In fact, in the 1970s South African engineers carved a tunnel through those mountains to bring water from the Orange River to the Fish River for irrigation in the Eastern Cape Karoo.

78. South Africa, *Final Report of the Drought Investigation Commission*, 51.

79. South Africa, *Final Report of the Drought Investigation Commission*, 52.

80. Kanthack, "Irrigation Development," 650.

81. Sutton, "Some Notes," 41.

82. South Africa, *Final Report of the Drought Investigation Commission*, 46.

83. "'The Great Thirstland': A Larger Egypt in South Africa," *Times of Natal*, 3 June 1920.

84. Evans, *Cultures of Violence*, 4.

85. South Africa, *Report of the Select Committee on Drought Distress Relief*, 64.

86. "The Forum," *Grocott's Mail,* 27 January 1925.

87. "Derf Gnik," letter to the editor, *Farmer's Weekly*, 9 June 1920, p. 1790.

88. G. P. Robarts to governor general, 14 March 1917; Ernest Long to governor general, 23 August 1918, NAR, GG vol. 1305, folder 35/22.

89. Leipoldt, "The Desert Shall Blossom," *Star*, 20 June 1921.

90. Weidner, *Fallacy*, 13.

91. Karlson, *Kalahari Problem* 6, 14.

92. August Karlson to High Commissioner of South Africa, September 19, 1923, NAR, GG vol. 1305, folder 35/50. See also Karlson, *Kalahari Problem.*

93. Ernest Long to governor general, 23 August 1918, NAR, GG vol. 1305, folder 35/28.

94. G. P. Robarts to governor general, 14 March 1917, NAR, GG vol. 1305, folder 35/22.

95. Van Sittert, "Nation-Building Knowledge," 97.

96. "The Waters of the Desert," *Cape Times*, 24 July 1920; "Kalahari Reclamation," *African World*, 28 August 1920.

97. "The Kalahari Dream," *Cape Times*, 26 July 1920.

98. Derf Gnik, letter to the editor, *Cape Times*, 30 July 1920.

99. G. H. Blenkinsop, "Work for the Unemployed," no newspaper title or date, Schwarz papers, Albany Museum.

100. William Jacobs, letter to the editor, *Farmer's Weekly*, 10 October 1923, p. 457.

101. "Schwarz Scheme Society Formed," *Farmer's Weekly*, 7 June 1933, p. 665.

102. "Derf Gnik," letter to the editor, *Cape Times,* 30 July 1920.

103. Reisner, *Cadillac Desert,* 114.

104. Long to governor general, 23 August 1918, NAR, GG vol. 1305, folder 35/28.

105. Ernest G. Long to governor general, 19 August 1918 and 23 August 1918, NAR, GG vol. 1305, folder 35/28.

106. August Karlson, *Times Engineering Supplement,* 2 September 1919.

107. Schwarz, *Kalahari,* 119, 129.

108. Van Reenen, "Resume,"179; "Droughts and Doorsteps," *Farmer's Weekly,* 6 October 1920, p. 783.

109. South African Senate, *Report from the Select Committee on Closer Land Settlement,* 27.

110. "South African Climate," *Star,* 20 July 1922; Schwarz, "The Kalahari Scheme as the Solution of the South African Drought Problem," 208; Schwarz, *Kalahari,* 45–46.

111. Schwarz, *Kalahari,* 42–43; Schwarz, "The Kalahari Scheme as the Solution," 220.

112. "Too Much State Irrigation!" *Farmer's Weekly,* 14 April 1915, p. 406.

113. South Africa, *Debates of the House of Assembly* 3 (1925): 1978.

114. Kanthack, letter to the editor, *Times Engineering Supplement,* February 1921.

115. Van Vuuren, *In the Footsteps of Giants,* 75.

116. "Best class" quote is from "Irrigation Development on the Sundays River," *Farmer's Weekly,* 17 March 1915, p. 112.

117. A. D. Lewis, "Irrigation in South Africa," *South Africa,* 26 May 1922.

118. Van Vuuren, *Footsteps,* 75–77.

119. Minnaar, "Graaff-Reinet," 63–65.

120. *Sunday Times,* 23 March 1924.

121. Forde, "Irrigation."

122. McKittrick, "Race and Rainmaking," 158. Farmers in the Transvaal, the Bathurst Farmer's Association, farmers in Victoria West, and citrus farmers along the Sundays sought to recruit Hatfield to South Africa.

123. Schwarz, letter to the editor, *Port Elizabeth Advertiser,* 29 August 1923.

124. "Derf Gnik," letter to the editor, *Farmer's Weekly,* 9 June 1920, 1790.

125. Hargreaves to G. M. Darell, 8 January 1927, NAR, GG vol. 2291, folder 11/162.

126. S. H. Boyle, letter to the editor, *Farmer's Weekly,* p. 833. "Dust" refers to the practice of seeding clouds by dispersing fine particulate, such as silver iodide or salt.

127. *Farmer's Weekly,* July 28, 1926, p. 1960. While *inyanga* depicted a specialist or professional in many of South Africa's indigenous languages (*inyanga yezulu,* or sky specialist, was one word for rainmaker), white South Africans gave it another, derogatory gloss—one that connoted primitivism.

128. Du Toit, *Report of the Kalahari Reconnaissance,* 49.

129. Schwarz, *Kalahari,* v.

130. Karlson's proposal can be found in NAR, PM vol. 1/2/102, folder 36/4A.

131. Schwarz, *Kalahari,* 32.

132. Schwarz, *Kalahari,* vi; 42.

133. Schwarz, *Kalahari,* 159–161.

134. Three months before he died, Ernest Glanville wrote of "Dr. Schwarz and his bold companions," and stated that the Namib and the Kalahari "threaten to join

hands." *Cape Argus*, 13 June 1925. Karlson claimed that Rider Haggard had expressed interest in his scheme: letter to the editor, *Journal of the African Society* 77, no. 20 (Oct. 1920): 77–79 (79). In 1935, Lawrence Green wrote approvingly of Schwarz's scheme in a chapter of *Great African Mysteries*, and stated that it was supported by "a distinguished group of scientists," 174–75.

135. *Schwarz*, "Kalahari and Its Possibilities," 10.

136. See, for example, Gessert, letter to the editor, *Farmer's Weekly*, 4 October 1922, n.p. Gessert not only seemed to carry no resentment that Schwarz had stolen his idea; he became one of Schwarz's most diehard supporters and amplifiers. Gessert initially focused on the increased rainfall that the Kalahari Scheme would provide; later, after a flurry of criticism of Schwarz's claim that his lakes would yield increase rainfall, Gessert switched gears. He continued to defend the claim of increased rainfall, but also insisted that the irrigation potential alone justified the project, since it promised agricultural stability.

137. "Solving the Riddle of the Lost Lakes," *Star*, 25 July 1925.

Chapter 6

1. Many of these articles were clipped and saved by Schwarz himself, who eventually hired a press clipping service. But, in keeping with his scattershot approach to citations in his own work, he pasted clippings haphazardly in several notebooks, and often failed to record the publication title and date.

2. R. M. Brown, "The Redemption of the Kalahari," 624; *Times of Natal*, 3 June 1920; *Diamond Fields Advertiser*, 5 June 1920; "Flooding the Kalahari,"*Queenstown Daily Representative and Free Press*, 19 December 1924; "Kan Zuid-Afrika's Regenval Vebeterd Worden," *De Zuid-Afrikaan*, 1 June 1920; "The Kalahari Dream," *Cape Times*, 26 July 1920.

3. *Bulawayo Chronicle*, 12 June 1920; Rolf Hartig, "Lahmseuche und Kalahari-Problem," *Mitteilungen der Farmwirtschafts-Gesellschaft*, 10 December 1920, pp. 687–93.

4. "Thirstland Redemption," *Farmer's Weekly*, 23 June 1920, p. 2013.

5. South Africa, *Debates of the House of Assembly* 3 (1925): 1986.

6. It was at this time that Schwarz began saving newspaper and magazine articles about his scheme.

7. Schwarz to Col. Amery, 9 July 1920, NAR, GG vol. 1305, folder 35/33. His request was ultimately denied.

8. Schwarz, *White Man's World*, 14–15.

9. William Jacobs, letter to the editor, *Farmer's Weekly*, 10 October 1923, p. 457.

10. W. G. Collins, letter to the editor, *Cape Times*, 11 September 1920.

11. D. S., letter to the editor, *Farmer's Weekly*, 10 August 1921, p. 2232.

12. Schwarz, "Kalahari Scheme as the Solution," 209.

13. Gold et al., "Control of Climate by Lakes," 175.

14. Smuts, "Science," 247.

15. "Professor Schwarz's Scheme," *Farmer's Weekly*, 13 October 1920, p. 961.

16. Akweenda, *International Law*, ch. 7.

17. Machado, *O Sul de Angola*.

18. Smuts to Foreign Office, 9 August 1920, NAR, GG vol. 165, folder 3/3182.

19. H. du Toit to Schwarz, 12 February 1921, NAR, LDB vol. 1365, folder 1318/12. Beinart, *Rise of Conservation*, deals with the Drought Commission in his

chapter on Heinrich du Toit. But historians have not recognized the connection between the commission and the popularity of Schwarz's scheme.

20. William Charles Scully, "The Desiccation of South Africa Part 1," *Cape Times*, 31 August 1922.

21. Brown, "Redemption of the Kalahari," 625.

22. Gregory, "Ancient River System," 539.

23. H. Paulsmeier, letter to the editor, *Farmer's Weekly*, 6 December 1922, n.p.

24. Schwarz, "Control of Climate," 171–72. Also Schwarz, letter to the editor, *Farmer's Weekly*, 24 August 1921, p. 2426.

25. Gold et al., "Control of Climate: Discussion," 175. To underscore how little was certain at the time, one discussant simultaneously supported Schwarz's meteorological claims and the Wegener hypothesis of continental drift—and was ridiculed not for his support of Schwarz but for his support of Wegener's hypothesis, which much of the scientific establishment of the time insisted was utterly implausible.

26. O'Donovan, letter to the editor, *Farmer's Weekly*, 14 November 1923; John Pons, letter to the editor, *Farmer's Weekly*, 5 December 1923 pp. 1264–65.

27. O'Donovan, letter to the editor, *Farmer's Weekly*, 14 November 1923, n.p.

28. "Kan Zuid-Afrika's Regenval Verbeterd Worden?," *De Zuid-Afrikaan*, 1 June 1920.

29. "Menace of Desiccation," *Cape Argus*, 9 June 1920.

30. "Die Bewässerung der Kalahari," *Allgemeine Zeitung*, 20 April 1925, p. 1.

31. Review in *Journal of the Royal African Society* 77, no. 20 (October 1920): 67–69.

32. Review of "The Kalahari; or, Thirstand Redemption," *Quarterly Journal of the Royal Meteorological Society* 47 (1921): 73.

33. "Reclaiming the Kalahari," *Cape Argus*, 18 September 1922.

34. "Südwestafrika und das Professor Schwarzsche Bewässerungsproblem," *Allgemeine Zeitung*, 20 April 1925.

35. W. G. Collins, letter to the editor, *Cape Times*, 11 September 1920.

36. South Africa, *Debates of the House of Assembly* 3 (1925): 1983.

37. J. R. Sutton, undated and uncited newspaper article, likely September 1925, Albany Museum, Schwarz papers.

38. "Thirstland Redemption: Reply to Dr. Sutton," *Diamond Fields Advertiser*, 25 April 1923. For information on the Rubidge family, including Walter, see Beinart, *Rise of Conservation*, ch. 9.

39. Philip Townshend, "Gospel of Tree-Planting," *Farmer's Weekly*, 31 August 1921, p. 2548.

40. South Africa, *Final Report of the Drought Investigation Commission*, 182.

41. "A Sufferer from Drought," letter to the editor, *Farmer's Weekly*, 23 February 1921, p. 3146.

42. Fred Nicholson to Government entomologist, 30 June 1921, Free State Archives Repository, CEN vol. 628; Ernest Galpin to chief, division of botany, 21 March 1927, Free State Archives Repository, CEN vol. 614.

43. "Tsama," letter to the editor, *Farmer's Weekly*, 19 February 1920, n.p.; H. Paulsmeier, letter to the editor, *Farmer's Weekly*, 6 December 1922, n.p.

44. William Jacobs, letter to the editor, *Farmer's Weekly*, 10 October 1923, p. 457.

45. South Africa, *Debates of the House of Assembly* 3 (1925): 1978.

46. Typescript of March 1918 Senate debate, p. 6, Schwarz papers, Albany Museum.

47. Schmokel, "Myth of the White Farmer."

48. "The Dry North." *Star*, November 11, 1919.

49. E. H. Louw of Karoo, *Debates of the House of Assembly* 3 (1925): 1981.

50. William Jacobs, letter to the editor, *Farmer's Weekly*, 10 October 1923, p. 457.

51. Robert McKinnon of Queenstown, letter to the editor, *Farmer's Weekly*, 14 July 1920, p. 2376.

52. Claude Lowe, letter to the editor, *Farmer's Weekly*, 23 February 1927, n.p.

53. John Pons letter to the editor, *Farmer's Weekly*, 19 January 1921, n.p.; "Dry Farmer," letter to the editor, *Farmer's Weekly*, 29 December 1920, p. 2231; Philip Townshend, letter to the editor, *Farmer's Weekly*, 24 December 1919, n.p.

54. Ernest G. Long to governor general, 23 August 1918, NAR GG vol. 1305, folder 35/28.

55. "Kalahari Scheme Advocated," *Star* (no newspaper or date given), Schwarz papers, Albany Museum.

56. Schwarz, "Kalahari Project in Relation to Distress," 322.

57. Weidner, *Fallacy*, 13.

58. A. Karlson to high commissioner, 19 September 1923, NAR, GG vol. 1305, folder 35/50.

59. Karlson, "The Lost Lakes of South Africa," NAR, PM vol. 1/2/102, folder 36/4A.

60. Schwarz, *Kalahari*, 156–57.

61. "Flooding the Kalahari," *Queenstown Daily Representative and Free Press*, 19 December 1924.

62. Karlson, "The Lost Lakes of South Africa," NAR, PM vol. 1/2/102, folder 36/4A.

63. Schwarz, "Kalahari Project in Relation to Distress," 322–23.

64. Schwarz, "Kalahari Project in Relation to Distress," 323.

65. Schwarz, *Kalahari*, 117; "Südwestafrika u. das Professor Schwarzsche Bewässerungsproblem," *Die Allgemeine Zeitung*, 20 April 1925.

66. W. G. Collins, letter to the editor, *Cape Times*, 11 September 1920.

67. Karlson, "The Lost Lakes of South Africa," NAR, PM vol. 1/2/102, folder 36/4A; Hyam, *Failure*, 104; David E. Torrance, "Britain, South Africa, and the High Commission Territories: An Old Controversy Revisited," *Historical Journal* 41, no. 3 (1998): 751–72.

68. Schwarz, *Kalahari*, 157.

69. Karlson, "The Lost Lakes of South Africa," 13, NAR, PM vol. 1/2/102, folder 36/4A.

70. Karlson, *Kalahari Problem*, 18.

71. Schwarz, *Kalahari*, 159–61.

72. Karlson, "The Lost Lakes of South Africa," 15, NAR, PM vol. 1/2/102, folder 36/4A.

73. Voyles, *Wastelanding*; Henni, "Nuclear Powers."

74. Schwarz, "Kalahari Project in Relation to Distress," 325.

75. Karlson, "The Lost Lakes of South Africa," 13, NAR, PM vol. 1/2/102, folder 36/4A.

76. W. G. Collins, letter to the editor, *Cape Times*, 11 September 1920.

77. "Flooding of the Kalahari," *Queenstown Daily Representative and Free Press*, 19 December 1924.

78. Schwarz, *Kalahari*, 159, used the "official estimate" of 27,000 heads of households.

79. Karlson, "The Lost Lakes of South Africa," 44, NAR, PM vol. 1/2/102, folder 36/4A.

80. Karlson, "The Lost Lakes of South Africa," NAR, PM vol. 1/2/102, folder 36/4A.

81. Karlson to high commissioner, 19 September 1923; Karlson, "The Lost Lakes of South Africa," NAR, PM vol. 1/2/102, folder 36/4A.

82. Hyam, *Failure*; Torrance, "Britain, South Africa."

83. Zollmann, "Negotiated Partition"; Posel, *Making of Apartheid.*

84. Barnhard, *Encountering Adamastor*, 47, 34.

85. Farini, *Through the Kalahari Desert.*

86. MacDonald, *Conquest of the Desert*, 13.

87. Schwarz, "Kalahari and Its Possibilities."

88. Hauptmann Friedrich von Erckert, "In der Kalahari," *Deutsche-Südwestafrikanischen Zeitung*, 5 December 1906, p. 5.

89. Karlson, *Kalahari Problem*, 17–18.

90. Karlson, "The Lost Lakes of South Africa," 11, NAR, PM vol. 1/2/102, folder 36/4A.

91. Schwarz, letter to the editor, *United Empire* 14, no. 10 (1923): 613.

92. "Fertile 'Deserts,'" *United Empire* 14, no. 10 (1923): 479.

93. "The Kalahari," *Windhoek Advertiser*, 13 August 1924, p. 2.

94. Irrigation Department Bulletin 4, NAN, SWAA vol. 598, folder A 66/1. The Reconnaissance Report has a chapter titled "The Nature of the Problem," but the idea of a problem recurs throughout the text.

95. "Fertile 'Deserts,'" *United Empire* 14, no. 8 (1923): 479.

96. Schwarz, *Kalahari*, 157.

97. Eduard Burchard, "Zur Ableitung des Kunene in die Etoschapfanne nach Prof. Schwarz," *Allgemeine Zeitung*, 15 June 1925 and 17 June 1925; see also McKittrick, "Making Rain, Making Maps."

98. F. T. Garbutt, letter to government secretary, Mafeking, 14 October 1920, and H. C. Weatherhilt, letter to government secretary, Mafeking, n.d., NAR, GG vol. 1305, folder 35/40.

99. Stigand to government secretary, Mafeking, 16 November 1920, NAR, GG vol. 1305, folder 35/40.

100. Stigand to government secretary, Mafeking, 16 November 1920, NAR, GG vol. 1305, folder 35/40.

101. R. Rothe to Outjo magistrate, 10 August 1925, NAN, SWAA vol. 598, folder A66/1.

102. South Africa, *Debates of the House of Assembly* 3 (1925): 1972.

103. Schwarz, *Kalahari*, 103.

104. G. H. Blenkinsop, "Work for the Unemployed," no newspaper title or date, Schwarz papers, Albany Museum.

105. Karlson, "The Lost Lakes of South Africa," 26, NAR, PM vol. 1/2/102, folder 36/4A.

106. Robert McKinnon letter to the editor, *Farmer's Weekly*, 14 July 1920, p. 2376.

107. F. Gessert, letter to the editor, *Farmer's Weekly*, 14 July 1920, p. 2377.

108. Markham, *South African Scene*, 402–3.

109. K. B. Quinan to Smuts, 28 May 1923, NAN, SWAA vol. 598, folder A66/1.

110. K. B. Quinan to Smuts, 28 May 1923, NAN, SWAA vol. 598, folder A66/1.

111. Director of irrigation to high commissioner for South West Africa, 27 February 1925, NAN, SWAA vol. 598, folder A66/1.

112. J. R. Sutton, "'Thirstland' Redemption: The Schwarz Theory Criticized," undated newspaper article, likely September 1925, Schwarz papers, Albany Museum.

113. PM Hertzog to Quinan, 15 September 1925, NAN, SWAA vol. 598, folder A66/1.; Lewis, preface to du Toit, *Report of the Kalahari Reconnaissance.*

114. Tilley, *Africa as Living Laboratory*; Jacobs, *Birders of Africa.*

115. Irrigation Department Bulletin 1, p. 3, NAR, GG vol. 1305, folder 35/59.

116. Lewis, preface to du Toit, *Report of the Kalahari Reconnaissance.*

117. F. H. Barber, "Is South Africa Drying Up?" *Agricultural Journal of the Cape of Good Hope*, vol. 36 (1910): 167.

118. McKittrick, "Making Rain."

119. Schwarz, *Kalahari*, 64.

120. Schwarz, *Kalahari*, 149.

121. Schwarz, *Kalahari*, 172.

122. Du Toit, "Kalahari and Some of Its Problems," 91.

123. Mukerji, *Impossible Engineering*, 44.

124. Levinkind, "Droughts in South Africa." Officially, the 1918 figures show 1 percent of the country as drought-stricken, a negligible amount. But also note that people may have experienced these as drought years because of the nature of the calculations; and that, incredibly, this 1941 piece still appears to use calendar years.

125. South Africa, *Debates of the House of Assembly* 3 (1925): 1974.

126. J. R. Sutton, "'Thirstland' Redemption: The Schwarz Theory Criticized," undated newspaper article, likely September 1925, Schwarz papers, Albany Museum.

127. Pim, *Financial and Economic Position of the Bechuanaland Protectorate*, 159, for rainfall totals by district. Yet *Die Weste*, which opposed Schwarz's scheme, argued, "We have had good rains this year, but that is not evidence that we should not take care of the future." *Die Weste*, 17 April 1925.

128. "A Professor's Unique Voyage," *Star*, 7 September 1925.

129. *Bechuanaland Protectorate Report 1925–26*, 25; Fleisch and Möhlig, *Kavango Peoples.*

130. Du Toit, *Report of the Kalahari Reconnaissance*, 10.

131. "S. African Irrigation Schemes," *Times* (London), 4 June 1925, p. 11.

132. Hahn to secretary for South West Africa, 27 April 1925, NAN, SWAA vol. 598, folder A66/1.

133. Yusoff, *Billion Black Anthropocenes*, 4.

134. Du Toit, Kalahari Reconnaissance diary, 62, University of Cape Town Libraries special collections, BC 722, Alex L. du Toit papers, L 1925.

135. Du Toit, Kalahari Reconnaissance diary, 47–48, University of Cape Town Libraries special collections, BC 722, Alex L. du Toit papers, L 1925.

136. Du Toit, Kalahari Reconnaissance diary, 65, University of Cape Town Libraries special collections, BC 722, Alex L. du Toit papers, L 1925.

137. Schwarz, Kalahari Expedition diary, 87, Albany Museum, Schwarz papers.

138. Du Toit, Kalahari Reconnaissance diary, 44, University of Cape Town Libraries special collections, BC 722, Alex L. du Toit papers, L 1925.

139. Du Toit, Kalahari Reconnaissance diary, 48–50, University of Cape Town Libraries special collections, BC 722, Alex L. du Toit papers, L 1925.

140. Du Toit, Kalahari Reconnaissance diary, 64–65, 69, University of Cape Town Libraries special collections, BC 722, Alex L. du Toit papers, L 1925.

141. Schwarz, Kalahari Expedition diary, 70, 73, 91, Albany Museum, Schwarz papers.

142. Schwarz, Kalahari Expedition diary, n.p., Albany Museum, Schwarz papers.

Chapter 7

1. Du Toit, *Report of the Kalahari Reconnaissance*, 54.
2. Du Toit, *Report of the Kalahari Reconnaissance*, 30–31.
3. Du Toit, *Report of the Kalahari Reconnaissance*, 56.
4. Jeffares, "Ngamiland Waterways Surveys of 1937," 1938.
5. Hyam, *Failure*, 106–9.
6. Bechuanaland Protectorate, *Report for 1925–26*, 9.
7. J. M. Moubray, letter to the editor, *Farmer's Weekly*, 26 January 1927, n.p..
8. "Verkenning van die Kalahari," *Die Landbouweekblad*, 18 August 1926, 761, 763.
9. "The Dry North," *Star*, 11 November 1919.
10. South Africa, *Debates of the House of Assembly* 3 (1925): 1981.
11. Irrigation Department Bulletin 3, NAN SWAA vol. 598, folder A66/1.
12. Du Toit, *Report of the Kalahari Reconnaissance*, 7–8, 49, 51.
13. A. Kern, letter to the editor, *Farmer's Weekly*, 29 March, 1933, p. 85.
14. "Farmer John," letter to the editor, *Farmer's Weekly*, 29 March 1933, p. 86.
15. Claude Lowe, letter to the editor, *Farmer's Weekly*, 23 February 1927, n.p.
16. L. A. Rose-Innes, letter to the editor, *Farmer's Weekly*, 5 January 1927, n.p.
17. L. A. Rose-Innes, letter to the editor, *Farmer's Weekly*, 16 February 1927, n.p.
18. A. Kern letter to the editor, *Farmer's Weekly*, 29 March 1933, p. 85.
19. "The Kalahari Report," *Star*, 26 July 1926. See also A. Kern, letter to the editor, *Farmer's Weekly*, 29 March 1933, p. 85.
20. "The 'Kalahari' Expedition," *Daily Dispatch* (East London), no date, Albany Museum, Schwarz papers; "Kalahari Scheme," *Grocott's Daily Mail*, 26 July 1926.
21. L. A. Rose-Innes, letter to the editor, *Farmer's Weekly*, n.p; see also Claude Lowe, letter to the editor, *Farmer's Weekly*, 23 February 1927, n.p.
22. P. F. le Roux, letter to the editor, *Farmer's Weekly*, 26 January 1927, n.p.
23. South Africa, *Debates of the House of Assembly* 8 (1927): 1439.
24. South Africa, *Debates of the House of Assembly* 8 (1927): 1435–37.
25. South Africa, *Debates of the House of Assembly* 8 (1927): 1447.
26. "De Grote Droogte," *De Kerkbode*, 2 March 1927.
27. "When the Kalahari Is Redeemed," *South Africa*, 13 January 1928.
28. Schwarz, "Eastern Kalahari Project."
29. "Kalahari Problem," *Star*, 8 July 1927.
30. "The Kalahari: Dr. du Toit Corrected," *Farmer's Weekly*, 24 August 1927, n.p.
31. Schwarz, *Kalahari and Its Native Races*, 151–207.
32. "Pioneer," letter to the editor, *Star*, 24 September 1927.
33. "The Kalahari," *Farmer's Gazette*, 11 November 1927.
34. H. C. de Wet, quoted in "Changes in the Rainfall: Cape Farmers Want Further Inquiry," *Star*, 23 July 1927.
35. A.R.L., "In Memoriam: The Late Professor E. H. L. Schwarz," *The Rhodian* 14 (1929): 3–4.
36. Gregory, "Prof. E. H. L. Schwarz," 101.

37. Prime Minister's Office, 21 May 1931, NAR, URU vol. 1207, minute number 1487.

38. Thompson, *Moisture and Farming*, 228.

39. Thompson, *Moisture and Farming*, 13–14.

40. "The Drought Dilemma," *Farmer's Weekly*, 15 February 1933, p. 1422; Temple Fyvie, letter to the editor, *Farmer's Weekly*, 22 March 1933, p. 27.

41. South Africa, *Debates of the House of Assembly* 55 (1946): 1084.

42. H. E. Bingham Boys, letter to the editor, *Farmer's Weekly*, 5 July 1933, p. 876.

43. By one measure, the export prices for agricultural products fell by 44 percent in two years, far outpacing declines in other sectors and compounding the effect of drought losses. Frankel, "Situation in South Africa," 95.

44. Ross et al., *Cambridge History Vol. 2*, 284.

45. Tischler, "Cultivating Race," ch. 7.

46. Ross et al., *Cambridge History Vol. 2*, 223.

47. Thompson, *Moisture and Farming in South Africa*, 94.

48. Van der Watt, "'It Is Drought,'" 169–70.

49. J. A. Manson, letter to the editor, *Farmer's Weekly*, 12 April 1933, p. 199.

50. Schonken, "State of the National Water Supply," 203.

51. Du Toit, "Some Considerations."

52. Richards, "Economic Revival," 628. Richards claimed in 1934 that government policies had become more rational—the same claim made fifteen years later by Ross. If the start of "drying out" and farmer distress was always moving into the more recent past, the discovery of rationality and wise use was always just beginning.

53. Leppan, *Agricultural Policy*, 39.

54. Leppan, *Agricultural Policy*, 45.

55. Leppan, *Agricultural Policy*, 37–38.

56. Leppan, *Agricultural Policy*, 45.

57. McDermott-Hughes, *Whiteness in Zimbabwe*.

58. Ross, *Soil Conservation*.

59. Van der Watt, "'It Is Drought,'" 42–43.

60. X Ray, "Drought Gets Home Again Despite Warnings," *Farmer's Weekly*, 11 December 1935, p. 1012.

61. P. F. Wall, "The Crime of Farming without Feed Reserves," *Farmer's Weekly*, 18 December 1935, p. 1084.

62. H. L. Simmons, "The Picture to Remember in Times of Abundance," *Farmer's Weekly*, 19 February 1936, p. 1739.

63. F. Nicholson, letter to the editor, *Farmer's Weekly*, 3 May 1933, p. 366.

64. Quoted in van der Watt, "'It Is Drought,'" 166–67.

65. Charles Hall, letter to the editor, *Farmer's Weekly*, 29 December 1937, pp. 1084-85; "Insuring against Drought," *Farmer's Weekly*, 29 December 1937, p. 1081.

66. J. A. van Niekerk, letter to the editor, *Farmer's Weekly*, 20 October 1943, pp. 220–21.

67. Leonard Flemming, letter to the editor, *Farmer's Weekly*, 22 March 1933, p. 29.

68. Leonard Flemming, letter to the editor, *Farmer's Weekly*, 22 March 1933, p. 29.

69. Nobbs to Machtig, 22 July 1931, NAUK, DO 35/393/4.

70. Downie to Harding, n.d., 1933, NAUK, DO 35/393/4.

71. Tischer, "Cultivating Race," ch. 7.

72. Temple Fyvie, letter to the editor, *Farmer's Weekly*, 22 March 1933, p. 27.

73. P.W. de Wet, letter to the editor, *Farmer's Weekly*, 15 February 1933, p. 1409.

74. Mrs. C.F. Middleton, letter to the editor, *Farmer's Weekly*, 15 February 1933, p. 1409.

75. A. Kern, letter to the editor, *Farmer's Weekly*, 29 March 1933, p. 85.

76. "Schwarz-plan Verg Ondersoek," *Die Volksblad*, 1 June 1933, p. 1.

77. Editorial, *Farmer's Weekly*, 21 June 1933, p. 761.

78. "Schwarz-plan Verg Ondersoek," *Die Volksblad*, 1 June 1933, p. 1. No record of this commission has been located, and neither Gessert nor Schwarz ever mentioned it.

79. C. Weidner, letter to the editor, *Farmer's Weekly*, 12 April 1933, p. 198.

80. Schonken, "State of the National Water Supply," 206.

81. E. G. Bryant, letter to the editor, *Farmer's Weekly*, 22 March 1933, p. 29.

82. South Africa, *Report of the Desert Encroachment Committee*, 16.

83. John Owen-Collett, letter to the editor, *Farmer's Weekly*, 12 July 1933, p. 944.

84. On the building of Kariba, see Tischler, *Light and Power*.

85. McKittrick, "Rainmaker Goes to Court."

86. N. E. Walton, letter to the editor, *Farmer's Weekly*, 28 March 1934, p. 97.

87. E. G. Bryant, letter to the editor, *Farmer's Weekly*, 20 June 1934, p. 898.

88. John Owen-Collett, letter to the editor, *Farmer's Weekly*, 23 May 1934, p. 627.

89. "Kalahari Society's Meeting," *Farmer's Weekly*, 6 June 1934, p. 799.

90. Ross et al., *Cambridge History Vol. 2*, 250; Jeeves and Crush, *White Farms*, 1.

91. Tayler, "'Our Poor.'"

92. Stebbing, "Encroaching Sahara," 515; Swift, "Desertification Narratives."

93. Sörgel, *Drei grossen "A,"* 101–3.

94. For a discussion of Sörgel's ideas, see Lehmann, *Desert Edens*.

95. Sörgel, *Atlantropa*, 70.

96. The Kalahari Scheme was integrated into Sörgel's vision despite his doubts that it would work—doubts apparently based on the report of the Kalahari Reconnaissance. Lehmann, *Desert Edens*, 69.

97. Sörgel and Siegwart, "Erschliessung Afrikas durch Binnenmeere."

98. Idriess, *Great Boomerang*, 247.

99. Timbury, *Battle for the Inland*, 22, 139. While the other schemes mentioned in this book had fallen out of public discussion by the early 1960s, the Bradfield Scheme continues to be the subject of political campaigns.

100. F. Gessert, letter to the editor, *Farmer's Weekly*, 29 March 1933, p. 87.

101. "Drying Up of Union: Harm Done by Man: Professor Obst's Views," *Rand Daily Mail*, 9 March 1933; also Korn and Martin, "Die jüngere geologische und klimatische Geschichte."

102. Lehmann, *Desert Edens*.

103. Obst, "Geomorphologische Forschungsreise."

104. Reuber et al., *Politische Geographien Europas*, 42; Obst, "Die Lebensräume der Weltvölker."

105. Korn, *Zwiegespräch in der Wüste*, 33–34. South African officials publicly distanced themselves from Obst's expedition, however.

106. "Dr. Obst and the Schwarz Theory," *Farmer's Weekly*, 17 April 1935, p. 382.

107. "New Kind of Brown Shirts Now," *Rand Daily Mail*, 12 September 1934, p. 12.

108. "Col. Reitz on Farm Problems," *Rand Daily Mail*, 12 September 1935, p. 8.

109. Obst, "Junge Krustenbewegungen," 452.

110. Obst, “Der Kampf gegen die ‘Austrocknung,’” 115–16.

111. E. Owen Wright, letter to the editor, *Farmer's Weekly*, 5 January 1938, pp. 1144–46.

112. A. de Villiers, letter to the editor, *Farmer's Weekly*, 9 February 1938, p. 1493.

113. John Owen-Collett, letter to the editor, *Farmer's Weekly*, 23 May 1934, p. 627.

114. A. de Villiers, letter to the editor, *Farmer's Weekly*, 9 February 1938, p. 1493.

115. Hyam, *Failure*, 144.

116. Hyam, *Failure*, 140–43.

117. “Col. Reitz's Kalahari Impressions,” *Farmer's Weekly*, 16 October 1935, p. 407.

118. Jeffares, “Report.” For details on Jeffares, see https://www.engineeringnews.co.za/article/jg-afrika-celebrates-100-years-of-excellence-2022-03-01-1. Accessed 18 October 2023.

119. Jeffares, “Report,” 41.

120. Jeffares, “Report,” 10–11.

121. Jeffares, “Report,” 63.

122. J. K. Phoks, letter to the editor, *Farmer's Weekly*, 3 May 1933, p. 366. The name is likely a pseudonym, given that no reference to this person has been found. The author was probably white and used the name as a kind of badge of authenticity.

123. “Swamp Drainage in Bechuanaland,” *Farmer's Weekly*, 28 July 1937, p. 1533.

124. Harlech to Machtig, 25 August 1945; high commissioner for South Africa to secretary of state for dominion affairs, 29 August 1945, NAUK, DO 35/1189.

125. Thompson, *Moisture and Farming*, 21, 28, 39–40.

126. Thompson, *Moisture and Farming*, 5.

127. Thompson, *Moisure and Farming*, 24.

128. Mackenzie, *Report on the Kalahari Expedition*, 1.

129. “Waters of the Kalahari,” *Times* (London), 25 August 1945, p. 5.

130. John Owen-Collett, letter to the editor, *Farmer's Weekly*, 12 July 1933, p. 944.

131. Du Toit, “Some Considerations,” 4.

132. Schumann and Thompson, *Study of South African Rainfall*, 35–36.

133. Thompson, *Moisture and Farming*, 102.

134. Thompson, *Moisture and Farming*, 97.

135. Thompson, *Moisture and Farming*, 100.

136. He also critiqued the much-lauded techniques of dry farming, which Heinrich du Toit and other progressive farmers had hailed as the salvation of South African agriculture. There was no point in creating a dry “soil mulch” on the surface of the land, Thompson said. Evaporation rates were so high that the water was gone before the farmer could get outside and cultivate. Thompson, *Moisture and Farming*, 175.

137. Thompson, *Moisture and Farming*, 16–17.

138. Quoted in Khan, “Soil Wars,” 442.

139. “Soil a Wasting Asset,” *Farmer's Weekly*, 23 October 1946, p. 95.

140. Quoted in van der Watt, “‘It Is Drought,’” 59, 62.

141. *Die Landbouweekblad*, 13 March 1940, p. 16; quoted in van der Watt, “‘It Is Drought,’” 60.

142. *Die Landbouweekblad*, 4 August 1943; quoted in van der Watt, “‘It Is Drought,’” 63.

143. Van der Watt, “‘It Is Drought,’” 63, 65–66.

144. H. Viedge, letter to the editor, *Farmer's Weekly*, 20 October 1943, p. 220.

145. Interestingly, this is almost precisely the length of time since Schwarz had

first identified a crisis, reflecting how the timeline kept shifting, although the sense of danger did not.

146. "An Indictment of South African Farming Methods," *Farmer's Weekly*, 17 November 1943, p. 412.

147. "Rain Maker?" *Time*, 8 November 1943, https://content.time.com/time/subscriber/article/0,33009,796267,00.html. Accessed 18 October 2023.

148. Harlech to C. R. Attlee, MP Dominions Office, 27 May 1943, NAUK, DO 35/1189.

149. Leppan, *Agricultural Policy*, 17–19.

150. Van der Watt, "'It is Drought,'" 73, notes that the Veld Trust blamed economic policies for farmers' lack of soil conservation strategies.

151. J. J. Theron, "Faulkner-Bewerings Geweeg en Verwerp," *Die Landbouweekblad*, 24 May 1944, p. 15; quoted in van der Watt, "'It Is Drought,'" 72.

152. Fick, *Abuse of the Soil*, 45. Emphasis mine.

153. Charles Hall, letter to the editor, *Farmer's Weekly*, 29 December 1937, pp. 1084–85.

154. Van der Watt, "'It Is Drought,'" 101.

155. Department of Agriculture and Forestry, *Reconstruction of Agriculture*, 6–7.

156. Dodson, *Soil Conservation Safari*; Delius and Schirmer, "Soil Conservation."

157. Department of Agriculture and Forestry, *Reconstruction of Agriculture*, 6–7.

158. Nicholls, *South African Native Policy*, 4.

159. "AGWK," letter to the editor, *Farmer's Weekly*, 19 April 1933, p. 254.

160. See, for example, H. Viege, letter to the editor, *Farmer's Weekly*, 3 November 1943, pp. 318–19; "South African," letter to the editor, *Farmer's Weekly*, 24 November 1943, p. 463; "Indigenous Tree," letter to the editor, *Farmer's Weekly*, 8 December 1943, p. 556.

161. Showers, *Imperial Gullies*; Jacks and Whyte, *Vanishing Lands*.

162. Delius and Schirmer, *Soil Conservation*, 735; Ross, *Soil Conservation*, 21; van der Watt, "'It Is Drought,'" 53.

163. "Icarus," letter to the editor, *Farmer's Weekly*, 31 October 1945, p. 443.

164. Daisy Schwarz, "'Schwarz of the Kalahari': The Man as I Knew Him," *Outspan*, 31 August 1945, pp. 13, 15, 49.

165. "The Kalahari Becomes a Practical Proposition," *Farmer's Weekly*, 19 June 1946, p. 1243.

Chapter 8

1. Mackenzie, *Report on the Kalahari Expedition*, 1, 20.

2. South Africa, *Debates of the House of Assembly,* 58 (1946): 8120.

3. Mackenzie, *Report on the Kalahari Expedition*, 2.

4. Mackenzie, *Report on the Kalahari Expedition,* 29.

5. Mackenzie, *Report on the Kalahari Expedition*, 16–19.

6. South Africa, *Debates of the House of Assembly* 58 (1946): 8089. Ironically, du Toit was an ardent opponent of the National Party.

7. South Africa, *Debates of the House of Assembly* 58 (1946): 8075–8106.

8. Oudtshoorn-boer, letter to the editor, *Die Landbouweekblad*, 7 February 1945, p. 17; G. J. van der Merwe, letter to the editor, *Die Landbouweekblad*, 14 March 1945, p. 34.

9. Marais, *Only the Refilling of the Kalahari Lakes*, 8, 12. The reference to republi-

canism reflects the National Party's commitment to cutting ties with the British Empire, something that was done in 1960 after white voters approved it in a referendum.

10. Kokot, *Investigation*, 56–57.

11. Kokot, *Investigation*, 64.

12. "Drought in Africa Worst in 100 Years," *New York Times*, 27 July 1949, p. 4.

13. W. F. Marais, letter to the editor, *Farmer's Weekly*, 29 March 1950, p. 14.

14. D. F. Kokot, "Kalahari Facts and Fantasies," *Farmer's Weekly*, 26 July 1950, p. 49.

15. John G. Collett to Harry Oppenheimer, 22 February 1969; John G. Collett, *Prevent Desert Encroachment with the Schwarz-Kalahari Scheme* (Cradock, South Africa: White and Boughton, 1949).

16. Booysen and Hofmeyr, *Save the Soil*, 1–4.

17. Booysen and Hofmeyr, *Save the Soil*, 33.

18. Booysen and Hofmeyr, *Save the Soil*, photos following p. 26.

19. South Africa, *Report of the Commission of Inquiry into European Occupancy of the Rural Areas*, 2–3, 50. Hugo, in "Frontier Farmers," 549–50, notes that by the late 1970s, the state saw the white farm population not only in terms of domestic security but in terms of securing the borders against Black liberation movements in neighboring countries and in the homelands. It offered subsidies to whites who occupied farms along these borders.

20. South Africa, *Report of the Commission of Inquiry into European Occupancy of the Rural Areas*, 58.

21. Foster, *Washed with Sun*, 242.

22. South Africa, *Report of the Commission of Inquiry into European Occupancy of the Rural Areas*, 5.

23. South Africa, *Debates of the House of Assembly* 55 (1946): 1502, 1522, 1526; 58 (1946): 8136.

24. Charles du Preez, letter to the editor, *Farmer's Weekly*, 14 November 1951, p. 18.

25. H. Holford, letter to the editor, *Farmer's Weekly*, 13 February 1952, p. 19.

26. Van Vuuren, *Footsteps of Giants*, 147.

27. Delius and Schirmer, "Soil Conservation," 736; Driver, "Anti-Erosion Policies."

28. South Africa, *Report of the Commission of Inquiry into European Occupancy of the Rural Areas.*

29. "Simplicitas," letter to the editor, *Rand Daily Mail*, 20 August 1954, p. 10; W. Kerr, letter to the editor, *Rand Daily Mail*, 15 April 1957, p. 4; A. van Heerden, "Kougha Water Can Augment Sundays River," *Farmer's Weekly*, 28 January 1959, pp. 27, 29.

30. Giliomee, *Afrikaners*; Dubow, *Apartheid.*

31. Dubow, *Apartheid*, 27.

32. David Mermelstein, "Anti-Apartheid Reader: The Struggle against White Racist Rule in South Africa," 97.

33. Roos, "South African History." It also had little support among Black South Africans. The Freedom Charter of 1955 asserted that the country belonged "to all who live in it, black and white," and this remained the policy of the ANC.

34. Quoted in Brits, "Voice of the 'People'?," 65.

35. Brits, "Voice of the 'People'?"

36. Dubow, *Apartheid*, 27, 30; Giliomee, *Afrikaners*, ch. 13.

37. Dubow, *Apartheid*, 112–13.

38. D. G. Steyn, Venterstad Farmers' Association chairman, letter to the editor, *Farmer's Weekly*, 12 November 1952, p. 21.

39. Simons, "Harnessing the Orange River."

40. Van Vuuren, *Footsteps of Giants*, 195.

41. Simons, "Harnessing the Orange River," 133.

42. Emmett and Hagg, "Politics of Water Management," 306.

43. Emmett and Hagg, "Politics of Water Management," 309–313.

44. Emmett and Hagg, "Politics of Water Management," 322.

45. Isaacman, "Cahora Bassa Dam," 104.

46. A series of dams was required due to the nature of the Kunene itself. A couple of them offered water for irrigation in Angola, but most were designed to ensure that the Ruacana power scheme at Ruacana Falls, at the border with South West Africa, would function. Work on the dams had to be abandoned upon Angolan independence in 1975, and some dams were subsequently damaged during the continuing war between South Africa and the army of the South West Africa People's Organization (SWAPO).

47. Quoted in World Council of Churches, *Cunene Dam Scheme*, 9.

Epilogue

1. Hipondoka, Kempf, and Jousse, "Paleo and Present Ecological Value of the Etosha Pan."

2. Nugent, "The Zambezi River."

3. McBride, Kruger, and Dyson, "Changes in Extreme Daily Rainfall Characteristics"; Karypidou, Katragkou, and Sobolowski, "Precipitation over Southern Africa."

4. Rohde et al., "Vegetation and Climate Change."

5. Haltinner and Sarathchandra, *Inside the World of Climate Change Skeptics*, 11.

6. "The Risks of Climate Engineering," *New York Times*, 12 February 2015; also Hamilton, *Earthmasters*, 168.

7. Kokot, *Investigation*, 94.

8. Hamilton, *Earthmasters*, 166.

9. Chukwumerije Okereke, "My Continent Is Not Your Giant Laboratory," *New York Times*, 18 April 2023.

10. Stephens et al., "Dangers of Mainstreaming Solar Geoengineering," 163.

11. Chukwumerije Okereke, "My Continent Is Not Your Giant Laboratory," *New York Times*, 18 April 2023.

BIBLIOGRAPHY

Adam, Heribert, and Hermann Giliomee. *Ethnic Power Mobilized: Can South Africa Change?* New Haven, CT: Yale University Press, 1979.

Adhikari, Mohamed. *Anatomy of a South African Genocide: The Extermination of the Cape San Peoples*. Athens: Ohio University Press, 2011.

Akweenda, Sackey. *International Law and the Protection of Namibia's Territorial Integrity*. New York: Springer, 1997.

Araripe, T. de Alancar. *Historia da provincia do Ceará, desde os tempos primitivos até 1850*. Recife, Brazil: Typographia do Jornal do Recife, 1867.

Anderson, Andrew A. *Twenty-Five Years in a Waggon in the Gold Regions of Africa*. London: Chapman and Hall, 1887.

Anderson, Katharine. "Looking at the Sky: The Visual Context of Victorial Meteorology." *British Journal for the History of Science* 36, no. 3 (September 2003): 301–32.

Anderson, Katharine. *Predicting the Weather: Victorians and the Science of Meteorology*. Chicago: University of Chicago Press, 2005.

Arrhenius, Svante. "The Fate of the Planets." *Scientific American Supplement*, no. 1866 (October 7, 1911): 238–39.

Asaka, Ikuko. *Tropical Freedom: Climate, Settler Colonialism, and Black Exclusion in the Age of Emancipation*. Durham, NC: Duke University Press, 2017.

Ashforth, Adam. *The Politics of Official Discourse in Twentieth-Century South Africa*. New York: Oxford University Press, 1990.

Ballantyne, Tony. *Orientalism and Race: Aryanism in the British Empire*. New York: Palgrave, 2002.

Barnard, W. S. *Encountering Adamastor: South Africa's Founding Geographers in Time and Place, Vol. 1*. Stellenbosch, South Africa: Sun Press, 2016.

Beattie, James. "Science, Religion, and Drought: Rainmaking Experiments and Prayers in North Otago, 1889–1911," in Beattie et al., eds., *Climate, Science, and Colonization*.

Beattie, James, Edward Melillo, and Emily O'Gorman, eds. *Eco-Cultural Networks and the British Empire: New Views on Environmental History*. London: Bloomsbury Academic, 2015.

Beattie, James, and Ruth Morgan. "Engineering Edens on This 'Rivered Earth'? A Review Article on Water Management and Hydro-Resilience in the British Empire, 1860s-1940s." *Environment and History* 23, no. 1 (February 2017): 39–63.

Beattie, James, Emily O'Gorman, and Matthew Henry, eds. *Climate, Science, and Colonization: Histories from Australia and New Zealand*. New York: Palgrave Macmillan, 2014.

Beinart, William. *The Rise of Conservation in South Africa: Settlers, Livestock, and the Environment, 1770–1950*. New York: Oxford University Press, 2003.

Beinart, William. *Twentieth-Century South Africa*. New York: Oxford University Press, 2001.

Beinart, William, Karen Brown, and Daniel Gilfoyle. “Experts and Expertise in Colonial Africa Reconsidered: Science and the Interpenetration of Knowledge.” *African Affairs* 108, no. 432 (2009): 413–33.

Beinart, William, and Peter Delius. “The Historical Context and Legacy of the Natives Land Act of 1913.” *Journal of Southern African Studies* 40, no. 4 (2014): 667–88.

Beinart, William, and Lotte Hughes. *Environment and Empire*. New York: Oxford University Press, 2007.

Belich, James. *Replenishing the Earth: The Settler Revolution and the Rise of the Angloworld*. New York: Oxford University Press, 2009.

Bell, Duncan. “John Stuart Mill on Colonies.” *Political Theory* 38, no. 1 (2010): 34–64.

Beningfield, Jennifer. *The Frightened Land: Land, Landscape and Politics in South Africa in the Twentieth Century*. London: Routledge, 2006.

Blanford, W. T. “On the Nature and Probable Origin of the Superficial Deposits in the Valleys and Deserts of Central Persia.” *Quarterly Journal of the Geological Society of London* 29, no. 1–2 (June 1873): 493–503.

Booysen, C. R., and S. J. G. Hofmeyr. *Save the Soil*. Cape Town: Juta and Co., 1955.

Bovill, E. William. “The Encroachment of the Sahara on the Sudan.” *Journal of the Royal African Society* 20, no. 79 (April 1921): 174–85.

Bovill, E. William. “The Encroachment of the Sahara on the Sudan (Concluded).” *Journal of the Royal African Society* 20, no. 80 (July 1921): 259–69.

Bowman, Isaiah. “The Pioneer Fringe.” *Foreign Affairs* 6, no. 1 (October 1927): 49–66.

Bozzoli, Belinda. *The Political Nature of a Ruling Class: Capital and Ideology in South Africa 1890–1933*. London: Routledge and Keegan Paul, 1981.

Bridgman, Jon. *The Revolt of the Hereros*. Berkeley: University of California Press, 1981.

Brincker, P. H. “Bemerkungen zu Bernsmanns Karte des Ovambolandes.” *Globus* 70, no. 5 (July 1896): 79–80.

Brits, J. P. “The Voice of the ‘People’? Memoranda Presented in 1947 to the Sauer Commission by ‘Knowledgeable’ Afrikaners.” *African Historical Review* 32, no. 1 (2000): 61–83.

Brown, John Croumbie. *Hydrology of South Africa; or, Details of the Former Hydrographic Condition of the Cape of Good Hope*. London: Henry S. King and Co, 1875.

Brown, R. M. “The Redemption of the Kalahari.” Review of *The Kalahari; or, Thirstland Redemption*, by E. H. L. Schwarz. *Geographical Review* 11, no. 4 (October 1921), 623–26.

Browne, John Hutton Balfour. *South Africa: A Glance at Current Conditions and Politics*. London: Longmans, Greene and Co., 1905.

Brownell, Josiah. *The Collapse of Rhodesia: Population Demographics and the Politics of Race*. London: I. B. Tauris, 2011.

Brückner, Eduard. “Über die Herkunft des Regens.” *Geographische Zeitschrift* 6 (1900): 89–96.

Bryce, James. *Impressions of South Africa*. London: Macmillan, 1897.

Byron, John Joseph. *Back to the Land! Notes for the Assistance of Settlers in South Africa*. N.P., 1913.

Cannon, William Austin. *General and Physiological Features of the Vegetation of the More Arid Portions of Southern Africa, with Notes on the Climatic Environment.* Washington: Carnegie Institution, 1924.

Carnegie, Dale. *Among the Matabele.* London: Religious Tract Society, 1894.

Catuneanu, O., H. Wopfner, P. G. Eriksson, B. Cairncross, B. S. Rubidge, R. M. H. Smith, and P. J. Hancox, eds., "The Karoo Basins of South-Central Africa." *Journal of African Earth Sciences* 43 (2005), 211–53.

Chapman, James. *Travels in the Interior of South Africa, Vol. 2.* London: Bell and Daldy, 1868.

Chetty, Suryakanthie. *"Africa Forms the Key": Alex Du Toit and the History of Continental Drift.* (Cham, Switzerland: Palgrave Macmillan, 2021.

Clynick, Tim. "Reformers, Rural Rehabilitation and Poor Whites on the Hartbeespoort Dam Irrigation Scheme under the Pact Government in South Africa, 1924–1929," in Alan H. Jeeves and Owen J. M. Kalinga, eds., *Communities at the Margin: Studies in Rural Society and Migration in Southern Africa, 1890–1980.* Pretoria: UNISA Press, 2002.

Coen, Deborah. *Climate in Motion: Science, Empire, and the Problem of Scale.* Chicago: University of Chicago Press, 2018.

Coetzee, J. M. *White Writing: On the Culture of Letters in South Africa.* New Haven, CT: Yale University Press, 1988.

Cole, Monica. *South Africa.* New York: E. P. Dutton, 1961.

Collett, John G. *Prevent Desert Encroachment with the Schwarz-Kalahari Scheme.* Cradock, South Africa: White and Boughton, 1949.

Connelly, Matthew, Matt Fay, Giulia Ferrini, Micki Kaufman, Will Leonard, Harrison Monsky, Ryan Musto, Taunton Paine, Nicholas Standish, and Lydia Walker. "'General, I Have Fought Just as Many Nuclear Wars as You Have': Forecasts, Future Scenarios, and the Politics of Armegeddon." *American Historical Review* 117, no. 5 (December 2012): 1431–60.

Conway, Erik M., and Naomi Oreskes. *Merchants of Doubt: How a Handful of Scientists Obscured the Truth on Issues from Tobacco Smoke to Global Warming.* New York: Bloomsbury Press, 2010.

Cory, H. T., and William P. Blake. *The Imperial Valley and the Salton Sink.* San Francisco: John J. Newbegin, 1915.

Creswell, Frederick H. P. "The Transvaal Labour Problem." *National Review* 40, no. 237 (November 1902): 444–56.

Crosby, Alfred. *Ecological Imperialism: The Biological Expansion of Europe, 900–1900.* New York: Cambridge University Press, 1986.

Cunniff, Roger L. "The Great Drought: Northeast Brazil, 1877–1880." PhD diss., University of Texas-Austin, 1970.

Dalrymple, G. Brent. *The Age of the Earth.* Palo Alto, CA: Stanford University Press, 1991.

Darwin, Charles R., and John W. Judd. *On the Structure and Distribution of Coral Reefs.* London: Ward, Lock, and Co., 1890.

Darwin, John. *The Empire Project: The Rise and Fall of the British World System, 1830–1970.* New York: Cambridge University Press, 2009.

Davie, Grace. *Poverty Knowledge in South Africa: A Social History of Human Science, 1855–2005.* New York: Cambridge University Press, 2015.

Davis, Diana K. *The Arid Lands: History, Power, Knowledge.* Cambridge, MA: MIT Press, 2015.

Davis, Diana K. "Restoring Roman Nature: French Identity and North African Environmental History." In Diana K. Davis and Edmund Burke III, eds., *Environmental Imaginaries of the Middle East and North Africa*. Athens: Ohio University Press, 2011.

Davis, Diana K. *Resurrecting the Granary of Rome: Environmental History and French Colonial Expansion in North Africa*. Athens: Ohio University Press, 2007.

Davis, Mike. "The Coming Desert." *New Left Review* 97 (January-February 2016): 23–43.

Delius, Peter, and Stefan Schirmer. "Soil Conservation in a Racially Ordered Society." *Journal of Southern African Studies* 26, no. 4 (December 2000): 719–42.

Denoon, Donald. *Settler Capitalism: The Dynamics of Development in the Southern Hemisphere*. New York: Oxford University Press, 1983.

Dodson, Belinda. "A Soil Conservation Safari: Hugh Bennett's 1944 Visit to South Africa." *Environment and History* 11 (2005): 35–54.

Doel, Ronald. Comments on H-Environment Roundtable Review of *Merchants of Doubt*, vol. 1, no. 2 (2011), 18–23.

Douglas, Kirsty. "'For the Sake of a Little Grass': A Comparative History of Settler Science and Environmental Limits in South Australia and the Great Plains." In *Climate, Science and Colonization*, edited by James Beattie, Emily O'Gorman, and Matthew Henry, 99–118. New York: Palgrave Macmillan, 2014.

Driver, Thackwray. "Anti-Erosion Policies in the Mountain Areas of Lesotho: The South African Connection." *Environment and History* 5, no. 1 (February 1999): 1–25.

D'Souza, Rohan. "Water in British India: The Making of a 'Colonial Hydrology.'" *History Compass* 4, no. 4 (July 2006): 621–28.

DuBois, W. E. B. "The Souls of White Folk," in *W. E. B. Du Bois: Writings*, 923–38. New York: Library of America, 1987.

Dubow, Saul. *A Commonwealth of Knowledge: Science, Sensibility, and White South Africa 1820–2000*. New York: Oxford University Press, 2006.

Dubow, Saul. *Racial Segregation and the Origins of Apartheid in South Africa, 1919–1936*. New York: St. Martin's Press, 1989.

Dunlap, Thomas. *Nature and the English Diaspora: Environment and History in the United States, Canada, Australia, and New Zealand*. New York: Cambridge University Press, 1999.

Du Toit, Alexander Logie. "The Kalahari and Some of Its Problems." *South African Journal of Science* 24 (1927): 88–101.

Du Toit, Alexander Logie. "Some Considerations upon Agriculture and Mining in South Africa." *South African Journal of Science* 31 (November 1934): 1–24.

Du Toit, Heinrich S. *Dry-Farming*. Potschefstroom, South Africa: Het Westen Printing Works, 1913.

Edwards, Paul. *A Vast Machine: Computer Models, Climate Data, and the Politics of Global Warming*. Cambridge, MA: MIT Press, 2010.

Emmett, Tony, and Gerard Hagg. "Politics of Water Management: The Case of the Orange River Development Project." In Meshack M. Khosa, ed., *Empowerment through Economic Transformation*. Durban: African Millennium Press, 2001: 299–328.

Endfield, Georgina, and David Nash. "Drought, Desiccation, and Discourse: Missionary Correspondence and Nineteenth-Century Climate Change in Central Southern Africa." *The Geographical Journal* 168, no. 1 (March 2002): 33–47.

Engerman, David C. "Introduction: Histories of the Future and the Futures of History." *American Historical Review* 117, no. 5 (2012): 1402–10.

Etherington, Norman. *The Great Treks: The Transformation of Southern Africa, 1815–1854*. New York: Longman, 2001.

Evans, Ivan. *Cultures of Violence: Lynching and Racial Killing in South Africa and the American South*. New York: Manchester University Press, 2010.

Fairhead, James, and Melissa Leach. *Misreading the African Landscape: Society and Ecology in a Forest-Savanna Mosaic*. Cambridge, UK: Cambridge University Press, 1996.

Farini, G. A. *Through the Kalahari Desert: A Narrative of a Journey with Gun, Camera, and Notebook to Lake N'Gami and Back*. London: Sampson Low, Martson, Searle & Rivington, 1886.

Fedorowich, E. K. "'Foredoomed to Failure': The Resettlement of British Ex-Servicemen in the Dominions 1914–1930." PhD diss., University of London, 1990.

Fick, J. C. *The Abuse of the Soil, Veld, and Water Resources of South Africa*. Cape Town: South African Interests Group, 1944.

Fleisch, Axel, and Wilhelm Möhlig. *The Kavango Peoples in the Past: Local Historiographies from Northern Namibia*. Cologne, Germany: Rüdigger Köppe Verlag, 2002.

Fleming, James Rodger. "Civilization, Climate, and Ozone: Ellsworth Huntington's 'Big' Views on Biophysics, Biocosmics, and Biocracy." In Martin Mahony and Samuel Randalls, eds., *Weather, Climate, and the Geographical Imagination: Placing Atmospheric Knowledges*. Pittsburgh, PA: University of Pittsburgh Press, 2020.

Fleming, James Rodger. *Fixing the Sky: The Checkered History of Weather and Climate Control*. New York: Columbia University Press, 2010.

Fleming, James Rodger. *Historical Perspectives on Climate Change*. New York: Oxford University Press, 2020.

Fleming, James Rodger. *Inventing Atmospheric Science: Bjerknes, Rossby, Wexler, and the Foundations of Modern Meteorology*. Cambridge, MA: MIT Press, 2016.

Forde, C. Daryll. "Irrigation in South Africa." *Geographical Journal* 65, no. 4 (April 1925): 342–49.

Foster, Jeremy. *Washed with Sun: Landscape and the Making of White South Africa*. Pittsburgh, PA: University of Pittsburgh Press, 2008.

Fournier, Luiz Mariano de Barros. *O Problema das seccas do nordeste*. Rio de Janeiro: Villas Boas, 1920.

Frankel, S. Herbert. "The Situation in South Africa." *Economic Journal* 43, no. 169 (March 1933): 93–107.

Geary, Daniel, Camilla Schofield, and Jennifer Sutton, eds. *Global White Nationalism: From Apartheid to Trump*. Manchester, UK: Manchester University Press, 2020.

Gessert, Ferdinand. "Auf der Flucht von Inachab zum Oranienfluss." *Globus* 57 no. 5 (2 February 1905): 79–80.

Gessert, Ferdinand. "Klimatische Folgen einer Kunene-Ableitung." *Globus* 71, no. 24 (19 June 1897): 391–92.

Gessert, Ferdinand. "Die Mutmasslichen Klimatischen Folgen einer Kunene-Ableitung." *Zeitschrift für Kolonialpolitik, Kolonialrecht und Kolonialwirtschaft* 1904:161–89.

Gessert, Ferdinand. "Reise längs Flussthäler des südwestlichen Gross-Namalandes." *Globus* 72, no. 12 (25 September 1897): 190–92.

Gessert, Ferdinand. "Der Seewind Deutsch-Südwestafrikas und seine Folgen." *Globus* 72, no. 19 (20 November 1897): 297–99.

Gessert, Ferdinand. "Über Rentabilität und Baukosten einer Kunene-Ableitung." *Globus* 85, no. 21 (2 June 1904): 338–41.

Gessert, Ferdinand. "Zur Aufforstungsfrage in Südwestafrika." *Globus* 85, no. 9 (3 March 1904): 134–36.

Giliomee, Hermann. *The Afrikaners: Biography of a People*. Cape Town: Tafelberg, 2003.

Gilmartin, David. "Imperial Rivers: Irrigation and British Visions of Empire." *Decentering Empire: Britain, India, and the Transcolonial World* (2006): 76–103.

Gold, E., F. E. Kanthack, J. W. Evans, and E. H. L. Schwarz. "The Control of Climate by Lakes: Discussion." *Geographical Journal* 57, no. 3 (March 1921): 174–81.

Gordin, Michael D., Helen Tilley, and Gyan Prakash, eds. *Utopia/Dystopia: Conditions of Historical Possibility*. Princeton, NJ: Princeton University Press, 2010.

Gordon-Cumming, R. George. *Five Years of a Hunter's Life in the Far Interior of South Africa, Vol. 1*. London: John Murray, 1850.

Goswami, Manu. "Imaginary Futures and Colonial Internationalisms." *American Historical Review* 117, no. 5 (December 2012): 1461–85.

Gray, Madi. "Race Ratios: The Politics of Population Control in South Africa," in L. Bonderstam and S. Bergstrom, eds., *Poverty and Population Control*. New York, Academic Press, 1980.

Green, Lawrence. *Great African Mysteries*. London: S. Paul and Co., 1935.

Green, Lawrence. *Thunder on the Blaauwberg*. Cape Town: Howard Timmins, 1966.

Gregory, J. W. "The Ancient River System of the Kalahari and the Possibility of Its Renewal," *Nature* 113 (12 April 1924): 539–40.

Gregory, J. W. *The Dead Heart of Australia: A Journey around Lake Eyre in the Summer of 1901–1902*. London: John Murray, 1906.

Gregory, J. W. "Prof. E. H. L. Schwarz," *Nature* 123 (19 January 1929): 100–101.

Gregory, J. W. "Professor Huntington on Climatic Pulsations." *Geographical Journal* 46, no. 4 (October 1915): 308–10.

Gregory, J. W. "The Reported Progressive Desiccation of the Earth." *Scientia* 17, no. 41 (1915): 328–44.

Grove, A. T. "The African Environment, Understood and Misunderstood," in Douglas Rimmer and Anthony Kirk-Green, eds., *The British Intellectual Engagement with Africa in the Twentieth Century* (London: Macmillan 2000), 179–206.

Grove, Richard. *Green Imperialism: Colonial Expansion, Tropical Island Edens and the Origins of Environmentalism, 1600–1860*. New York: Cambridge University Press, 1996.

Grove, Richard. "Scottish Missionaries, Evangelical Discourses, and the Origins of Conservation Thinking in Southern Africa 1820–1900." *Journal of Southern African Studies* 15, no. 2 (January 1989): 163–87.

Hall, Anthony. *Drought and Irrigation in North-East Brazil*. New York: Cambridge University Press, 1978.

Haltinner, Kristin, and Dilshani Sarathchandra. *Inside the World of Climate Change Skeptics*. Seattle: University of Washington Press, 2023

Hamilton, Carolyn. *Terrific Majesty: The Powers of Shaka Zulu and the Limits of Historical Invention*. Cambridge, MA: Harvard University Press, 1998.

Hamilton, Clive. *Earthmasters: The Dawn of the Age of Climate Engineering*. New Haven, CT: Yale University Press, 2014.

Hancock, W. H., and J. van der Poel. *Selections from the Smuts Papers*, vols. 1 and 2. New York: Cambridge University Press, 1966.

Headlam, Cecil, ed. *The Milner Papers: South Africa 1899–1905, Vol. 2.* London: Cassel and Company, 1933.

Heffernan, Michael J. "Bringing the Desert to Bloom: French Ambitions in the Sahara Desert during the Late Nineteenth Century: The Strange Case of '*la Mer intérieure.*'" In Denis Cosgrove and Geoff Petts, eds., *Water, Engineering, and Landscape: Water Control and Landscape Transformation in the Modern Period*, 94–114. London: Belhaven Press 1990.

Henni, Samia. "Nuclear Powers," *Architectural Review*, 1 June 2022: 64–68.

Herschensohn, Allan C. *A Great South Africa.* Johannesburg: Unionist Party, 1911.

Hertzog, J. B. M. *The Segregation Problem: General Hertzog's Solution.* Cape Town: Nationalist Party of South Africa, 1926.

Higginson, John. *Collective Violence and the Agrarian Origins of South African Apartheid, 1900–1948.* New York: Cambridge University Press, 2015.

Hipondoka, Martin H. T., Jürgen Kempf, and Helene Jousse. "Paleo and Present Ecological Value of the Eosha Pan, Namibia: An Integrative Review." *Journal of the Namibian Scientific Society* 61 (2013): 67–85.

Hodson, Arnold W. *Trekking the Great Thirst: Sport and Travel in the Kalahari Desert.* Johannesburg: Africana Book Society, 1977 (orig. London: T. Fisher Unwin, 1912).

Hoffman, M. T., S. W. Todd, Z. N. Ntshona, and S. D. Turner. "Land Degradation in South Africa." Unpublished final report. Cape Town: National Botanical Institute, 1999. http://www.pcu.uct.ac.za/pcu/resources/databases/landdegrade. Accessed 18 October 2023.

Holdich, Thomas, Dr. Blanford, Martin Conway, Prof. Seeley, Douglas Freshfield, J. S. Flett, H. R. Mill, Dr. Evans, and Prince Kropotkin. "The Desiccation of Eur-Asia: Discussion." *Geographical Journal* 23, no. 6 (June 1904): 734–41.

Hough, Franklin B. *Report upon Forestry*. Washington: US Government Printing Office, 1878.

Hugo, Pierre. "Frontier Farmers in South Africa." *African Affairs* 87, no. 349 (October 1988): 537–52.

Humboldt, Alexander von. *Ansichten der Natur mit wissenschaftlichen Erläuterungen, Vol. 1.* First edition. Tübingen, Germany: J. G. Cotta'schen Büchhandlung, 1808. Accessed through archive.org, 2 September 2020.

Humboldt, Alexander von. *Ansichten der Natur mit wissenschaftlichen Erläuterungen, Vol. 1.* Third edition. Stuttgart, Germany: J. G. Cotta'scher Verlag, 1849.

Hutchins, D. E. *Cycles of Drought and Good Seasons in South Africa.* Wynberg, South Africa: The Times, 1889.

Hyam, Ronald. *The Failure of South African Expansion, 1908–1948.* New York: Africana Publishing, 1972.

Hyslop, Jon. "The Imperial Working Class Makes Itself White: White Laborism in Britain, Australia, and South Africa." *Journal of Historical Sociology* 12, no. 4 (December 1999): 398–421.

Idriess, Ion L. *The Great Boomerang.* Sydney: Angus and Robertson, 1942.

Isaacman, Allen. "Cahora Bassa Dam and the Delusion of Development," *Daedalus* 150, no. 4 (Fall 2021): 103–23.

Jacks, G. V., and R. O. Whyte. *Vanishing Lands: A World Survey of Soil Erosion.* New York: Doubleday, 1939.

Jackson, Thomas Jefferson. *The Cause of Earthquakes, Mountain Formation, and Kindred Phenomena Connected with the Physics of the Earth.* Philadelphia: N.P., 1907. Reprinted in *Proceedings of the American Philosophical Society* 45, 1906.

Jacobs, Nancy. *Birders of Africa: History of a Network.* New Haven, CT: Yale University Press, 2016.

Jasanoff, Sheila. "Future Imperfect." In Sheila Jasanoff and Sang Hyun-Kim, eds., *Dreamscapes of Modernity: Sociotechnical Imaginaries and the Fabrication of Power.* Chicago: University of Chicago Press, 2015.

Jeeves, Alan, and Jonathan Crush. *White Farms, Black Labor: The State and Agrarian Change in South Africa, 1910–1950.* Portsmouth, NH: Heinemann, 1997.

Jeffares, J. L. S. "Report of the Ngamiland Waterways Surveys of 1937." Unpublished manuscript, 1938.

Kalb, Martin. *Environing Empire: Nature, Infrastructure, and the Making of German Southwest Africa.* New York: Berghahn Books, 2022.

Kanthack, F. E. "Irrigation Development in the Cape Colony." *Agricultural Journal of the Cape of Good Hope* 34, no. 6 (1909): 645–57.

Kanthack, F. E. "Notes on the Kunene River, Southern Angola." *Geographical Journal* 57, no. 5 (May 1921): 321–36.

Karlson, August. *The Kalahari Problem.* Johannesburg: Argus, 1919.

Karypidou, Maria Chara, Eleni Katragkou, and Stefan Pieter Sobolowski. "Precipitation over Southern Africa: Is There Consensus among Global Climate Models (GCMs), Regional Climate Models (RCMs) and Observational Data?" *Geoscientific Model Development* 15 (2022): 3387–3404.

Keegan, Timothy. "Gender, Degeneration and Sexual Danger: Imagining Race and Class in South Africa, ca. 1912." *Journal of Southern African Studies* 27, no. 3 (2001): 459–77.

Keegan, Timothy. *Rural Transformations in Industrializing South Africa: The Southern Highveld to 1914.* Johannesburg: Ravan, 1986.

Kennedy, Dane. *Islands of White: Settler Society and Culture in Kenya and Southern Rhodesia.* Durham, NC: Duke University Press, 1987.

Khan, Farieda. "Soil Wars: The Role of the African National Soil Conservation Association in South Africa, 1953–1959." *Environmental History* 2, no. 4 (October 1997): 439–59.

Kienetz, Alvin. "Nineteenth-Century South West Africa as a German Settlement Colony." PhD diss., University of Minnesota, 1976.

Koorts, Lindie. "'The Black Peril Would Not Exist if It Were Not for a White Peril That Is a Hundred Times Greater': D. F. Malan's Fluidity on Poor Whiteism and Race in the Pre-Apartheid Era, 1912–1939." *South African Historical Journal* 65, no. 4 (2013): 555–76.

Korn, Hermann. *Zwiegespräch in der Wüste: Briefe und Aquarelle aus dem Exil 1935–1946.* Göttingen, Germany, and Windhoek, Namibia: Klaus Hess Verlag, 1996.

Korn, Hermann, and Henno Martin. "Die jüngere geologische und klimatische Geschichte." *Zentralblatt für Mineralogie, Geologie, und Paläontologie* 11 (1937): 456–73.

Koselleck, Reinhart. *Futures Past: On the Semantics of Historical Time.* Trans. Keith Tribe. Cambridge, MA: MIT Press, 1990.

Krikler, Jeremy. "Re-Thinking Race and Class in South Africa: Some Ways Forward." In Wulf D. Hund, Jeremy Krikler, and David Roediger, eds., *Wages of Whiteness and Racist Symbolic Capital.* Berlin: Lit Verlag, 2010.

Krikler, Jeremy. *White Rising: The 1922 Insurrection and Racial Killing in South Africa.* Manchester, UK: Manchester University Press, 2005.

Kropotkin, Prince. "The Desiccation of Eur-Asia." *Geographical Journal* 23, no. 6 (June 1904): 722–34.

Kutzleb, Charles Robert. "Rain Follows the Plow: The History of an Idea." PhD diss., University of Colorado, 1968.

Laflin, P. *The Salton Sea: California's Overlooked Treasure.* Indio, CA: Coachella Valley Historical Society, 1995.

Lake, Marilyn. *Progressive New World: How Settler Colonialism and Transpacific Exchange Shaped American Reform.* Cambridge, MA: Harvard University Press, 2019.

Lake, Marilyn. "White Man's Country: The Trans-National History of a National Project." *Australian Historical Studies* 122 (2003): 346–53.

Lake, Marilyn, and Henry Reynolds. *Drawing the Global Colour Line: White Men's Countries and the International Challenge of Racial Equality.* Cambridge, UK: Cambridge University Press, 2008.

Lang, Lis. *White, Poor and Angry: White Working Class Families in Johannesburg.* Burlington, VT: Ashgate, 2003.

Lau, Brigitte, and Christel Stern. *Namibian Water Resources and Their Management.* Windhoek: National Archives of Namibia, 1990.

Lave, Rebecca. "Introduction to Book Review Forum." *AAG Review of Books* 7, no. 1 (2 January 2019): 35–46.

Legg, Stephen. "Debating the Climatological Role of Forests in Australia, 1827–1949: A Survey of the Popular Press." In James Beattie et al., eds., *Climate, Science, and Colonization.*

Lehmann, Philipp. "Average Rainfall and the Play of Colors: Colonial Experience and Global Climate Data." *Studies in History and Philosophy of Science* 70 (August 2018): 38–49.

Lehmann, Philipp. "Infinite Power to Change the World: Hydroelectricity and Engineered Climate Change in the Atlantropa Project." *American Historical Review* 121, no. 1 (February 2016): 70–100.

Leppan, Hubert Dudley. *Agricultural Policy in South Africa.* Johannesburg: Central News Agency, 1931.

Lester, Alan. "Imperial Circuits and Networks: Geographies of the British Empire: Imperial Circuits and Networks." *History Compass* 4, no. 1 (January 2006): 124–41.

Leutwein, Theodor. *Elf Jahre Gouverneur in Deutsch-Südwestafrika.* Berlin: Ernst Siegfried Mittler and Son, 1906.

Levinkin, L. "Droughts in South Africa: A Preliminary Study of their Extent, Severity, and Frequency of Occurrence since 1904." *Farming in South Africa,* March 1941: 84–87.

Lewis, Stephen., *The Economics of Apartheid.* New York: Council on Foreign Relations, 1990.

Li, Tania Murray. "Beyond 'the State' and Failed Schemes." *American Anthropologist* 107, no. 3 (2005): 383–94.

Limerick, Patricia. *Desert Passages: Encounters with the American Deserts.* Albuquerque: University of New Mexico Press, 1985.

Linton, Jamie. *What Is Water? The History of a Modern Abstraction.* Vancouver: University of British Columbia Press, 2010.

Livingstone, David. *Missionary Travels and Researches in South Africa.* New York: Harper and Brothers, 1858.

Lowell, Percival. *Mars as the Abode of Life*. New York: Macmillan, 1908.

Lvovitch, M. I. "World Water Balance (General Report)." In IASH/UNESCO/WMO, *World Water Balance: Proceedings of the Reading Symposium July 1970, Vol. 1*. Gentbrugge, Belgium: IASH, 1972: 401–15.

Macdonald, Allan John. *Trade, Politics and Christianity in Africa and the East*. New York: Longmans, 1916.

MacDonald, William. *The Conquest of the Desert*. London: T. Werner Laurie, 1913.

MacDonald, William. *The Settler and South Africa*. London: Union Castle Line, 1913.

Machado, Carlos. *O Sul de Angola e as Aguas de Rio Cunene*. Lisbon: Oficinas Gráficas da Biblioteca Nacional, 1922.

Mackenzie, Donald. *The Flooding of the Sahara: An Account of the Proposed Plan for Opening Central Africa to Commerce and Civilization from the North-West Coast*. London: S. Low, Marston, Searle and Rivington, 1877.

MacMillan, William M. *The South African Agrarian Problem and Its Historical Development*. Witwatersrand, South Africa: Central News Agency, 1919.

Mamdani, Mahmood. *Citizen and Subject: Contemporary Africa and the Legacy of Late Colonialism*. Princeton, NJ: Princeton University Press, 1986.

Marais, W. F. *Only the Refilling of the Kalahari Lakes Can Halt the Advancing Desert in Southern Africa*. Cradock, South Africa: White and Boughton, 1948.

Markham, Violet R. *The South African Scene*. London: Smith, Elder and Co., 1913.

Marsh, George Perkins. *The Earth as Modified by Human Action: A Last Revision of 'Man and Nature.'* New York: C. Scribner, 1898 [1874].

Marsh, George Perkins. *Man and Nature; or, Physical Geography as Modified by Human Action*. New York: C. Scribner, 1864.

Martel, E.-A. *Nouveau Traité des Eaux Souterraines*. Paris: Librairie Octave Doin, 1921.

Mauldin, Erin Stewart. *Unredeemed Land: An Environmental History of Civil War and Emancipation in the Cotton South*. New York: Oxford University Press, 2018.

McBride, Charlotte M., Andries C. Kruger, and Liesl Dyson, "Changes in Extreme Daily Rainfall Characteristics in South Africa, 1921–2020," *Weather and Climate Extremes* 38 (December 2022); https://doi.org/10.1016/j.wace.2022.100517.

McCray, W. Patrick. *The Visioneers: How a Group of Elite Scientists Pursued Space Colonies, Nanotechnologies, and a Limitless Future*. Princeton, NJ: Princeton University Press, 2017.

McDermott-Hughes, David. *Whiteness in Zimbabwe: Race, Landscape, and the Problem of Belonging*. New York: Palgrave Macmillan, 2010.

McKittrick, Meredith. "Capricious Tyrants and Persecuted Subjects: Reading between the lines of Missionary Records in Precolonial Northern Namibia." In Toyin Falola and Christian Jennings, eds., *Sources and Methods in African History: Spoken, Written, Unearthed*, 219–38. Rochester, NY: Rochester University Press, 2003.

McKittrick, Meredith. "Making Rain, Making Maps: Competing Geographics of Water and Power in Southwestern Africa." *Journal of African History* 58, no. 2 (2017): 187–212.

McKittrick, Meredith. "Race and Rainmaking in 20th-Century Southern Africa." In Martin Mahoney and Sam Randalls, eds., *Weather, Climate and the Geographical Imagination*, 152–67. Pittsburgh, PA: Pittsburgh University Press, 2020.

McKittrick, Meredith. "The Rainmaker Goes to Court: Death, Dependency, and the Production of Colonial Ignorance." *Journal of Namibian Studies* 26 (2019): 51–70.

McKittrick, Meredith. "Talking about the Weather: Settler Vernaculars and Climate Anxieties in Early Twentieth-Century South Africa." *Environmental History* 23 (2018): 3–27.

McKittrick, Meredith. "Theories of 'Reprecipitation' and Climate Change in the Settler Colonial World." *History of Meteorology* 8 (2017): 74–94.

McKittrick, Meredith. *To Dwell Secure: Generation, Christianity and Colonialism in Namibia*. Portsmouth, NH: Heinemann, 2002.

Mermelstein, David. *The Anti-Apartheid Reader: The Struggle against White Rule in South Africa*. New York: Grove Press, 1987.

Minnaar, A. de V. "Graaff-Reinet and the Great Depression (1929–1933)." MA thesis, Rhodes University, 1978.

Moffat, Robert. *Missionary Labors and Scenes in Southern Africa*. New York: Carter, 1888.

Moloney, Alfred. *Sketch of the Forestry of West Africa with Particular Reference to Its Present Principal Commercial Products*. London: S. Low, Marston, Searle, and Rivington, 1887.

Money, Duncan, and Danelle van Zyl-Hermann, *Rethinking White Societies in Southern Africa 1930s–1990s*. London: Routledge, 2020.

Morrell, Robert. *From Boys to Gentlemen: Settler Masculinity in Colonial Natal, 1880–1920*. New York: Cambridge University Press, 2001.

Mukerji, Chandra. *Impossible Engineering: Technology and Territoriality on the Canal du Midi*. Princeton, NJ: Princeton University Press, 2009.

Murray, John. "On the Total Annual Rainfall of the Land of the Globe, and the Relation of Rainfall to the Annual Discharge of Rivers." *Scottish Geographical Magazine* 3 (1887): 65–77.

"Must Humanity Perish of Thirst? The Possible Desiccation of the Earth through the Depredations of Underground Watercourses." *Scientific American Monthly* 4 (October 1921): 305–7.

Nicholls, George Heaton. "South African Native Policy." *African Affairs* 44, no. 175 (April 1945): 73–80.

Nicholson, Sharon. "The ITCZ and the Seasonal Cycle over Equatorial Africa." *Bulletin of the American Meteorological Society* 99, no. 2 (February 2018): 337–48.

Nitsche, Georg. "Aus dem Deutsch-Südwestafrikanischen Schutzgebiete: Der Ursprung der jährlichen Überschwemmung in Ovamboland." *Mitteilungen aus den Deutschen Schutzgebieten* 26 (1913).

Nugent, Chris. "The Zambezi River: Tectonism, Climate Change, and Drainage Evolution." *Paleogeography, Paleoclimatology, Paleoecology* 78, nos. 1–2 (1990): 55–69.

O'Brien, Chris. "Imported Understandings: Calendars, Weather and Climate." In *Climate, Science and Colonization*, edited by James Beattie, Emily O'Gorman, and Matthew Henry, 195–212. New York: Palgrave Macmillan, 2014.

Obst, Erich. "Geomorphologische Forschungsreise von Prof. Dr. E. Obst und Dr. K. Kayser in Südafrika." *Geographische Wockenschrift* 3, no. 40 (28 October 1935): 963–65.

Obst, Erich. "Junge Krustenbewegungen in Südafrika und ihre klimatischen Folgen." *Forschungen und Fortschritte* 12, no. 35/36 (December 1936): 449–52.

Obst, Erich. "Der Kampf gegen die 'Austrocknung' in Afrika." *Deutsche Kolonial-zeitung* 50, no. 3 (1938): 112–16.

Obst, Erich. "Die Lebensräume der Weltvölker." *Zeitschrift für Politik* 29, no.1/2 (January-February 1939): 1–10.

O'Gorman, Emily. "'Soothsaying' or 'Science'? H. C. Russell, Meteorology, and Environmental Knowledge of Rivers in Colonial Australia." In *Climate, Science and Colonization*, edited by James Beattie, Emily O'Gorman, and Matthew Henry, 174–94. New York: Palgrave Macmillan, 2014.

Oswell, William. *William Cotton Oswell, Hunter and Explorer, Vol. 1*. London: Heinemann, 1900.

Passarge, Siegfried. *Die Kalahari: Versuch einder physisch-geographischen Darstellung der Sandfelder des südafrikanischen Beckens*. Berlin: Dietrich Reimer, 1904.

Penn, Nigel. "The British and the 'Bushmen': The Massacre of the Cape San, 1795 to 1828." *Journal of Genocide Research* 15, no. 2 (2013): 183–200.

Penn, Nigel. *The Forgotten Frontier*. Athens: Ohio University Press, 2005.

Phillips, Sarah T. "Lessons from the Dust Bowl: Dryland Agriculture and Soil Erosion in the United States and South Africa, 1900–1950." *Environmental History* 4, no. 2 (1999): 245–66.

Plummer, F. E. *Rainfall and Farming in the Transvaal, Part 1*. Pretoria: Transvaal University College, 1927.

Posel, Deborah. *The Making of Apartheid 1948–1961*. Oxford, UK: Clarendon Press, 1991.

Posel, Deborah. "The Meaning of Apartheid before 1948: Conflicting Interests and Forces within the Afrikaner Nationalist Alliance." *Journal of Southern African Studies* 14, no. 1 (1987): 123–39.

Potten, D. H. "Aspects of the Recent History of Ngamiland," *Botswana Notes and Records* 8 (1976): 63–86.

Powell, John Wesley. "Trees on Arid Lands." *Science* 12, no. 297 (October 1888).

Powell, Miles. *Vanishing America: Species Extinction, Racial Peril, and the Origins of Conservation*. Cambridge, MA: Harvard, 2016.

Prada Samper, José Manuel de. "The Forgotten Killing Fields: 'San' Genocide and Louis Anthing's Mission to Bushmanland, 1892–1863," *Historia* 57, no. 1 (2012): 172–87.

Price, R. R. M. *Catalogue of Publications of the Geological Survey*. Pretoria: Council for Geoscience, 2001.

Rafferty, John. *Oceans and Oceanography*. New York: Britannica Educational Publishing, in association with Rosen Educational Services, 2011.

Reid, Percy C. "Journeys in the Linyanti Region. *Geographical Journal* 17, no. 6 (June 1901): 573–85.

Reisner, Marc. *Cadillac Desert: The American West and Its Disappearing Water*. New York: Penguin Books, 1993.

Rennell, James. *The Geographical System of Herodotus Examined and Explained*. London: W. Bulmer, 1800.

Reuber, Paul, Anke Struver, and Gunter Wolkersdorfer, eds. *Politische Geographien Europas: Annäherungen an ein umstrittenes Konstrukt*. Münster, Germany: Lit, 2005.

Richards, C. S. "Economic Revival in South Africa." *Economic Journal* 44, no. 176 (1934): 616–30.

Roe, William. "Notes on a Form of Irrigation Well Suited to a Large Portion of the Colony, and Especially the Karroo and Inland Plateaux." *Agricultural Journal of the Cape of Good Hope* 10, no. 1 (1897): 25–27.

Rohde, Richard F., M. Timm Hoffman, Ian Durbach, Zander Venter, and Sam Jack. "Vegetation and Climate Change in the Pro-Namib and Namib Desert Based on Repeat Photography: Insights into Climate Trends." *Journal of Arid Environments* 165 (April 2018): 119–31.

Rolle, Andrew. *John Charles Frémont: Character as Destiny*. Norman: University of Oklahoma Press, 1991.

Roos, Neil. "South African History and Subaltern Historiography; Ideas for a Radical History of White Folk." *International Review of Social History* 61, no. 1 (April 2016): 117–50.

Ross, J. C. *Soil Conservation in South Africa: A Review of the Problem and Developments to Date*. South Africa Department of Agriculture, 1963.

Ross, Robert, Anne Kelk Mager, and Bill Nasson, eds. *The Cambridge History of South Africa, Vol. 2, 1885–1994*. Cambridge, UK: Cambridge University Press, 2011.

Roudaire, François Elie. *Rapport à M. le Ministre de l'Instruction Publique sur la Mission des Chotts: Études Relatives au Projet de Mer Intérieure*. Paris: Imprimerie Nationale, 1877.

Sayre, Nathan. *The Politics of Scale: A History of Rangeland Science*. Chicago: University of Chicago, 2017.

Schinz, Hans. *Deutsch-Südwest-Afrika*. Oldenburg, Germany: Schulzesche Hof-Buchhandlung, 1891.

Schmidt, Wilhelm R., ed. *Albert Schmidt: Ein bergischer Baumesiter*. Erfurt, Germany: Sutton Verlag, 2008.

Schmokel, Wolfe. "The Myth of the White Farmer: Commercial Agriculture in Namibia." *International Journal of African Historical Studies* 18, no. 1 (1985): 93–108.

Schnegg, Michael. "The Life of Winds: Knowing the Namibian Weather from Someplace and from Noplace." *American Anthropologist* 121, no. 4 (2019): 830–44.

Schonken, J. D. "The State of the National Water Supply." *South African Journal of Science* 27 (1930): 201–12.

Schulz, Aurel, and August Hammar. *The New Africa*. London: Heinemann, 1897.

Schumann, T. E. W., and W. R. Thompson. *A Study of South African Rainfall: Secular Variations and Agricultural Aspects*. Pretoria: J. L. van Schaik, 1934.

Schwarz, Bill. *The White Man's World*. New York: Oxford University Press, 2011.

Schwarz, E. H. L. *Causal Geology*. London: Black and Son, 1910.

Schwarz, E. H. L. "The Control of Climate by Lakes." *Geographical Journal* 57, no. 3 (March 1921): 166–74.

Schwarz, E. H. L. "The Desiccation of Africa: The Cause and the Remedy." *South African Journal of Science* 15, no. 1 (1918): 139–90.

Schwarz, E. H. L. "The Eastern Kalahari Project." *South African Mining and Engineering Journal*, 3 December 1927: 353–54.

Schwarz, E. H. L. "The Former Land Connection between Africa and South America." *Journal of Geology* 14, no. 2 (1906): 81–90.

Schwarz, E. H. L. *The Kalahari and Its Native Races*. London: H. F. and G. Witherby, 1928.

Schwarz, E. H. L. "The Kalahari and Its Possibilities." *Journal of the African Society* 77, no. 20 (October 1921): 1–12.

Schwarz, E. H. L. *The Kalahari; or, Thirstland Redemption*. Cape Town: T. Maskew Miller, 1920.

Schwarz, E. H. L. "The Kalahari Project in Relation to Distress in South Africa." *Matériaux pour létude des calamites* 1, no. 4 (1924) : 291–331.

Schwarz, E. H. L. "The Kalahari Scheme as the Solution of the South African Drought Problem." In *Report of the 21st Annual Meeting of the South African Association for the Advancement of Science, Bloemfontein* 1923, 208–22. Johannesburg: South African Association for the Advancement of Science, 1924.

Schwarz, E. H. L. "The Lost Land of Agulhas." In *Report of the Twelfth Annual Meeting of the South African Association for the Advancement of Science,* 169–79. Cape Town: South African Association for the Advancement of Science, 1915.

Schwarz, E. H. L. "The Menace to Ovamboland." *United Empire* 10, no. 5 (May 1919): 253–55.

Schwarz, E. H. L. "The Rivers of Cape Colony." *Geographical Journal* 27, no. 3 (March 1906): 265–79.

Schwarz, E. H. L. "The Three Paleozoic Ice-Ages of South Africa." *Journal of Geology* 14, no. 8 (1906): 683–91.

Scully, William Charles. *Between Sun and Sand: A Tale of an African Desert.* Cape Town: J. C. Juta, 1898.

Seekings, Jeremy. "'Not a Single White Person Should Be Allowed to Go Under': *Swartgevaar* and the Origins of South Africa's Welfare State, 1924–1929." *Journal of African History* 48 (2007): 375–94.

Shaw, Paul. "The Desiccation of Lake Ngami: An Historical Perspective." *Geographical Journal* 151, no. 3 (November 1985): 318–26.

Showers, Kate. *Imperial Gullies: Soil Erosion and Conservation in Lesotho.* Athens, Ohio: Ohio University Press, 2005.

Sim, J. M. "The Modification of South African Rainfall." In *Report of the Fourteenth Annual Meeting of the South African Association for the Advancement of* Science, 318–26. Cape Town: South African Association for the Advancement of Science, 1917.

Simons, H. J. "Harnessing the Orange River." In W. M. Warren and N. Rubin, eds., *Dams in* Africa, 128–45. London: Frank Cass, 1968.

Smuts, Jan Christian. "Science in South Africa." *Nature* 116 (15 August 1925): 245–49.

Sörgel, Herman. *Atlantropa.* Zürich: Fretz and Wasmuth, 1932.

Sörgel, Herman. *Die drei grossen "A": Gross Deutschland und Italienisches Imperium, die Pfeiler Atlantropas.* Munich: Piloty & Loehle, 1938.

Sörgel, Herman, and Bruno Siegwart. "Erschliessung Afrikas durch Binnenmeere: Saharabewässerung durch Mittelmeersenkung." *Beilage zum Baumeister* 3 (1939): 37–39.

South African Archival Records Natal, Vol. 1. Cape Town: Cape Times, 1958.

Spies, F. J. du T., D. W. Krüger, and J. J. Oberholster, eds. *Die Hertzogtoesprake, Vols. 1 and 2.* Johannesburg: Perskor, 1977.

Stehr, Nico, ed. *Eduard Brückner: The Sources and Consequences of Climate Change and Climate Variability in Historical Times.* Dordrecht, Netherlands: Kluwer, 2000.

Stebbing, E. P. "The Encroaching Sahara: The Threat to the West African Colonies." *Geographical Journal* 85, no. 6 (June 1935): 506–19.

Steinbach, Daniel Rouven. "Carved Out of Nature: Identity and Environment in German Colonial Africa." In Christina Folke Ax, Niels Brimnes, Niklas Thode Jensen, and Karen Oslund, eds., *Cultivating the Colonies: Colonial States and their Environmental* Legacies, 47–77. Athens: Ohio University Press, 2011.

Stephens, Jennie, Prakash Kashwan, Duncan McLaren, and Kevin Surprise. "The Dangers of Mainstreaming Solar Geoengineering: A Critique of the National Academics Report." *Environmental Politics* 32, no. 1 (2023): 157–66.

Stigand, A. G. "Ngamiland." *Geographical Journal* 62, no. 6 (1923): 401–19.

Stoler, Ann Laura. *Along the Archival Grain: Epistemic Anxieties and Colonial Common Sense*. Princeton, NJ: Princeton University Press, 2009.

Stovall, Tyler. *White Freedom: The Racial History of an Idea*. Princeton, NJ: Princeton University Press, 2021.

Sutton, J. R. "Some Notes on Rainfall and Run-Off in South Africa." *South African Geographical Journal* 5 (December 1922): 41–44.

Swart, Sandra. "'Bushveld Magic' and 'Miracle Doctors': An Exploration of Eugène Marais and C. Louis Leipoldt's Experiences in the Waterberg, South Africa, 1906–1917." *Journal of African History* 45, no. 2 (2004): 237–55.

Swart, Sandra. "The 'Five Shilling Rebellion': Rural White Male Anxiety and the 1914 Boer Rebellion." *South African Historical Journal* 56, no. 1 (2006): 88–102.

Sweeney, Kevin Z. "Wither the Fruited Plain: The Long Expedition and the Description of the 'Great American Desert.'" *Great Plains Quarterly* 25, no. 2 (Spring 2005): 105–17.

Swift, Jeremy. "Desertification Narratives, Winners and Losers." In Melissa Leach and Robin Mearns, eds., *The Lie of the Land: Challenging Received Wisdom on the African* Environment, 73–90. Oxford, UK: James Currey, 1996.

Tate, Ralph. "Inaugural Address by Professor Ralph Tate." *Report of the Fifth Meeting of the Australasian Association for the Advancement of Science, September 1893.* Sydney, 1894.

Tate, Ralph. "Post-Miocene Climate in South Australia." *Transactions and Proceedings and Report of the Royal Society of South Australia* 8 (1884–85): 49–59.

Tayler, Judith. "'Our Poor': The Politicization of the Poor White Problem, 1932–42." *African Historical Review* 24, no. 1 (1992): 40–65.

Thompson, W. R. *Moisture and Farming in South Africa*. Johannesburg: Central News Agency, 1936.

Tilley, Helen. *Africa as Living Laboratory: Empire, Development, and the Problem of Scientific Knowledge, 1870–1950*. Chicago: University of Chicago Press, 2011.

Timbury, F. R. V. *The Battle for the Inland: The Case for the Bradfield and Idriess Plans*. Sydney: Angus and Robertson, 1944.

Tischler, Julia. "Cultivating Race." Unpublished manuscript.

Tischer, Julia. *Light and Power for a Multiracial Nation: The Kariba Dam Scheme in the Central African Federation*. Houndmills, Basingstoke, UK: Palgrave Macmillan, 2013.

Torrance, David E. "Britain, South Africa, and the High Commission Territories: An Old Controversy Revisited." *Historical Journal* 41, no. 3 (1998): 751–72.

Towner, E. T. "Lake Eyre and its Tributaries." *Queensland Geographical Journal* 43 (1955): 65–95.

Valencius, Conevery Bolton. *The Health of the Country: How American Settlers Understood Themselves and Their Land*. New York: Basic Books, 2002.

VanderPost, Cornelis. "Early Maps of Ngamiland and the Okavango Delta." *Botswana Notes and Records* 37 (2005):196–207.

Van der Watt, Susanna Maria Elizabeth. "'It Is Drought, Locusts, Depression . . . and the Lord Knows What Else': A Socio-Environmental History of White Agriculture in the Union of South Africa." MA thesis, University of Stellenbosch, 2009.

Van Duin, Pieter. "Artisans and Trade Unions in the Cape Town Building Industry, 1900–1924." In Wilmot G. James and Mary Simons, eds., *Class, Caste, and Color: A Social and Economic History of the South African Western Cape*, 95–110. New Brunswick, NJ: Transaction Publishers, 1992.

Van Onselen, Charles. *The Seed Is Mine: The Life of Kas Maine, a South African Sharecropper, 1894–1985*. New York: Hill and Wang, 1997.

Van Reenen, R. J. "A Resume of the Drought Problem in the Union of South Africa." In *Report of the 21st Annual Meeting of the South African Association for the Advancement of Science*, 178–92. Johannesburg: SAAAS, 1923.

Van Sittert, Lance. "Nation-Building Knowledge: Dutch Indigenous Knowledge and the Invention of White South Africanism, 1890–1909." In David M. Gordon and Shepherd Krech, eds., *Indigenous Knowledge and the Environment in Africa and North America*, 94–109. Athens: Ohio University Press, 2012.

Van Sittert, Lance. "The Supernatural State: Water Divining and the Cape Underground Water Rush, 1891–1910." *Journal of Social History* 37, no. 4 (Summer 2004): 915–37.

Van Vuuren, Lani. *In the Footsteps of Giants: Exploring the History of South Africa's Large Dams*. Gezina, South Africa: Water Research Commission, 2012.

Veracini, Lorenzo. "The Imagined Geographies of Settler Colonialism." In Tracey Banivanua Mar and Penelope Edmunds, eds., *Making Settler Colonial Space*, 179–97. New York: Palgrave Macmillan, 2010.

Veracini, Lorenzo. *Settler Colonialism: A Theoretical Overview*. New York: Palgrave Macmillan, 2010.

Veracini, Lorenzo. "'Settler Colonialism': Career of a Concept." *Journal of Imperial and Commonwealth History* 41, no. 2 (2013): 313–33.

Von François, Curt. "Bericht von Hauptmann C. v. François über seine Reise nach dem Okavango-Fluss." *Mitteilungen von Forschungsreisenden und Gelehrten aus den deutschten Schutzgebieten* 4 (1891): 205–12.

Von Weber, O. "Ferdinand Gessert, Farmer in Sandverhaar." *Afrikanischer Heimatkalender* 1972, 31–38.

Voyles, Traci Brynne. *Wastelanding: Legacies of Uranium Mining in Navajo Country*. Minneapolis: University of Minnesota Press, 2015.

Wallace, Marion. *A History of Namibia from the Beginning to 1990*. New York: Oxford University Press, 2014.

Walther, Johannes. *Das Gesetz der Wüstenbildung in Gegenwart und Vorzeit*. Leipzig, Germany: Quelle and Meyer, 1912 (orig. 1900).

Ward, Robert De Courcy. *Climate, Considered Especially in Relation to Man*. New York: G. P. Putnam's Sons, 1918.

Weart, Spencer. *The Discovery of Global Warming*. Cambridge, MA: Harvard University Press, 2008.

Weidner, Carl. "Lost Lakes of South Africa." *United Empire* 10, no. 1 (January 1919): 26–29.

Weidner, Carl. *The Fallacy of Schwarz's Kalahari Rain-Making Magic*. Cape Town: Salesian Press, 1925.

Weiner, Douglas R. *A Little Corner of Freedom: Russian Nature Protection from Stalin to Gorbachev*. Berkeley: University of California Press, 1999.

Weisiger, Martha. *Dreaming of Sheep in Navajo Country*. Seattle: University of Washington Press, 2009.

Wellington, John. *South West Africa and Its Human Issues*. Oxford, UK: Clarendon, 1967.

Widney, Joseph. "The Colorado Desert." *Overland Monthly* 10 (January 1873): 44–50.

Wilson, James Fox. "On the Progressing Desiccation of the Basin of the Orange River in Southern Africa." *Proceedings of the Royal Geographical Society of London* 9, no. 3 (1864–65): 106–9.
Wilson, James Fox. "Water Supply in the Basin of the River Orange, or Gariep, South Africa." *Journal of the Royal Geographical Society of London* 35 (1865): 106–29.
Woeikov, A. "Der Einfluss der Wälder auf das Klima." *Petermanns geographische Mitteilungen* 31 (1885): 81–87.
Wooding, Robert. "Populate, Parch, and Panic: Two Centuries of Dreaming about Nation-Building in Inland Australia." In John Butcher, ed., *Australia under Construction: Nation-Building: Past, Present and Future*, 57–70. Canberra: Australian National University, 2008.
World Council of Churches. *Cunene Dam Scheme and the Struggle for the Liberation of Southern Africa*. Geneva: World Council of Churches, 1971.
Worster, Donald. *Rivers of Empire: Water, Aridity, and the Growth of the American West.* Oxford, UK: Oxford University Press, 1992 (orig. 1985).
Wright, A. W. "Grahamstown of Today." In T. Sheffield, *The Story of the Settlement, with a Sketch of Grahamstown as It Was and Grahamstown as It Is*. Grahamstown, South Africa: Grocott and Sherry, 1912.
Wroebel, David. *Promised Lands: Promotion, Memory, and the Creation of the American West.* Lawrence: University Press of Kansas, 2002.
Wybergh, W. "Imperial Organization and the Colour Question." *Contemporary Review* 91 (June 1907), 805–15.
Yusoff, Kathryn. *A Billion Black Anthropocenes or None*. Minneapolis: University of Minnesota Press, 2018.
Zilberstein, Anya. *A Temperate Empire: Making Climate Change in Early America*. New York: Oxford University Press, 2016.
Zimmerman, A. *Alabama in Africa: Booker T. Washington, the German Empire, and the Globalization of the New South*. Princeton, NJ: Princeton University Press, 2010.
Zollman, Jakob. "Negotiated Partition in South Africa: An Idea and Its History (1920s–1980s)." *South African Historical Journal* 73, no. 2 (2021): 406–34.

Published Government Reports

Bechuanaland Protectorate Report for 1925–1926. London: HM Stationery Office, 1926.
Cape of Good Hope, First Annual Report of the Geological Commission. Cape Town: W. A. Richards and Sons, 1896.
Kokot, D. F. *An Investigation into the Evidence Bearing on Recent Climatic Changes over Southern Africa*. Pretoria: Government Printers, 1948.
Mackenzie, L. A. *Report on the Kalahari Expedition 1945*. Pretoria: Government Printers, 1946.
Pim, Alan. *Financial and Economic Position of the Bechuanaland Protectorate*. London: HM Stationery Office, 1933.
South Africa, *Debates of the House of Assembly*.
South Africa. *Final Report of the Drought Investigation Commission*. Cape Town: Government Printers, 1923.
South Africa. *Report from the Select Committee on Closer Land Settlement.* Cape Town: Cape Times, 1911.
South Africa. *Report from the Select Committee on Droughts, Rainfall, and Soil Erosion.* Pretoria: Union of South Africa, 1914.

South Africa. *Report of the Commission of Inquiry into European Occupancy of the Rural Areas*. Pretoria: Government Printer, 1960.

South Africa. *Report of the Desert Encroachment Committee*. Pretoria: Government Printer, 1951.

South Africa. *Report of the Select Committee on Drought Distress Relief*. Cape Town: Cape Times, 1916.

South Africa. *Report of the Select Committee on European Employment and Labor Conditions*. Cape Town: Cape Times, 1913.

South Africa. *South African Native Affairs Commission*, Vols. 1–9. Cape Town: Cape Times, 1904–5.

South Africa Department of Agriculture. *Report for the Period 31st May 1910 to 31st December 1911*, Cape Town: Government Printers, 1913.

South Africa Department of Agriculture and Forestry. *Reconstruction of Agriculture: Report of the Reconstruction Committee of the Department of Agriculture and Forestry*. Pretoria: Government Printer, 1945.

South Africa Department of Irrigation. *Report of the Kalahari Reconnaissance of 1925*. Pretoria: Government Printing and Stationery Office, 1926.

Newspapers and Magazines

Agricultural Journal of the Cape of Good Hope

Allgemeine Zeitung

Deutsch-Südwestafrikanischen Zeitung

Farmer's Weekly

Die Landbouweekblad

Nature

Rand Daily Mail

South African Agricultural Journal

Star (Johannesburg)

United Empire

Windhoek Advertiser

Archival Collections

Cape Town Archives Repository, Cape Town, South Africa

Alex Logie du Toit papers, Special Collections, University of Cape Town, Cape Town, South Africa

Free State Archives Repository, Bloemfontein, South Africa

Imperial College Archives, London, United Kingdom

National Archives of Namibia, Windhoek, Namibia (NAN)

National Archives Repository, Pretoria, South Africa (NAR)

National Archives of the United Kingdom, Richmond, United Kingdom (NAUK)

E. H. L. Schwarz papers, Albany Museum, Grahamstown, South Africa

William Charles Scully papers: Fonds A1312, Wits University Historical Papers Research Archive, Johannesburg, South Africa

INDEX

Page numbers in italics refer to illustrations.

Abdication of the White Man, The (Schumann), 240–41
Act of Union, 183
Africa, 10, 28, 30, 34, 44, 59, 66, 69, 79, 164, 176, 184, 186, 189, 212, 214, 248; German empire in, 213; "lost lakes," 47; political protest in, 216; rainfall, from land's surface, 71; surface water, rare in, 137–38; water, and white man's land, 244; white men in, and language of duty, 178; yeoman-style racial democracy in, goal of, 129. *See also* southern Africa
African National Congress (ANC), 216, 240
African nationalism, 240
Afrikaners, 24–25, 57, 98, 103, 111, 118–20, 125–27, 131, 134, 150, 169, 200, 238; British, as threat to culture, 109, 116; "Dorsland trekkers," 257n56; farmers, 142; language and culture, securing of, 239–40; as lower-class whites, 106; nationalists, and link between *volk* and land, 128; "nomadic trekker," 95; "poor white problem," and Afrikaner nationalism growth, 117; as poor whites, 122; precarity of, 129; Schwarz scheme, opposition to, 230; soil conservation, as safeguard of *volk*, 222; soil erosion, as threat to, 222. *See also* Boers
Agassiz, Louis, 24
agrarianism: agrarian mystique, 126; as gendered, 127; romanticizing of, 126
Albany Museum, 58
Algeria, 26, 41
amaXhosa people, 58
American conservation movement, white racial dominance, and perceived threats to, 6
American Geological Society, 166
Andes Mountains, 70
Andrew Carnegie Corporation, 120
Anglo-Boer War. *See* South African War
Angola, 31, 45, 59–60, 155, 170–71, 182–83, 185–86, 209, 257n56, 289n46
Angola–South West Africa Boundary Commission, 170–71, 184
apartheid, 14–15, 231, 232–33, 245; as survival plan, 239; white minority, domination of Black majority, 240
Argentina, 76, 184
aridity, 8, 44, 56, 63–64, 69–70, 73, 91–92, 101, 200, 229–30; rainfall and, 20, 66; as white preoccupation, 226
arid lands, 9, 135, 137; as abnormal, settler views of, 190; colonial roots and, 11–12; deserts as empty, perception of, 180–81; irrigation in, 151; "redemption," as term for, 14; transformation, in need of, 102
Arizona, 66
Asia, 12, 28–30, 43, 48, 231
Aswan Dam, 137
Atlantropa, 212; as pacifist project, 213
Australasian Association for the Advancement of Science, 260n65

Australia, 2, 7–10, 12–13, 17–18, 25–26, 28–29, 40–41, 43, 58, 60, 64, 66–67, 79, 107, 124, 136–37, 140, 160, 201–2, 231, 256n25; drought-proofing interior, 212; White Australia policy, 212–13; as "white man's country," 108

baaskap, 240
"back to the land" movement, 20, 125, 127, 130, 202; poor whites, and independent farming, 128; as transnational and racial, 126
Basutoland, 19, 105, 107, 179, 183, 213, 215
BaTawana, 51
Bathurst Farmer's Association, 277n122
Battle of Blood River, 103
Bechuanaland Protectorate, 19, 39, 86, 107, 118, 156–57, 168, 178, 181–83, 185–86, 190, 193, 195–96, 198–99, 206–7, 215–17, 229; white farmers in, 90. *See also* Botswana
Beinart, William, 99, 106
Belich, James, 7–9, 26
Black farmers, 12, 49, 96, 102, 105, 112, 119, 133, 142, 157, 224–25, 234, *235*, 274n30
Black liberation movements, 288n19
Black people, displacement and disenfranchisement of, 3
Black poverty, 123
Black sharecropping, 112–13, 130, 132, 149
Blanford, William Thomas, 29–30, 40; elevation of land theory, 64
Boers, 50, 129, 140; farming, as antithesis of progressive farming, 141. *See also* Afrikaners
botany, 5
Botha, Louis, 57
Botletle River, 40, 51, 180, 189, 192, 196
Botswana, 2, 31, 34–35, 39, 44, 218, *219*, 257n56. *See also* Bechuanaland Protectorate
Bowman, Isaiah, 8, 12, 15, 241
Bradfield, Edward (E. R.), 110, 151
Brazil, 2, 41, 70; reservoir building, 68
British Empire, 1, 25, 61, 288n9; hydrology and, 136–37; and imperialism, 138–39
Brown, John Croumbie, 31–32, 35–36, 49
Brückner, Eduard, 12, 76, 264n67; "Brückner's law," 70; rain, on land-based origins of, 69–70, 75, 93
Bryce, James, 258n58
Byron, J. J., 128, 198

California, 12, 66, 76, 77, 159, 202; Salton Sea, 172
Canada, 26, 29, 140, 256n25
Cape Colony, 24, 26, 56–57, 64, 147
Cape Geological Society, 53
Cape Geological Survey, 25, 57, 61, 188
Cape of Good Hope, 24
Cape Province, 29, 170, 214; as white man's country, 242
Cape Province Agricultural Association, 214
Cape Town (South Africa), 30, 38, 53, 71, 84, 139, 155; "Day Zero," 246–47
capitalism, 245; industrial, 3; modern, 15
Carey Act, 168
Carnegie Commission of Investigation on the Poor White Problem, 120, 124
Challenger expedition, 68
Chamberlin, Thomas, 59
Chapman, James, 42, 51–52
China, 30, 77, 199, 247
Chobe River, 18, *35*, 42, 76, 79, 158, 189, 193
climate, 108; determinism, 109
climate change, 4, 69, 79, 114, 169–71, 220, 246–47; moral hazard and, 248
climate science, 12
climate skepticism, 21, 247
climatic seasonality, 11
climatology, 5, 10, 12, 23–24
closer settlement, 129–30, 157, 202–3; as generative, 152
Coen, Deborah, 12
Coetzee, J. M., 17, 74, 128, 134
cohabitation, 134–35
Collett, John G., 206, 232

colonialism, geoengineering as echo of, 248
Colorado Desert, 28, 54–55, 157; Salton Sea, 41
Congress of the Cape Province Agricultural Association, 170
Conradie, J. H., 174, 198
Conroy, Andrew Meintjes, Conroy expedition, 229–31, 236
continental drift, 57–58, 279n25
Conway, Erik M., 4
Corstorphine, George, 25–26, 57
Creswell, Frederic, 110, 193
Crosby, Alfred, "neo-Europes," 7
Crush, Jonathan, 139
Cuvelai: Basin, 46; plain, 47, 209, 243–44; River, *33*, 185, 191, 209

Dark Continent, 1
Darwin, Charles, 23–24
Davis, Diana K., 11–12, 63–64. 263n40
decolonization, 240
deforestation, 49, 64, 66, 71, 94, 211, 261n100; desertification, link to, 11–12, 62–63, 65
de Jager, Andries, 52–53, 90, 159, 190, 198, 262n113
Department of Irrigation, 188–89
desertification, 64–65; deforestation, links to, 11–12, 62–63, 65
deserts, as empty, perception of, 180–81
desiccation, 52–53, 68, 75–76, 78–80, 88, 100–101, 103, 105, 164, 173–74, 214, 220; behavior of farmers, as cause of, 98; Black South Africans, as cause of, 219; deforestation, links to, 11–12, 62–63; geology, as to blame for, 218, 219; geopolitical importance of, 213; global anxieties over, 211; human agency, as cause of, 64–65; and mismanagement of land, 64–65; and white innocence, 231; whites' existence, as threat to, 128
de Wet, P. J., 87, 91–92, 176, 207
Die Landbouweekblad (Agricultural Weekly) (newspaper), 98–99, 103, 141, 169, 204, 230
Dingane's Day (Day of the Covenant), 103
disequilibrium ecology, 9
Division of Soil and Veld Conservation, 227
Doel, Ronald, 4–5
Dreyer, Thomas, 207
drought, 9, 62–65, 79, 81–84, 87, 91, 98–99, 106, 118, 120–22, 135, 158, 171–72, 176, 183, 190, 194, 198, 200, 214, 248, 282n124, 284n43; as calamity, 201; as divine displeasure, 267n41; drought-proof farms, 148–50; end of, 209; global shift in agriculture, coinciding with, 102, 144–45; and landless whites, 105; and "poor whites," 10; as ruin of white men, 181; and white farmers, 1–2, 204; white population, and decline in, 115, 201
Drought Commission, 85, 87, 90, 92, 99, 103–4, 113–17, 121, 137–38, 140, 145–46, 152–53, 158, 170–72, 174, 176, 220–21; and white settlers, 100–102, 225
Drought Distress Relief Committee, 61, 146
Drought Investigation Commission. *See* Drought Commission
"drying out," 42, 80, 85, 88, 94, 114, 123, 134, 136, 138, 153, 179, 214, 233, 246–47, 257n42, 284n52; existential language and, 221
drylands, 8–9, 11, 15, 17, 23, 26–27, 29–30, 34–35, 55, 62, 68, 70, 81, 104, 141, 259n16; race, effect on, 48
Du Bois, W. E. B., 6
Dubow, Saul, 17, 239–40, 273n147
Dust Bowl, 149, 211
Dutch East India Company, 30
Dutch Reformed Church, Kakamas irrigation scheme, 121, 124, 146
du Toit, Alexander Logie, 162, 188, 190–93, 198–99, 208–9, 216, 220, 230, 232, 272n94; continental drift theory, as proponent of, 57–58; Schwarz scheme, rejection of, 195–97, 218

du Toit, Heinrich, 99–100, 102, 106, 113, 140–41, 146–48, 171, 174, 199, 223, 286n136; on soil erosion, 103–4; on "white civilization," and environmental ills, 101

East Africa, 47, 75, 77, 137, 199, 231
East Coast Fever, 142
Eastern Cape, 2, 49, 58–59, 87, 90, 92, 96, 142, 144–45, 161, 178, 232, 241, 276n77
East Indies, 27
ecology, 5
Edwards, Paul, 11, 264n67
efundja (flood), 46
Egypt, 7, 73–74, 76–77, 137, 159
environmental conservation, 3
erasure: conceptual, 134; discursive, 17; imaginative, 13
Eswatini, 19, 34. *See also* Swaziland
Etosha Pan, 46, 75, 169–70, 184–86, 191, 244–45
Europe, 7, 11, 18, 29, 41, 48, 58, 62, 64, 69, 73, 79, 199, 202, 212, 214, 231, 247; rainfall, generated from land-based moisture idea, 65–66; rural depopulation, 126
Evans, Ivan, 134, 154
evaporation, 65, 68, 75, 77, 153, 194; and dam-building, 60, 155, 195, 203; and drylands, 27, 31, 86; and farmers, 86, 159; lack of data about, 69, 86, 137, 216, 221; purported links to rainfall, 64, 66, 69–70, 94–95, 190, 195
evapotranspiration, 8–9, 27

Farini, Guillermo, 183
farmers, 237; agricultural modernization, critique of, 143; and antierosion works, 202; and artificial rainfall, 158, 160–62; backveld, 141–42, 144, 147, 224–26; "backward," 144, 146–47; Boer, 141; as bulwark, against Black majority, 234; "checkbook farmers," 159; as collectivity, 101; and degeneration, 222; and drought losses, 146–47; and drought-proof farms, 148–49, 158, 202, 226; and dry farming, 148–50, 158; endless expansion, goal of, 204; and irrigation, 149–50, 152, 158–59, 162; and livestock disease, 142; and livestock farms, 148; and progressive farming, 143, 146, 204; progressivism and backwardness, adopting language of, 142; rainfall, conservation of, 158; and scientific farming, 92, 126, 132, 140–45, 204, 233–34; and soil erosion, 143, 145, 223–24; and state intervention, 210–11; and storing water, 202. *See also* Black farmers; white farmers
Farmer's Weekly (newspaper), 98–99, 102, 119, 134, 139–45, 147–49, 159, 161–62, 169, 199–201, 203–4, 206–7, 230; and agricultural fertility, 202; Kalahari fantasy, resurrection of, 227–28; as Schwarz skeptics, 167, 227
Fick, J. C., 224
FitzPatrick, Percy, 160
folk hydrology, 39–40
foraging communities, genocide of, 30
Forde, C. Daryll, 160
Foster, Jeremy, 13, 235
Fournier, Luiz Mariano de Barros, 70, 262n11
France, 25–26, 41, 199, 222
Frederiks, Cornelius, 78
Frederiks, Paul, 71, 78
Freedom Charter, 288n33
Frémont, Charles John, 66
Frere, Bartle, 108
"frontier wars," 58

Geldenhuys, Frans, 98–99, 141
geoengineering, 248
Geological Commission of the Cape of Good Hope, 24
German Colonial Society, 72–73
Germany, 23, 25–26, 71–72, 74, 77–78, 89, 103, 129, 170–71, 208, 214, 223, 258n69; *Blut und Boden* policy, 126; geographical expansion, necessity for, 213

Gessert, Ferdinand, 71–72, *73*, 74–75, 91, 103, 129, 137, 187, 213, 232, 263n47, 278n136; afforestation, encouraging of, 77; as colonial optimist, 73, 78; dam building, proposal of, 76
Gilbert, Grove Karl, 75
Giliomee, Hermann, 109, 115, 117, 120, 239
Glanville, Ernest, 164, 277n134
"global color line," 3, 6
Global South, 248
global white solidarity, "white man's lands," quest to secure, 6
Globus (magazine), 74, 77
Gobi Desert, 75
Grand Apartheid, 8, 21–22, 131, 183
Great Britain, 23, 43, 56–59, 103, 117–19, 129, 137, 160, 195–96, 217, 222
Great Depression, 127, 139, 200, 226
Great Drought, 87–88, 127, 201–2, *203*, 204, 206, 208, 213, 219–20, 226
Great Replacement Theory, 248, 249
Green, Lawrence, 164, 277n134
Gregory, John Walter, 29, 43–44, 53, 172, 200

Haggard, Rider, 277n134
Hahn, C. H. L., 191
Hall, Charles, 161–62, 204, 224
Hamilton, Clive, 247
Hartbeesport Dam, 124, 137–38, 155, 163, 201, 227
Hartig, Rolf, 167
Harvard University, 248
Hatfield, Charles, 7, 160–62, 277n122
Hedin, Sven, 30
Heimat (homeland), 26, 72
Herero people, 77–78, 103, 180
Hertzog, J. B., 111–12, 115–16, 120, 131–32, 139, 188, 195–96, 210, 273n142
Hitler, Adolf, 213–14
Homestead Act, 28
Homo sapiens, 58
Hughes, David McDermott, 13; "hydrology of hope," 202–3
Humboldt, Alexander von, 12, 65
Huntington, Ellsworth, 268n89
hydroengineering: and Black labor, 179; nature, and conquering of, 137; and segregation, 179–80; social engineering, as inseparable from, 178
hydrology, 5, 136–37, 218, 232–33

identity politics, of knowledge, 249
Idriess, Ion, 212
imperialism, 5, 63, 150–51; British, 138–39; and white men, in opposition to colonized subjects, 6
India, 58, 97, 136–37, 159, 247, 256n14
Indigenous knowledge systems, 4–5, 249; Indigenous agency, 255n10; west wind, 267n62
Indigenous populations, 8, 77–78, 88–90, 107–8; ceasing to matter, 157; elimination of, 104. *See also* Native Americans
Intertropical Convergence Zone, 11, 93
inyanga, 277n127

Jeeves, Alan H., 139
Jeffares, John L. S., 52, 216–17
Johannesburg (South Africa), 24, 95, 118, 137, 139, 209, 236; as boomtown, 25
Judd, John Wesley, 23, 53, 56, 70

Kalahari, The; or, Thirstland Redemption (Schwarz), 14, 163–66, 171, 184
Kalahari and Its Native Races, The (Schwarz), 199
Kalahari Desert, 1, 5, 10, 35, 38, 41, 44–45, 48–51, 53, 65, 71, 78, 80, 155, 172–73, 176–79, 187, 197–98, 200, 212, 221, 246; aerial survey of, 216; and atomic power, 238; diamond deposits in, 199; drier climate of, 42; expeditions to, 229–30, 232, 236; flooding of, 92, 206, 208, 210; grasslands, 28, 184; hydrology of, 218; as imagined place, 183; irrigation, 199; legends about, 183–84; paradox of, 184; ranching, 238; redeeming of, for white settlers, 218; redemption of, 232; and white settlers, 181; as wilderness, 183–84

Kalahari Development Company, 168
"Kalahari problem," 184
Kalahari Problem, The (Karlson), 184
Kalahari redemption, 169, 206; commitment to, 209
Kalahari Reconnaissance, 188, 193, 195–96, 198, 207–9, 285n96
Kalahari Scheme, 3, 14, 21, 99, 104, 136, 156, 162, 167, 169, 173, 176–78, 180, 188, 200, 206, 213–15, 218, 222, 231–33, 236, 258n10, 278n136, 285n96; excitement over, 185; as expansive, 186; inspiration for, 54–55; Kalahari fantasy, as resurrection of, 227–28; poor whites, guarantee of good living to, 164; public enthusiasm for, 154, 171; racial demographics, and shifting of, 164. *See also* Kalahari Thirstland Redemption Scheme; Schwarz scheme
Kalahari Thirstland Redemption Scheme, 13, 17, 20–21, 23, 60, 71, 121, 242, 256n14. *See also* Kalahari Scheme; Schwarz scheme
Kalahari Thirstland Redemption Society, 209–10, 220; creation of, 207. *See also* Schwarz Kalahari Society
Kanthack, Francis, 47, 64–65, 142, 147, 152–53, 158–61, 170–74, 184–86, 196, 232
Kariba Dam, 138, 209, 243
Karlson, August, 111, 113, 137, 155, 157–58, 163, 178–82, 184, 187, 277n134
Karoo, 7, 30, 33, 53, 55–56, 59, 82, 84, 89–92, 122, 129, 135, 140, 152, 155, 161, 167–68, 170, 187, 201, 206, 215
Katombora rapids, 195, 208–9, 238, 243, 246
Kavango River, 34, *35*, 42, 45, 47, 51, 162, 178, 188, 209–10, 213, 232. *See also* Okavango Delta
Kayser, Kurt, 213
Keegan, Timothy, 109, 144
Kenya, 126
Kern, Adolph, 207–8
Khama, Tshekedi, 216
Khoekhoe pastoralists, 30
knowledge: experiential 92–93; experts, 92; identity politics of, 249; Indigenous, 4–5, 249, 255n10, 267n62; Native, and white geology, 192; scientific, 143; transnational, and white experts, 5; white popular, 7; white vernacular, 81
Kokot, Daniel, 231–32, 238, 248
Korn, Hermann, 214
Koselleck, Reinhart, 13
Krikler, Jeremy, 15–16, 109, 115
Kropotkin, Peter, 43, 63–64, 75–76
Kuhn, Alexander, 72–73
Kunene River, 38, 158, 171–72, 188, 191, 213, 229, 243–46, 265n93, 289n46; colonial survey of, 184; diverting of, 76, 181, 244

Lake, Marilyn, 6, 9–10
Lake Eyre, 43
Lake Ngami, 33–34, 36, 38, 42, 45, 47, 176, 187, 217, 245–46; disappearance of, 48, 50–53, 65, 262n111; as "lost lake," 65, 79; and papyrus rafts, 50–52, 195, 261n103; restoration of, 216; shrinking of, 44
Lake Tritonis, 29
Land Bank. *See* National Land and Agricultural Bank
land degradation, 211, 219
landless whites: droughts, vulnerable to, 105, 118–19; and white poverty, 121
Lands Department, 118–20, 271n83
Landsdiens (youth organization), 222
League of Nations, 59–60, 89, 177, 258n69
Lehmann, Philipp, 14, 267n55
Leipoldt, Johann, 155
Lemuria, 59, 262n11
Leppan, Hubert, 202, 218
Lesotho, 19. *See also* Basutoland
Lesseps, Ferdinand de, 67
Lewis, Alfred, 151, 160
liberty, racializing of, 6
Limerick, Patricia, "wish-fulfillment geography," 29
Linton, Jamie, 29
Linyanti River. *See* Chobe River

Livingstone, David, 33–36, 38, 41–42, 45, 48, 52, 75, 191, 230
Long, Ernest, 155, 157, 178
Lugard, Frederick, 44

MacDonald, William, 39, 133, 141, 145, 149, 152, 182–83
Machado, Carlos Roma, 171
Mackenzie, Donald, 41
MacMillan, William Miller, 114–15
Makgadikgadi plain, *36*, 40, 47, 167, 189–90, 193, 245–46
Malan, Daniel François, 110, 113–14, 239
Malan, M. L., 123
Malawi, 107
Malherbe, E., 120–21, 124
Mamdani, 269n8
Manning, Charles, 89
Marais, Wilfred, 231–32, 238
Marchand, B. P. J., 121, 123–24, 146, 154
Markham, Charles, 187–88, 213
Markham, Violet, 113, 116, 187
Mars, 42–43, 59
Marsh, George Perkins, 63, 66–67, 264n56, 269n107
Martel, Édouard-Alfred, 41
Martin, Henno, 214
Mbambangandu (rainmaker), 209–10
Merensky, Hans, 213
Merriman, John X., 117, 122, 124, 147
Mesopotamia, 137, 156
meteorology, 5, 10
Midland Farmers and Wool Growers Association, 220
Milner, Alfred, 109, 117, 126, 138–39, 140, 142
Milsom, Fred, 118–20
missionaries, 49, 51
Mitchell, Thomas, 29
modernity, 81, 154
Moffat, Robert, 32, 35, 42
Money, Duncan, 124
moral hazard, 248
Mozambique, 107, 155, 213, 243
Murray, John, 68, 264n66; rain, on land-based origins of, 69–70, 93

Nama people, 71–72, 78, 103
Namib Desert, 31, 71, 75, 91, 170, 246
Namibia, 2, 19, 30–32, 39, 45, 188, 218, *219*, 257n56; Great Namaqualand, 71. *See also* South West Africa
Natal, 26, 82–83, 105, 110–11, 201
National Academies of Science, Engineering, and Medicine (NASEM), 248
National Land and Agricultural Bank, 122, 139, 147–48, 202
National Socialists, 213
National Society for Agriculture, 70
National Veld Trust, 221–24
Native Americans, absence of trees, held responsible for, 63. *See also* Indigenous populations
Natives Land Act, 83, 111–12, 119, 125, 130, 178, 216
Naudé, J. J., 117, 124
New Zealand, 124, 140
Ngamiland Waterways Surveys, 216
Nicholls, George Heaton, 225
Nitsche, Georg, 47
North Africa, 2, 41, 48, 63, 67
North America, 1, 8, 17–19, 28, 43, 48, 58, 64, 66, 75, 107–8, 124, 161, 231; rural depopulation, 126
Northern Rhodesia, 107, 182–83, 186
Nyasaland, 107

Obst, Erich, expedition of, 213–14
oceanography, 68
Okavango Delta, 33–34, *35*, 38, 45, 51–52, 179–81, 185, 189–90, 216, 229
Oppenheimer, Harry, 232
Orange Free State, 24, 56–57, 82, 84, 88, 90–92, 98–99, 105–6, 112, 114–15, 123, 144, 157, 167, 186, 201, 206–7, 234, *235*, 237
Orange River Development Project, 241–42; Gariep Dam, *243*
Oreskes, Naomi, 4
Origin of Species, The (Darwin), 23–24
Oswell, William Cotton, 40
Outspan (magazine), 227
Ovamboland, 32, 45, 76, 83, 89, 179–80, 184, 191, 209, 213, 244
Owen-Collett, John, 206–7, 209–10, 215, 220, 232

Pan Africanist Congress, 240
Panama Canal, 156–57, 187–88
Passarge, Siegfried, 38, 41, 51, 53, 66, 69, 71, 75–77, 191, 261n102
pastoralism, 224
Paulsmeier, Heinrich, 79, 172, 232
Pearson, Charles, 9–10, 108, 241
Persia, 29, 40, 64
plaasroman (farm novel), 201
planetary desiccation, 43, 56, 79, 268n89; astronomy in, role of, 260n78
platteland, 236; racial demographics of, 234–35; white depopulation of, as threat to white civilization, 234
"pluvial periods," 43–44, 47
"poor whiteism," 106, 121, 135; racial order, jeopardizing of, 117
"poor white problem," 104, 114, 116–17, 119, 122, 126, 139, 195–96; "back to the land," 106; and racial pride, 125; response to, 106; solution to, 211
poor whites, 120–22, 129–30, 146, 164, 181, 237; "back to the land" movement, 128; as danger to state, 117; and independent farming, 128; manual labor, resistance to performing, 124–25; poor man and, distinction between, 123–24; racial hierarchy of, 125; racial privilege, 124
population removal, 131
Porfirio Díaz, José de la Cruz, 66, 264n55
Portugal, 170–71, 243–44
Posel, Deborah, 131
Powell, John Wesley, 7, 63, 68, 75
Powell, Miles, 6–7
Pretoria (South Africa), 111, 142–43, 161
progressivism, 6, 96, 99, 142

Queenstown Farmer's Association, 96
Quinan, Kenneth, 187–88, 213

racial anxiety, 249
racial privilege, 124; and responsibility, 101
racial "replacement," fears of, 21–22
racism, 4, 16, 239; scientific, 63
"Rand Revolt," 115
Reitz, Denys, 214, 217–18
Rennell, James, 29, 65
reprecipitation, 68
Rhodes, Cecil, 25–26
Rhodesia, 118, 157, 210
Rhodes University College, 58–59
Robarts, G. P., 155–56
Rogers, Arthur, 25–26, 57, 61–62, 258n12
Roos, Neil, 16, 239
Ross, J. C., 284n52
Rotherforth, C. H., 119–20
Roudaire, François, 67, 70, 75, 264n57
Royal Colonial Institute, 168, 199
Royal Geographical Society, 43, 48, 170, 172, 197
rural white precarity, and landlessness, 117–19

Sahara Desert, 63–65, 75, 79, 231; and white settlement, 211–12
Schinz, Hans, 38
Schmidt, Albert, 76–77, 79
Schoeman, Johannes Hendrik, 122, 135, 141–42
Schonken, J. D., 202, 208
Schumann, Theodor, 220–21, 223, 240–41
Schwarz, Bill, 10, 169
Schwarz, Daisy, 207, 227
Schwarz, Ernest, 2, 5, 14–15, 18–20, 26, 30–31, 33, 39, 47, 49, 53–54, 57, 58, 60–62, 64, 70–71, 77, 79, 83–84, 90–91, 100, 106–7, 114–15, 125, 127–29, 133, 136, 152, 154–59, 162, 165–66, 169, 171–74, 176–77, 181–82, 184, 187, 195–98, 208, 210, 212, 217–18, 237, 243–45, 248, 256n14, 258n10, 258n12, 277n134, 278n1, 287n145; African and Dutch place names, invoking of, 89; African land rights, protection of, 178–80, 186, 193; background of, 23–24; and climate change, 170; congenial presence of, 163; death of, 200, 231; doodle of, *163*; as "English Imperialist," 167;

experiential knowledge and universal science, and bridging gap between, 93; Galileo, comparison to, 227; on Kalahari, as "imagined place," 183; Kalahari Reconnassiance, participation in, 188–94; as lionized, 223; "lost lakes," 185–86, 190, 193; on native rights, 168; Observatory House, 59; opponents of, 231–32; popularity of, 188; as populist, 164; as "Prophet of the Kalahari," 228; public support, cultivating of, 168; racially charged language against supporters of, 175; on "river capture," 163–64; *Scientific African*, as editor of, 25; settler discourse, embrace of, 52; speculation of, 199; supporters of, 206–7, 209, 213–14, 219–21, 231, 233, 241, 247, 278n136, 279n25; surveying geology of Karoo, 56; as techno-optimist, 241; unorthodox views, embracing of, 59; water running to waste, idea of, 137; white innocence, protecting of, 178

Schwarz Kalahari Society, 209, 231. *See also* Kalahari Thirstland Redemption Society

Schwarz scheme, 3, 7, 10, 13, 17–18, 20–21, 38, 44, 52, 56, 59, 80, 98, 110–11, 116, 121, 130, 135, 155, 166–68, 170–71, 177–78, *182*, 184–85, 187, 202, 206–7, 209–10, 213, 215–17, *219*, 220, 222, 229, 238, 248, 277n134, 278n6; African land rights, protection of, 186; Afrikaner opposition to, 230; "back to the land" concept, 126; as cautionary tale, 249; climate effects, estimation of, 172; core beliefs of, 233; geology for desiccation, blame of, 218–19; geopolitics, shaping of, 218; as historical oddity, 232; hydro-utopianism of, 238; as identical to Gessert's version, 76; inspiration for, 54–55; irrigation, shared imaginary of, 236; irrigation potential, emphasis on, 165; Kalahari Reconnaissance, investigation of, 188–98; as low-tech, 158; meaning of, change in, 208; opposition to, 236–37, 282n127; popularity of, 158, 162, 183, 200; progressive farming, bound up with, 146, 176; publicity campaign, 162; radical partition, proposal of, 240; rejection of, 195–97, 218, 223, 231; salinization, problem of, 186; skeptics of, 172–75; and soil conservation, 219; supporters of, 230, 232, 236–37, 242; techno-optimism of, 154, 247; as unrealistic, 156; westerly wind, and reigning in, 92; white fragility, effect on, 181; and white innocence, 232–33; and white settlers, 106–7. *See also* Kalahari Scheme; Kalahari Thirstland Redemption Scheme

science denialism, 4

Scientific African (magazine), 25

scientific agriculture, 3, 141

scientific farming, 92, 126, 132, 140–42, 144–45, 204, 233–34

Scotland, 68. *See also* Great Britain

Scully, William Charles, 32–33, 36, 52, 89–90, 92–93, 97, 101, 133, 164, 171, 174–75, 259n32

segregation, 3, 14–15, 21, 113, 123, 131, 133–34, 238–39, 273n147; and hydrological engineering, 179–80; as spatial concept, 19; and white supremacy, 180

Select Committee on Closer Land Settlement, 111, 133

Select Committee on Drought Distress Relief, 121–22, 124, 147

Select Committee on Droughts, Rainfall, and Soil Erosion, 84, 95

Select Committee on European Employment and Labor Conditions, 125

sertão, 26, 28, 62–63, 70, 79, 262n11

settler colonialism, 2–3, 9, 17–19, 21, 29, 32, 60, 72, 102, 107, 110, 237, 240, 247, 249, 256n25; conceptual erasure, as key aspect of, 134; as demographic replacement, 104; environment and, relationship of, 7; explosion of, 7–8; Indigenous population, effacement and replacement of, 8, 104; and vision of green lands for white men, 4

settler imaginaries, 134
Settler Revolution, 26–27
Sharpeville Massacre, 240–41
Siegwart, Bruno, 212
Sim, James, 49, 94
Smuts, Jan, 57, 103, 109–10, 117, 120, 156, 167–68, 170–71, 177, 183, 188
soil erosion, 82, 95–96, 103–4, 133–34, 143, 152, 208, 211, 221, 223, 226, 233; Afrikaner people, as threat to, 222; evils of, 141, 145; as natural occurrence, 97
Sörgel, Herman, 211–12, 285n96
South Africa, 2–3, 7, 12, 19, 23–24, 31–33, 35–36, 39–40, 48, 58, 60, 78–79, 82, 89, 91, 105, 114, 116, 118, 137, 143, 147, 155, 165, 167, 170, 182, 187–88, 190, 199, 210, 217–18, 222, 224–25, 229–30, 237, 242, 244–45, 247, 270n27, 272n94, 277n122, 277n127, 286n136; absentee landlords, 111; Afrikaners, as majority of voters in, 57; agrarian imagination, 141; agrarian life, changes in, as threat to, 126; as "agricultural country," 127–28; agropastoralist societies in, 108; apartheid, 14–15, 231, 239–40; aridity, and white settlement, 233; arid lands, 135; artificial rain, fascination with, 160–62; "back to the land" movement, 20, 106, 125–28, 130, 202; as big game preserve, 133; Black competition in, 114, 120–21; Black erasure, 134; Black farmers, 119, 132–33; Black farmers, as threat to white man's land, 215; Black future, fear of, 201–2; Black invisibility, 133–34, 234; Black labor, 104, 107, 130–31, 153, 226, 235, 239–40; Black majority, 109, 234, 238, 240–41, 249; Black majority, as invisible, 138, 180; Black majority, as "problem," 108; Black resistance, 240; Black South Africans, relocated to homelands, 226; Black "specter," facing of, 181–82; "Black spots," elimination of, 216; Black tenants, 111–13; boundaries of, 183; British Commonwealth, withdrawal from, 240; Britons, as minority among white population, 109–10; "checkbook farmers," 159; cities, growth of, 125; climate, changes in, 65; closer settlement, 129–30, 152, 157; cognitive dissonance, 17; cohabitation in, as out of the question, 134–35; colonial nationalism and scientific rationality, 25; countryside, white-Black relations in, 125; deforestation, 94; demographics, of plantation colony, 110; desiccation of, 56, 100–101, 103; desiccation of, as threat to rural whites, 105, 213; drought in, 99–100, 106, 144–45, 201–2, 208, 248, 282n124; dry farming in, 148–49; "drying out," 1, 80, 85, 90, 94, 102, 114, 153–54, 173, 176–77, 213–14, 221, 233, 246; election of 1948, as watershed, 238; emerging identity, 25; encroaching desert, fear of, 102; ethnic favoritism, in rural land policies, 119; existential issues, of whites in, 249; expert and popular opinion on "drying out" in, gulf between, 80–81; farming, as national virtue, 127; floods in, 98; fragility, of white population, 181; Grand Apartheid and, as white man's land, 8, 21–22; hydrology of, 60–62, 64; imaginative erasure of Black South Africans, 13; irrigation, 152, 158–60, 162, 164, 178–80, 209, 276n77; irrigation, as way to secure white dominance, 151, 157; Kalahari expedition to, 195, 196–98; as majority-Black country, 107, 124–25; minority status of white population, as existential threat, 18; "national suicide," 103–4; natural resources in, waste of, 138; pass laws, 240; platteland, 234–36; "poor white problem," 104, 106, 113, 117, 119, 125–26, 139, 195–96, 211; poor whites, 10, 15, 20, 104, 117, 181; protectorates, lack of control over, 215–16; rainfall, and soil erosion, 223; rainfall, and white depopulation,

221; rainfall in, 81, 83–87, 92–93, 97–98, 148–49, 173–75, 208, 211, 220–21, 231, 233, 246; redemption of, 151, 227; as republic, 240; rural whites, 16, 125–26; scarcity, perceptions of, 249; scientific elite, 5, 11, 142, 173–74; and scientific farming, 126, 132, 140; segregation, 131, 238–39; as settler colony, 107, 110; settler explosion, 26–27; soil erosion in, "evil" of, 141, 211; South West Africa, claim over, 243; and South West Africa, war between, 289n46; and Sundays River irrigation scheme, 159–60; "swamping" in, fear of, 109–10; Swartberg mountains area, 25, 53–54; territorial expansion, support of, 183, 215; wasted water, ideology of, 60–62, 153, 194, 233; water, as necessary for industrial and urban expansion in, 136; watersheds, 60; as white, vision of, 13; "white civilization," as tenuous state, 104; white countryside, vision of, 130; white farmers, 49–50, 57, 107, 125, 130–33, 138–40, 153, 157–58; white farmers, and redemption of, 200; white fears, 102, 110, 114; "white genocide" in, 248, 250; white identity, and water, 136; white immigration, 18, 20, 106–7, 141; white immigration, encouraging of, 20, 107, 113, 129, 215; white innocence in, "drying out," 233; as white man's country, as spatial and racial idea, 125; "White Man's Country," Black labor and idea of, 181; white man's land, as egalitarian rural order in, 126; and white man's land, farming in, 127; "white man's land," status as, 15, 106, 108–12, 114–15, 117, 125–27, 132, 134–35, 140, 157–58, 215, 239–40; white population, 110–11, 113–15, 119, 157, 181, 202–3; white poverty, 15, 18, 117; white prestige in, concerns over, 106; white rural depopulation, 113–14, 120–21; whites, as embattled minority, 109, 115; white settlement, 19, 49, 63, 153, 164, 166, 178–80, 223, 233; whites' existential fears, 135; white stability, as responsibility of state, 176; white supremacy, 239, 241; white survival in, fears about, 115, 194; World War II, entry into, 211

South African Agricultural Journal (journal), 93–95

South African Agricultural Union, 199

South African Association for the Advancement of Science, 2, 94, 170, 220, 242; creation of, 258n10; drought symposium, 158

South African College, 25–26

South African Museum, 25–26

South African Native Affairs Commission, 108, 116, 131

South African Philosophical Society, 25, 108, *163*

South African Republic. *See* Transvaal

South African War, 1, 56, 84, 87–88, 103, 106, 109–11, 117, 134, 150; rural precarity, exacerbation of, 105

South America, 43, 58–59, 231; "Great Lake," 70

southern Africa, 10, 23, 30, 38–39, 48, 58, 60, 66, 74, 76, 79, 82, 89–91, 101, 104, 113, 155, 163–64, 166, 182, 191, 209, 213, 227, 231; aridity of, 70–71; Black majority, as threat to white survival, 103; as climate change hotspot, 246; desiccation of, 47, 80; drying up of, 44, 138, 187; drylands, 17; flooding of, 209; ghost landscapes, 41; Indigenous populations, as ceasing to matter, 157; missionaries in, 49; and rinderpest epizootic, 53; water sources, as squandered, 61; white man's land, transformation into, 165; and white settlers, 209

Southern Rhodesia, 19, 26, 107, 182–83, 226, 243. *See also* Zimbabwe

South West Africa, 19, 20, 26, 33–34, 59, 75–76, 79–80, 85, 89, 91, 97, 103, 107, 126, 144, *150*, *151*, 155, 157–58, 164–65, 167, 170–73, 182–86,

South West Africa (*continued*) 188, 190–91, 206–8, 229; afforestation in, 77; as African California, 73; colonial economy, 45; dam building in, 73–74; flooding in, 210; Germany's overseas empire, as part of, 71–72; Indigenous Herero population and German farmers in, war between, 77–78; and landless whites, 129; liberation movement, 243; South Africa, war with, 289n46; and white farmers, 72, 78. *See also* Namibia
South West Africa People's Organization (SWAPO), 289n46
Stals, Jacobus, 114, 127–28
Staples-Cooke, F. G., 119–20
Stebbing, Edward Percy, 211
Stephen Long Expedition, 28
Stewart, Charles, 161
Stigand, Almar, 185–86, 261n102
Stoler, Ann Laura, 100
Stovall, Tyler, 6
Suez Canal, 67, 156, 187–88
Sundays River irrigation scheme, 159–61
Supan, Alexander, 69–70
Sutton, John Richards, 153, 174–75, 188
Swart, Sandra, 7, 105
swartgevaar (Black peril), 117
Swaziland, 19, 107, 155, 183. *See also* Eswatini

Tanganyika, 75
Tarka Farmers' Association, 138
Tate, Ralph, 40, 66, 260n65
Theal, George McCall, 225
Thompson, W. R., 218, 220–21, 286n136
Timbury, Frederick, 212–13
Tischler, Julia, 96, 256nn17–18
Transvaal, 24–25, 56–57, 81, 84, 92, 105, 112, 114, 119, 135, 138, 144, 150, 155, 159, 167, 176–78, 201, 204, 213, 237, 257n56, 277n122; Sabi Reserve, 87–88
Transvaal Mining Industry Commission, 110
trekboers, 30, 132

Uitlanders, 25
Union of South Africa. *See* South Africa
United Nations (UN), 240, 243
United States, 6, 26, 29, 48, 72–73, 76, 99, 113, 141, 158–60, 222, 256n14, 256n25; arid lands of, 7, 63, 75, 137, 247; artificial rain in, 223; Great Plains of, 9, 12, 28, 101; "redemption," environmental and racial meaning in, 14. *See also* US South
University of Cape Town, 25–26
urbanization, 116; urban poverty, visibility of, 117
US Bureau of Reclamation, 151, 155, 157
US South, 124; segregation, enforcement of, 113
Utah, Great Salt Lake, 75, 172

van der Watt, Susanna Maria Elizabeth, 142
van Hoepen, Egbert, 207
van Onselen, Charles, 112
van Reenen, Reenen, 158
van Sittert, Lance, 156
van Zyl-Hermann, Danelle, 124
Veracini, Lorenzo, 8; conceptual erasure, 134
Victoria Falls, 45, 47, 186–87, 208–9
virga, 90–91, 93
Vollmer, Edmund, 72
Von François, Curt, 34
von Trotha, Lothar, 77–78

Walther, Johannes, 69–70
Washington, Booker T., 124
waste, 120, 195; desert, 61, 184; meanings of, 60–61; as moral issue, 138; of natural resources, 138; nonengineered rivers, description for natural flow of, 61; and resource neglect, 60; "running to waste," 54–55, 60–62, 136–38, 145, 152–53, 187, 194, 218, 233, 237; sterile, 28, 41–42; of water, 60, 223; waterless, 33, 76, 145
Wegener, Alfred, 57–58, 279n25
Weidner, Carl, 50–51, 87, 129, 137, 149, 155–56, 163, 175, 178, 182–83, 208, 241
West Africa, 2, 63, 211
West Indies, 27
white farmers, 18, 30, 70, 78, 83, 86–87,

90, 92, 106, 107, 113–14, 147, 149, 157–58, 178, 200–201, 204, 206, 209, 216, 246, 274n30; in debt, 139–40; "drying out," 153; environmental ills, push back against responsibility for, 102; failure to modernize, 143; as key to survival, of white civilization, 233; and scientific farming, 144, 233–34; state largesse toward, 237–38, 241–42; and white innocence, 231
white fears, 2, 14, 114; of extinction, 3, 10, 20, 102, 110, 194
white identity: class lines, fractured along, 119, 124, 130; construction of, 3; loss of, 201–2; science and modernity, as cornerstone of, 154; water, as crucial to, 136
white innocence, 226, 231–32; and "drying out," 233
white nationalism, 3, 5, 21; resurgence of, 248
whiteness, 124; "backveld" farmers', questioning of, 147; global politics of, 2, 6; transnational, elitist bent of, 6; and white aspirations, 3; and white entitlement, 6; and "White Geology," 191–92; and white landlessness, 15, 118–19; and white men's lands, 9, 13–15; white prestige, and economic impoverishment, as threat to, 116; and white racial anxieties, 248–49; and white responsibility, 49; white superiority, and myth of, 10
white poverty, 103, 125, 208, 239; alarm over, 117; causes of, 120; and landlessness, 121; as pathology, 123; and racial privilege, 124; and soil conservation, 222; solution to, in cities, 211; state, as solution to, 154; and white rural prosperity, as responsibility of state, 237
white precarity, 122, 129, 158; and climate change, 114
white settlement, 7, 9, 60, 63, 78–80, 102, 137–38, 153, 164, 166, 184, 193; and aridity, 233; and environmental degradation, 225; nature's equilibrium, disrupting of, 48–49; water, fear of disappearing, 40
white supremacy, 3, 7, 19, 113, 224, 239, 241–43; myth of, 10; and segregation, 180; and white innocence, 232–33; white poverty, as threat to, 106, 117; white prosperity and, relationship between, 115; white survival, equated to, 240
white survival, 239; white supremacy, equated to, 240
white tenants (*bywoners*), 105
Widney, Joseph, 28, 66, 70, 264n55
Willcocks, William, 137
Wilson, James Fox, 48–50, 261n96
Wirsing, H. F., 117–18
Witbooi, Hendrik, 78
Woeikov, Alexander, 69, 263n47
Wolfe, Patrick, 256n25
Wolmarans, Andries D. W., 128
Wooding, Robert, 10
World War I, 118, 160, 241, 258n69
World War II, 200, 211, 230, 241
Worster, Donald, 9–10, 150–51

Xhosa territories, 58

Yusoff, Kathryn, 191–92

Zambesi River, 34, 40, 45, 75, 138, 182, 189, 191–93, 198, 210, 229, 238, 245–46; damming of, 76, 79, 185, 195, 209, 243; diverting of, 187, 195, 206, 208, 231, 243; engineering of, 188, 242–43
Zambia, 107. *See also* Northern Rhodesia
Zimbabwe, 2, 13, 19, 221. *See also* Southern Rhodesia
Zollman, Jakob, 134
Zulu kingdom, 103, 108